CONTEMPORARY
ENVIRONMENTAL
ISSUES
FIFTH EDITION
Michael C. Slattery

Kendall Hunt
publishing company

D0165821

Front cover image info:

Fuglefjorden Iceberg, 79 degrees 45.5' N, 011 degrees 29.5' E,

Digital photograph (Adam Fung, 2017)

Adam Fung (M.F.A. Painting, University of Notre Dame, B.F.A. Painting, Western Washington University) is Associate Professor of Art at Texas Christian University. In 2016, Adam participated in The Arctic Circle Artist Residency and Solstice Expedition, a program whose mission is to introduce artists and scientists to the planet's North (www.thearcticcircle.com). Immersed in the Arctic environment for three weeks aboard a sailing ship, Fung used a drone, cameras, and other video equipment to collect imagery from around Svalbard, the Arctic Ocean, and the polar ice cap. The icebergs and glaciers captured in his photographs and paintings no longer exist as depicted, having shifted in the duration of a year of winter's snow and summer's melt. The larger question at hand is in what form will these places exist, if at all, in the coming years, decades, or millenniums.

More information can be found at www.adamfung.com. © Kendall Hunt Publishing Company

*Proceeds from the sale of this book go toward supporting TCU's Rhino Initiative, specifically rhino protection and rescue. Please see www.ensc.tcu.edu/rhino for more information.

Kendall Hunt
publishing company

www.kendallhunt.com
Send all inquiries to:
4050 Westmark Drive
Dubuque, IA 52004-1840

PAK ISBN: 978-1-5249-8073-3
Text Alone ISBN: 978-1-5249-7961-4

Printed in the United States of America

For my son Liam, again
Isn't this a marvelous planet?

Contents

Preface

Our relationship with the Earth is changing at an unprecedented rate. The pace of change is accelerating not only from our advancing technology but also from world population growth, economic growth, and increasingly frequent collisions between expanding human demands and the limits of the Earth's natural systems. Scientists now say we are in a new stage of the Earth's history, the Anthropocene, when we humans have become the globe's principal agent of change. We frequently hear that current global consumption levels could result in ecosystem collapse by the middle of the century and that environmental catastrophe looms ahead unless we make major changes soon.

Whether or not current human pressure on the Earth's ecosystems threatens our future as a species, one thing seems certain: we cannot continue to consume at northern levels indefinitely. To begin a shift toward a "sustainable society," significant action is required now on a range of issues. Fortunately, people are capable of changing their behaviors and values. Often these changes stem from exposure to new information or experiences. I have seen this first hand in Costa Rica, where I take students from my university on a three-week trip each spring. In Costa Rica, they get to see widespread environmental degradation (deforestation, soil erosion, overgrazed landscapes, etc.) juxtaposed against pristine rainforests teaming with wildlife. Many begin to grasp, for the first time, the extent of our impact on the Earth's ecosystems, as well as the challenges we face in trying to strike a balance between development and conservation.

I believe that everyone should be exposed to such information and experiences, because it develops a level of environmental literacy that is necessary for dealing with the challenges of the 21st century. The sad truth, however, is that environmental literacy is almost nonexistent in formal education. At most universities, undergraduates are required to take a set of common courses—chemistry, physics, calculus, English, etc.,—that make up a "core" of the degree plan. Hardly anywhere do you see a required course in global environmental issues, although issues related to the environment affect each of us in our daily lives. I believe that every citizen should become fluent in the principles of environmental science, demonstrating a working knowledge of the basic grammar and underlying concepts of environmental wisdom. This book has been written with this goal in mind.

Why am I writing yet another book on the environment when there are several currently on the market? In short, I believe that this book is different. Most of the traditional introductory college texts are, in my opinion, too broad, attempt to do too much, and have become "environmental encyclopedias" laden with too many facts about the environment. I have used several textbooks in my introductory course over the past decade, and during this time, I have become increasingly frustrated with students having to spend well in excess of $100 for these books when we can realistically only cover half of the material in a typical 15-week semester course. A lot of the time, the best value students get out of a book is the money they receive when they sell it back to the bookstore! What I really wanted to create is a book that presents a candid analysis of the major environmental issues the world currently faces: one that is inexpensive, informative, and makes students *think* about how the environment affects their lives and how their actions affect the environment. I always half-jokingly said I wanted a book that students could read in the bathtub! Well, the water may get a little cold, but I certainly hope that I have written a book that you would *want* to keep after the course is over!

The issues covered in this book are among the most pressing we face: population growth (and food), energy, atmospheric pollution, ozone depletion, climate change, deforestation, biodiversity loss, soil degradation, and water quantity and quality. I chose the topics after surveying the environmental studies faculty at my university, asking them a simple question: What environmental issues do we want our students—all our students, and not just environmental science majors—to be conversant in when they walk across that commencement stage? Interestingly, there was almost unanimous agreement among the faculty on the issues, with one or two personal interest topics emerging. It is not an exhaustive list, nor is it meant to be. Rather, my approach to writing the book was to cover these key issues and cover them well, providing the current state of scientific knowledge, yet written at a level that is digestible by the non-science major. However, adequate solutions to environmental problems also require well-informed ethical, aesthetic, political, and cultural perspectives, in addition to basic science and economics. I have attempted to weave these perspectives throughout the text. The following questions, in particular, are ones to think about as you read the material:

1. What (if any) are the ethical responsibilities of humans relating to the natural environment?
2. What is the role of science in the environmental debate?
3. Is scientific research value-neutral?
4. Does nature have intrinsic value?
5. Are there ethical principles that constrain how we use resources and modify our environment?
6. How do we achieve a balance between human values and interests and our obligations and responsibilities to nature?

My intention is that, upon completion of this book, students will have developed: (1) an understanding of the complexity of the delicately balanced processes that shape the natural world, (2) an understanding of the need to make informed and responsible decisions with regard to the development of the Earth, and (3) an appreciation of the notion of humans as the dominant species and an understanding of the consequences of human-induced changes to the environment.

A NOTE ABOUT THE DIAGRAMS AND DATA

The majority of the diagrams in this book are conceptual figures and data-rich graphs from the scientific literature that have been simplified and redrawn to *tell the story*. The illustrations convey important ideas, such as the relationship between carbon dioxide concentration and global temperature projections or among the amount of land required to feed humanity, population growth, and rates of soil loss. Some illustrations may seem light hearted, but they have all been chosen specifically to either inform or simply inspire. Of particular importance to me is that the graphs are current. In all cases, I have gone back to the original journal articles or relevant reports and redrawn the graphs so that all are standardized and as up-to-date as possible. You may also notice that the data and references are from some of the world's leading scientific journals, predominantly *Science* and *Nature*, as these publish the most current and thoroughly reviewed scientific research. I have also included URLs where appropriate so that you can keep track of the latest data and trends.

ACKNOWLEDGMENTS

I am extremely grateful to the following people, each of whom helped in some way in getting this fifth edition completed. To my colleagues at Texas Christian University for their support and friendship over the years, especially Ken Morgan, Richard Hanson, Nowell Donovan, Helge Alsleben, John Holbrook, Becky Johnson, Leo Newland, Phil Hartman, Ray Drenner, Magnus Rittby, Tory Bennett, Art Busbey, Kristi Argenbright, Stephanie Sunico, and Pete McKone. I thank Lauren Geffert for her keen eye and critical editing, and Paul Gares and Thad Wasklewicz at East Carolina University, for their friendship and asking questions no one else dare ask. My thanks to Amanda Smith and Lara McCombie at Kendall Hunt for being so easy to work with.

I am forever grateful to Jan Wade for being the greatest mom to my son, and to my students who make driving in to school on a Monday morning well worth the effort. To my three special ticos in the cloud forests of Costa Rica, Gustavo Orozco, Ronald Calvo, and, especially, Gustavo Abarca, all extraordinary guides and true friends who inspire me to be more careful with the beauty that surrounds us. I am forever changed by the work and environmental ethic of Dr. Will Fowlds, a recent addition to my life doing extraordinary work on rhino conservation in South Africa. A special thanks to David Hardwick for everything. Finally, to my wife Lauren, you make it all seem possible. I love you T. M. D.

Welcome to the Anthropocene

INTRODUCTION

This is Earth at night lit up by human activity (Figure 1.1). This iconic view of our planet is both mesmerizing and troubling. All of the activity occurs within a few miles of atmosphere and a few meters of biosphere, sitting on top of a few inches of soil. How much activity can this shell sustain before we seriously start disrupting its ecological functioning?

Figure 1.1 This composite image (which has become a popular poster) shows a global view of Earth at night, compiled from over 400 satellite images by NASA scientists. Image credit: NASA/NOAA (www. earthobservatory.nasa.gov).

In 2002, Nobel Prize-winning chemist, Paul Crutzen,[1] published a landmark paper titled "Geology of Mankind–The Anthropocene" in which he proposed a new geological age, one in which humans have been *the* major influence on climate and the environment.[2] In this paper, Crutzen suggests that the Anthropocene started in the latter part of the eighteenth century, essentially at the onset of the Industrial Revolution.[3] Analysis of air from this period trapped in polar ice clearly shows the beginning of growing global concentrations of carbon dioxide and methane. This corresponds with the invention of the steam engine and ever-increasing industrial development of the Industrial Revolution.

While scientists may disagree as to the exact timing of the onset of the Anthropocene, there is little doubt that the most intense human impacts on the environment began around the mid-20th century, which we now refer to as the Great Acceleration (Figure 1.2). Human activity appears large enough to have had a substantial effect on planetary functioning, including climate change, land cover change, overharvesting of resources, and more. The reason we focus on the mid-20th century as a key moment in time is because of a range of socio-economic variables and environmental indicators that began to take off following the redirection of wartime economies to mass production and mass industrialization. The Industrial Revolution then continued to spread outside of its European core to newly independent countries in Asia and Africa.[4]

[1] Paul Crutzen was one of the chemists who shared the 1995 Nobel Prize in Chemistry for his work on ozone depletion.

[2] Source: https://www.nature.com/articles/415023a

[3] The era known as the Industrial Revolution was a period from about 1760 onwards in which fundamental changes occurred in agriculture, textile and metal manufacture, transportation, etc. in England.

[4] See Mali (2017), *Annual Review of Environment and Resources,* Vol. 42, p. 77–104, for an excellent overview of the current debate surrounding the concept of the Anthropocene.

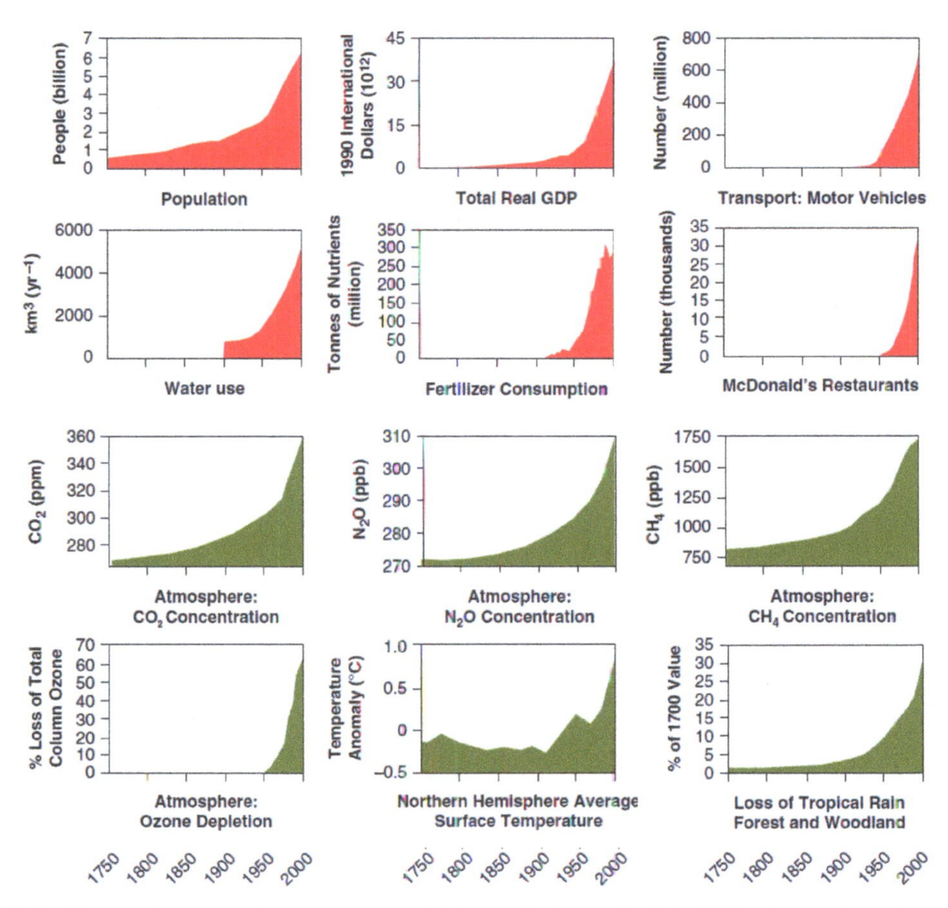

Figure 1.2 The increasing rates of change in human activity since the beginning of the Industrial Revolution and global scale changes in the Earth system as a result of the dramatic increase in human activity. In the diagrams, red represents economic development and consumption indices while green represents ecological impacts. Significant increases in rates of change occur around the 1950s in each case and illustrate how the past 70 years have been a period of dramatic and unprecedented change in human history (Source: Steffen et al. (2011), *Philosophical Transactions of the Royal Society A*, Vol. 369: p. 842–867).

In reality, humans have been altering the planet ever since the dawn of agriculture (Figure 1.3). In fact, some scientists argue that the expansion of farming at about 10,000 years before present, after the end of the last ice age, is key in terms of human impact on the environment. Others suggest the subsequent spread of mining civilizations or the post-1500 conquest of the Americas should be given equal consideration in terms of the onset of widespread environmental damage.[5] Either way, we are in unchartered territory when it comes to human alteration of the environment, especially our impact on the climate and the functioning of the biosphere. What we are experiencing now is not "the norm" or simply an extension of what we have been doing over the last 10,000 years.

[5] Some scientists have proposed that we designate 05:29:45 CDT on July 16, 1945 as the official onset of the Anthropocene, following the first atmospheric detonation of a nuclear device. (Source: Monastersky (2015), *Nature,* doi:10.1038/nature.2015.16739).

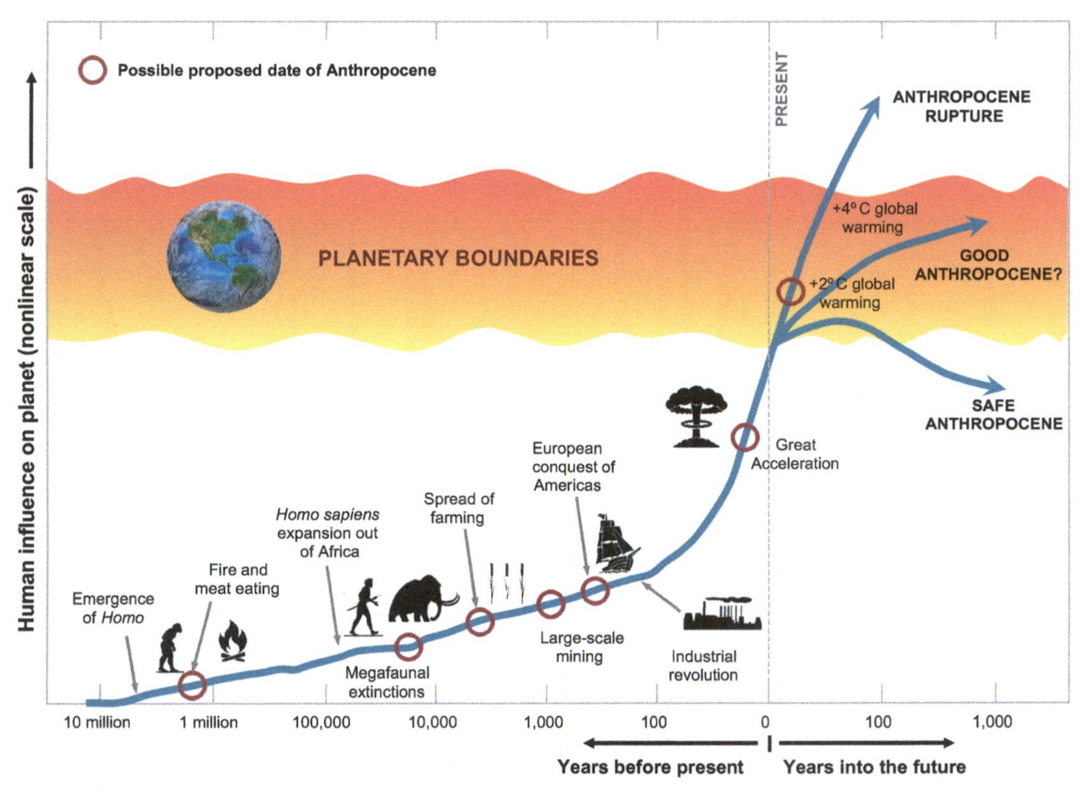

Figure 1.3 A timeline of increasing human influence on the Earth system, with some proposed start dates for the Anthropocene highlighted (Source: Mali (2017), *Annual Review of Environment and* Resources, Vol. 42, p. 77–104).

If we accept that we have entered the Anthropocene, then two of the overarching questions are: What happens now? And can we get out of it? These are very difficult questions to answer. The primary reason is that we have yet to determine with any degree of certainty whether there are fundamental limits to how much disruption human activity can place on our planet before irreversible change occurs (see Box 1.1 for short discussion on scientific uncertainty). If there are such limits and we are able to identify and measure them, then what is our best strategy to stay within these so-called sustainable planetary boundaries?

Take biodiversity, a topic we will explore in detail in Chapter 8. Ecologists argue that we are now firmly in the **sixth mass extinction**, a period when we are losing species at unprecedented rates. Some 25% of all mammals on the planet are under threat from human activity, meaning that they are likely soon to become **endangered species** where extinction is imminent (Figure 1.4). The situation for amphibians is even worse: 41% of those are threatened. Can we afford to lose them? Would the planet continue to function if we did?

BOX 1.1

Scientific Uncertainty and Confidence

Scientists treat uncertainty as a given, a characteristic of all information that must be honestly acknowledged and communicated. Climate Change is probably the best and most highly publicized example of a high-stakes, high-uncertainty problem. Predictions about future warming trends are frequently criticized because they are exactly that, *predictions*, made on imperfect computer models with imperfect data. That should not be surprising. After all, we cannot obtain measurable data about future events. We cannot know for certain what temperatures will be like in 2100 because 2100 will not happen for another 80 years! Instead, we use past information to construct a simulation model that produces data about a hypothesized future. These simulations are only as good as our model assumptions, and our estimates are valid only as long as future conditions are similar to the conditions used to build the model. So, when it comes to deciding how to implement climate policies based on predictions, what consequences are we willing to risk?

Most environmental regulations, particularly those in the U.S., demand certainty. When scientists are pressured to supply this nonexistent commodity, there is not only frustration and poor communication, but also mixed messages in the media. Because of uncertainty, political and economic interest groups often manipulate environmental issues toward their own agendas. No one should blame policymakers for wanting to make unambiguous, defensible decisions. After all, regulations are much easier to write and enforce if they can be stated in absolute, certain terms. However, to use science rationally to make policy decisions, we must deal with uncertainty.

One way scientists deal with uncertainty is to place some level of *confidence* in their data and predictions. The problem is that there is always some sort of mismatch between what scientists say about how certain they are and what the general public thinks an expert means. Here, the term *confidence* has very specific, statistical meaning. For example, when scientists say that it is "extremely likely" that humans are mostly to blame for rising global temperatures during the latter half of the 20th century (Chapter 6), what they really mean is that they are 95% certain of that conclusion. In science, 95% certainty is considered the gold standard because there simply is no such thing as absolute proof. Scientists are "virtually certain" (which equates to a 99% confidence) about many fundamental processes and laws (gravity is a good example), but there are many factors that are not fully understood or under control. Thus, some level of uncertainty will always exist. It is frustrating that there are always people (and special interest groups) who think that when scientists say they are uncertain, that means do nothing. This reasoning is illogical. After all, we are fairly certain that we won't have a car accident every time we drive, yet we still buy car insurance just to be sure. Simply stated, we never should use uncertainty as justification for inaction.

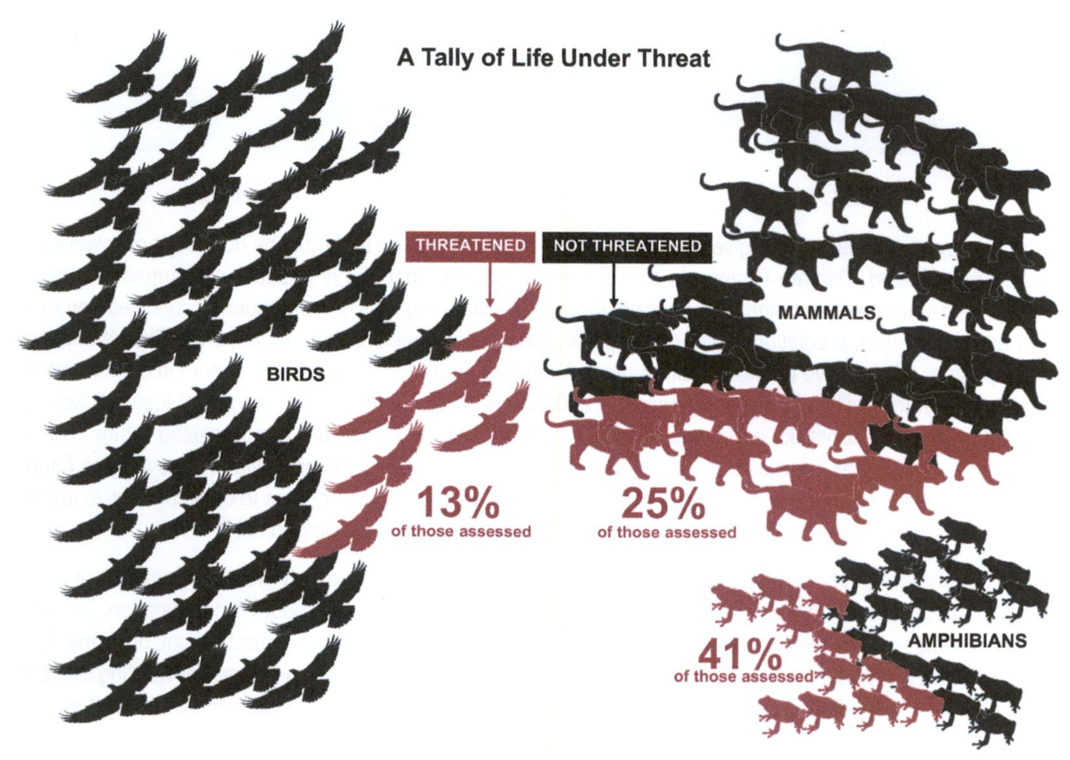

Figure 1.4 The International Union for the Conservation of Nature has evaluated 52,205 species for their ability to survive. 13% of birds, 25% of mammals, and 41% of amphibians around the globe are considered high risks for extinction in the wild. Because most known species of birds, mammals, and amphibians have been evaluated, scientists are confident about the percentage of each group that is threatened (Source: www.iucn.org).

MORALITY AND THE GREAT ACCELERATION

The environmental effects of the Great Acceleration are the focus of this book. These effects are clearly visible at the global scale: higher levels of CO_2 in the atmosphere, rising air and ocean temperatures, rising sea levels, biodiversity loss, soil erosion, increasing consumption of freshwater supplies, and so on. The Great Acceleration is the most profound and rapid shift in the human–environment relationship that the Earth has experienced, and it has happened largely within one human lifetime at a scale and speed that is truly remarkable.

The primary ethical challenge we face in protecting and improving Earth's environmental health is the human tendency to modify land to benefit human development with disregard for the value of nature. Consider, for example, a farmer draining a **wetland** to cultivate more land. This may seem like a small intrusion on nature when compared to the 660,000 hectares of Amazon rainforest cleared for agriculture and cattle ranching in 2017.[6] Yet both choices trade the **intrinsic value** of

[6] This is equivalent in area to 1.23 million football fields.

biodiversity for the financial benefits of exploiting the land's **instrumental value.** These are two very important terms, so let us explore these in a bit more detail.

The field of environmental ethics makes a distinction between instrumental value and intrinsic value. The instrumental value is the value as *means* to further some other ends, whereas the intrinsic value is the inherent value, regardless of whether it also is useful as means to other ends (i.e., it is an end in and of itself). For instance, a certain wild plant may have instrumental value because it provides the ingredients for some medicine. However, some may see the plant simply as beautiful and having value in and of itself, independent of its prospects for furthering medical aid. This is the plant's intrinsic value. Do we not have a moral duty to protect, or at least refrain from damaging, those things that have intrinsic value?

These may seem like lofty philosophical questions, but I want you to think about the question of value as it relates to the environment as you read this book and go about your daily life. In particular, think about where you stand with regard to the environment and some of the choices you make that affect it. Ask yourself why you make the choices you do. Do not feel bad if your answer does not feel quite right. This is not about laying blame or instilling guilt. The truth is we all create change in our environment, both good and bad. Our relationship with the natural world is a dynamic and interactive one–it always has been.

Some scholars and thinkers have argued that the roots of our ecological crisis stem from the perception that nature is an endless resource and that humans have dominion over it. We call this exploitative view of the natural world **anthropocentrism** (or human-centeredness). Whether or not you subscribe to this view, the reality is that our ability to destroy and exploit the environment is higher now than it ever has been in human history. The technological and economic forces that provide valuable material benefits are also producing the very conditions that undermine the planet's ability to sustain life. How, then, do we decide what is worth protecting and what can be exploited?

PUTTING A PRICE ON NATURE AND ITS SERVICES

Think for a moment about tropical rainforests. These are the world's richest biodiversity resource. They also provide important environmental services for the planet. For instance, they function as temperature regulators by absorbing and storing carbon dioxide, helping to mitigate the impacts of a changing climate. Of course, they also generate the oxygen that sustains life. Yet no one pays for these services. They are free.

In 1997, a group of **ecological economists**, led by Robert Costanza at the University of Maryland, argued that if the importance of nature's free benefits could be adequately quantified in economic terms, policy decisions could "better reflect the value of ecosystem services and natural capital".[7] Drawing upon earlier studies that aimed at estimating the value of a wide variety of ecosystem goods and services, from waste assimilation and the renewal of soil fertility to crop pollination

[7] Source: Costanza, R. et al. (1997), *Nature*, Vol. 387: p. 253–260.

and splendid scenery, the research team estimated the economic value of the entire biosphere to be somewhere between $16 and $54 trillion dollars per year at that time. Its average value, according to this group of scientists, is about $33 trillion per year.

This estimate attracted enormous public attention at the time. In feature stories with titles such as "How Much Is Nature Worth? For You, $33 Trillion," and "What Has Mother Nature Done for You Lately?" dozens of newspapers and magazines, including the *New York Times* and *Newsweek*, covered the Costanza study. The details of how the economists and scientists came up with this figure lie well beyond the scope of this book, but understanding the notion of ecosystem services is critically important in the context of conservation and preservation.

The world's economies clearly depend on the ecological life-support systems nature provides. However, we almost never give any weight in policy decisions to such services because we have failed to capture adequately their value *in economic terms*. Of course, actually assigning a dollar value to the services provided by a particular ecosystem is extremely difficult and controversial. For example, look at Figure 1.5. How much is this scenic river in central Costa Rica truly worth? You may be able to estimate the value of the outputs from this ecosystem, such as the fish in the river or the gravel along the banks, or the costs of *replicating* the ecosystem, or, at least, parts of it. Still, how would you price the aesthetic, artistic, educational, spiritual, and scientific benefits people find in natural places as the Sarapiqui River?

Figure 1.5 The Sarapiqui River in Costa Rica (Photo: Mike Slattery).

Some scholars argue that valuation of ecosystems is either impossible or unwise and that we cannot place a value on the intangible attributes such as environmental aesthetics or long-term ecological benefits. Others argue that we have to rethink how we deal with the environment and that, unless we put a price on it, we will not know what is truly worth saving. Either way, the debate serves an important purpose, namely to help us recognize the *free* benefits ecosystems provide and hopefully prompt us to defend these systems from relentless exploitation and destruction.

THE NEED FOR ENVIRONMENTAL STEWARDSHIP

One of the goals in writing this book is to instill in you an appreciation of **environmental stewardship**, which is an increasingly important hallmark of the educated and responsible citizen. The concept arises from the recognition that the dynamics of modern society have placed limits on the growth that our planet can support (those sustainable planetary boundaries referred to earlier). Responsible citizenship requires that we identify these limits and factor them into ethical decision-making by societal leadership.

Merriam-Webster's Dictionary defines a *steward* as "one who acts as a supervisor or administrator, as of finances and property, for another or others." Under this definition, everyone is a steward of Earth. We do not own the environment. No one does. Yet, if we view ourselves as caretakers of the environment, then, surely, it is our responsibility to manage Earth's resources to the best of our abilities, securing their availability for our use and for the use of future generations. Every day, we make countless choices that affect our environment. Thinking of ourselves as a stewards of this environment will help reduce negative impacts, securing a healthier, better world beginning right now and into the future.

What makes a decision informed and responsible? What makes an action right or wrong? Questions such as these are the focus of the discipline known as **environmental ethics**, the branch of philosophy that studies the moral relationship of human beings to the environment. There are many ethical decisions that humans make with respect to the environment. For example:

- Should we continue to clear-cut forests for the sake of human consumption?
- Should we continue to make gasoline-powered vehicles, depleting fossil fuel resources, when the technology exists to create low or zero emission vehicles?
- Should we continue to dam rivers for water supply for non-essential uses, such as lawn irrigation, knowing the detrimental impacts dams can have?
- Is it ethical for humans to cause the extinction of a species for the perceived or real convenience of humanity?

Consider a mining company that has performed open-pit mining in some previously unspoiled area (Figure 1.6). Does the company have a moral obligation to restore the landform and surface ecology? What is the value of a human-restored environment compared with the original natural one? When we begin thinking carefully about environmental ethics, we quickly realize that diverse

Figure 1.6 An open pit mine at Pickering Knob, West Virginia (Photo courtesy of iLovemountains.org).

points of view exist on almost every topic and that the notion of *value* is, ultimately, subjective. This book will present many alternative viewpoints that will challenge you to identify what you truly believe about the natural world and your place in it. I hope that as you read, you will begin to understand the basis and need for making informed and responsible decisions with regard to our dynamic, yet fragile environment. Now, let us get started.

2

Human Population Growth

INTRODUCTION

Humans have been a remarkably successful species (Figure 2.1). For millennia, the human population has been expanding, migrating, consuming, and civilizing. For the past two centuries, we have also been industrializing and developing, and our relationship with the Earth and its ecosystems is changing. One of the world's most influential and recognized scientists, Harvard biologist Edward O. Wilson, sees the 21st century as "the century of the environment," a time when humans will either celebrate and preserve the diversity of life on Earth or simply destroy it. According to Wilson:

> *It's obvious that the key problem facing humanity in the coming century is how to bring a better quality of life—for 8 billion or more people—without wrecking the environment entirely in the attempt.*

Some people believe that there are too many people on Earth, arguing that overpopulation is the primary factor impacting the planet's ability to cope but is that really the case? What does that term "overpopulation" really mean, given that the entire world's population can fit quite

Figure 2.1 Humans have been a remarkably successful species and the global population is projected to surpass 9 billion by the middle of the century (copyright: Steve Greenberg). Steven Greenberg, *Seattle Post-Intelligence*, 1993. Used with permission.

Figure 2.2 These makeshift homes on the bank of a river in Mumbai, India, are typical of the city's slum areas. Note, however, the satellite dish on one of them (center) © iStock.com/Gordon Dixon.

comfortable into Norway, albeit in a Manhattan-like environment? And how much of what we see as environmental stress is a result of a large population versus other lifestyle factors, such as how we choose to live and how we produce, consume, and waste our resources? The global population figure of 7.6 billion,[1] and the literature about so-called overpopulation, inevitably leads us to look to the economically disadvantaged regions of the world where there are generally high populations and plenty of examples of catastrophic environmental degradation. Scenes of starvation in dusty, overgrazed lands, and vast numbers of people living in **acute poverty** among piles of garbage and open sewers seem fairly commonplace in our media (Figure 2.2). Certainly, there are many who are economically disadvantaged, but they tend to consume significantly fewer resources than we do in America. For example, it is estimated that the average American will drain as many resources as 35 Indians and consume 53 times more goods and services than someone from China.[2] Nevertheless, a key question is simply this: Is limiting population growth a key factor, possibly *the* key factor, in protecting the global environment? But before we address that, let us examine some of the principles underlying population growth.

[1] How big is 7 billion? It is hard to make sense of such a giant figure, but 7 billion steps would take you 133 times around the world, for a total of 3.3 million miles!

[2] Source: www.sierraclub.org, 2012.

FUNDAMENTALS OF POPULATION GROWTH

Population growth is the change in a population over time. The term can technically refer to any species, but we almost always use it in reference to humans, specifically to the growth of the population of the world. A population's **rate of natural increase** (r) is the **crude birth rate** (b) minus the **crude death rate** (d):

$$r = \frac{\text{births } (b) - \text{deaths } (d)}{\text{sample size (or population at the start of the year)}} \tag{2.1}$$

where b *is* the number of births in a year (estimated to be 3.85 million in the United States in 2017)[3] and d the number of deaths in a year (2.72 million in the United States in 2017).[4] Thus, the rate of natural increase in the United States is currently around 0.35% (assuming a population of 324.7 million on January 1, 2017).[5] The second half of the 20th century saw great declines in death rates in many countries around the world due primarily to increase in health care (e.g., better nutrition and immunizations) and sanitation (septic infrastructure in urban areas). This reduction of death rates can drastically affect population growth as much as, or even more than, birth rate alone.

When looking specifically at a country, the total population is also affected by immigration (people entering) and emigration (people leaving). Therefore, they must be included when calculating the total **population growth rate** (pgr), as follows:

$$pgr = \frac{[(\text{births} - \text{deaths}) + (\text{immigration} - \text{emigration})]}{\text{sample size}} \tag{2.2}$$

For example, from 2011–2015, the U.S. population increased naturally (i.e., births minus deaths) by 6.8 million people; however, an additional 3.34 million people immigrated during this time (including legal and non-legal immigration), meaning that almost one third of the growth in the U.S. population was due to people entering the country from another country, currently, a controversial topic (Figure 2.3).[6] However, applying the crude death rate to a whole population can be misleading. For example, the number of deaths can be higher for developed nations than in less developed countries, despite standards of health being better in developed countries. This is the result of the **age structure** of a population, a topic we will examine toward the end of this chapter.

The **doubling time** (Td) is the period of time required for a population to double in size. When the growth rate is constant, the quantity undergoes **exponential** (or geometric) **growth** and has a constant doubling time that can be calculated directly from the growth rate by dividing 70 by the percentage growth rate:

$$Td = \frac{70}{r} \tag{2.3}$$

[3] Source: National Center for Health Statistics
[4] Source: Centers for Disease Control and Prevention
[5] Source: https://www.census.gov/popclock/
[6] Source: www.census.gov and migrationpolicy.org

Figure 2.3 Immigration around the world has become an extremely controversial topic, from migration into Western Europe to President Trump's famous wall along the Mexico-U.S. border. © Shutterstock.com.

For example, given a population growth rate of 0.35% in the United States in early 2018, the population should double in about 200 years (i.e., 70/0.35). Thus, if the growth rate in the United States were to remain constant, the population will double from its current roughly 328.5 million to about 657 million by 2218, with the emphasis on *if the growth rate remains constant.* Examining the doubling time can give a more intuitive sense of the long-term impact of growth than simply viewing the percentage growth rate (Figure 2.4). For example, with an annual growth rate of 3.8% (which does not really seem that high), the doubling time for a population is just 18.4 years. A doubling time of 70 years corresponds to a growth rate of 1%.

The numerator in the equation (2.3) is taken from the so-called **rule of 70**, a useful rule of thumb that explains the time periods involved in exponential growth at a constant rate. Exponential growth occurs when the growth rate of a function is always proportional to the function's current size. This implies that the larger the quantity gets, the faster it grows. Anything that grows by the same percentage every year is growing exponentially. If this still seems confusing, a well-known story, said to have originated in Persia, may help. The story tells of a courtier who presented a rather beautiful chess set to the king. In return, all the courtier asked is that the king give him one grain of rice for the first square, two grains for the second square, four grains (or double again) for the third, and so on. The king must have been somewhat mathematically challenged for he agreed. The result? The eighth square required 128 grains while the twelfth took more than one pound. Eventually, every grain of rice in the kingdom had been used. Even today, the total world rice production would not be enough to meet the amount required for the final square of the chessboard!

The *simple* exponential model described here is sometimes called the **Malthusian growth model**, named after the Reverend Thomas Malthus (1766–1834), who authored *An Essay on the Principle*

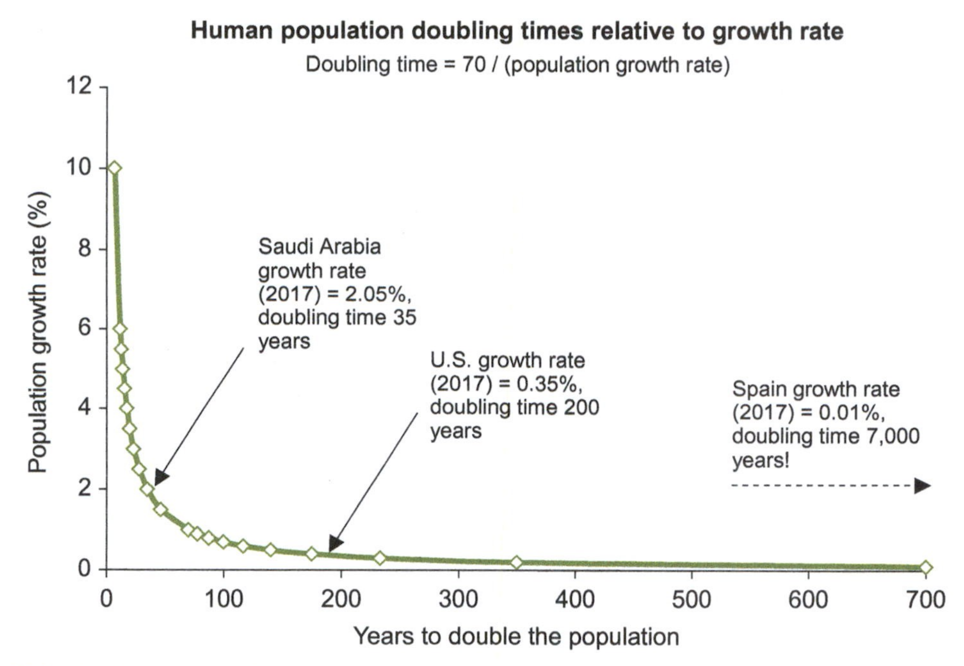

Figure 2.4 Doubling times relative to the rate of natural increase.

of Population, one of the earliest and most influential books on population. The Malthus principle of population was based on the idea that population, if unchecked, increases at an exponential rate (i.e., 1, 2, 4, 8, 16, etc.), whereas the food supply grows at an **arithmetic** (or linear) rate (i.e., 1, 2, 3, 4, etc., as shown in Figure 2.5). Malthus hypothesized that the planet would ultimately return to subsistence-level conditions as a result of agricultural or economic production eventually being outstripped by growth in population, the so-called Malthusian catastrophe.

Nothing can grow at a constant rate indefinitely. Even under optimum environmental conditions, an organism's **biotic potential** will be restricted by **environmental resistance** (Figure 2.6). These regulating factors may be decreasing oxygen supply, low food supply, disease, predators, and limited space, among others. When environmental resistance is paired with biotic potential, a population will reach its **carrying capacity**, and the growth curve will follow an S-shape (known as a logistic curve), as shown in Figure 2.6. Here, the initial stage of growth is approximately exponential; then, as environmental resistance increases, the growth slows and eventually stops.

Carrying capacity refers to the level of population that can be supported for organisms in an environment, given the amount of food, habitat, water, and so on present. It is the number of individuals an environment can support without significant negative impacts to the organism and its environment. It is possible for a species to exceed its carrying capacity temporarily, referred to as **overshoot**, until mass fatalities occur as shortages in water, food, and other resources, take effect. This dieback may be a relatively gradual correction back toward the carrying capacity (Figure 2.7) and can also be catastrophic. A catastrophic dieback is more devastating for a population because

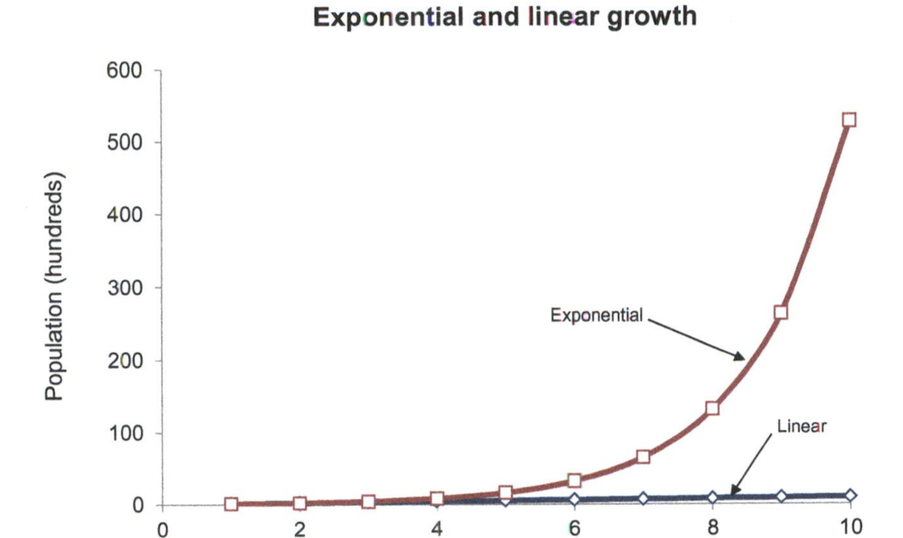

Figure 2.5 Exponential (i.e., geometric) and linear growth curves. Note how exponential growth "gains speed" and how unstoppable it is, once it gains momentum.

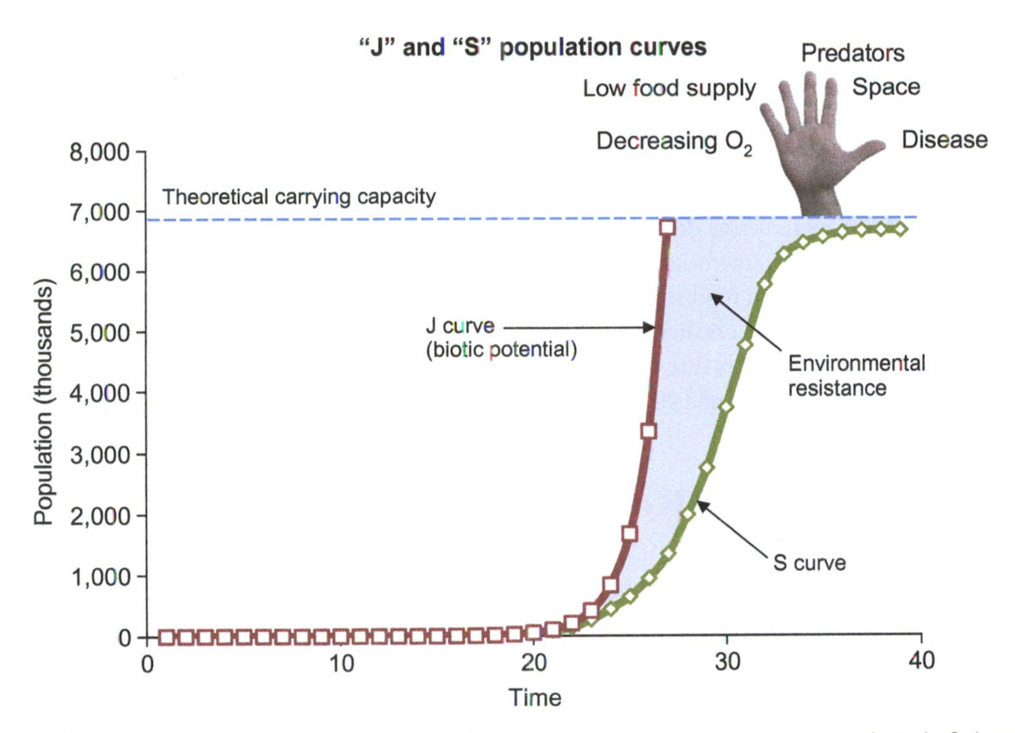

Figure 2.6 J and S population curves. Environmental resistance causes a population's growth curve to approximate the S-shape as numbers approach the carrying capacity.

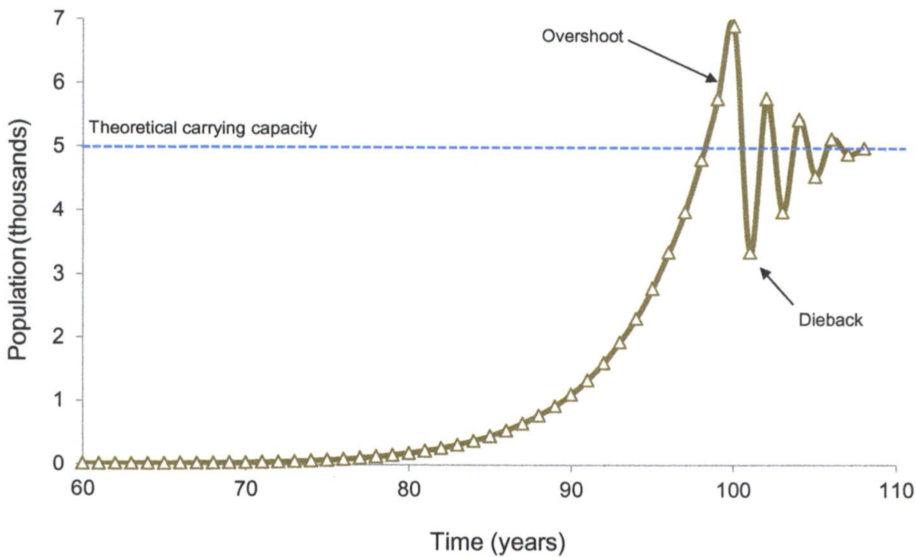

Figure 2.7 Population overshoot and dieback, leading to oscillations around the carrying capacity.

it results in mass deaths and stresses for the entire species. The population of the species after the dieback generally falls far below the carrying capacity in an overcorrection. Such was the case on Isle Royale, the largest island in Lake Superior and a U.S. National Park in Michigan. The island, just over 45 miles long and 9 miles wide at its widest point, is well known for its moose and wolf populations and is among the world's best studied predator–prey relationships (see http://www.isleroyalewolf.org/). Moose have only been present on the island since about 1900 and wolves since about 1950. The first five decades of moose occupation on Isle Royale, which occurred in the absence of wolves, were characterized by significant fluctuations. The absence of predation allowed population explosions, with food shortages caused by unchecked moose herbivory and catastrophic winters driving population crashes. Now, the two species appear to be in balance (Figure 2.8), punctuated at times by crashes due to other factors (e.g., the wolf crash of 1980–1982 to infectious disease and the moose die off of 1995 to severe winter).

HUMAN POPULATIONS

The Current Situation

In 2017, our global population surpassed 7.5 billion, and this figure continues to grow at rates that were unprecedented prior to the 20th century. Based on the data shown in Figure 2.9, you can see that the world's population has been growing exponentially since the industrial revolution. Historical population figures, in terms of when each billion milestone was met, are given in Table 2.1. These numbers show that the world's population has tripled in 72 years and doubled in 38 years up to the

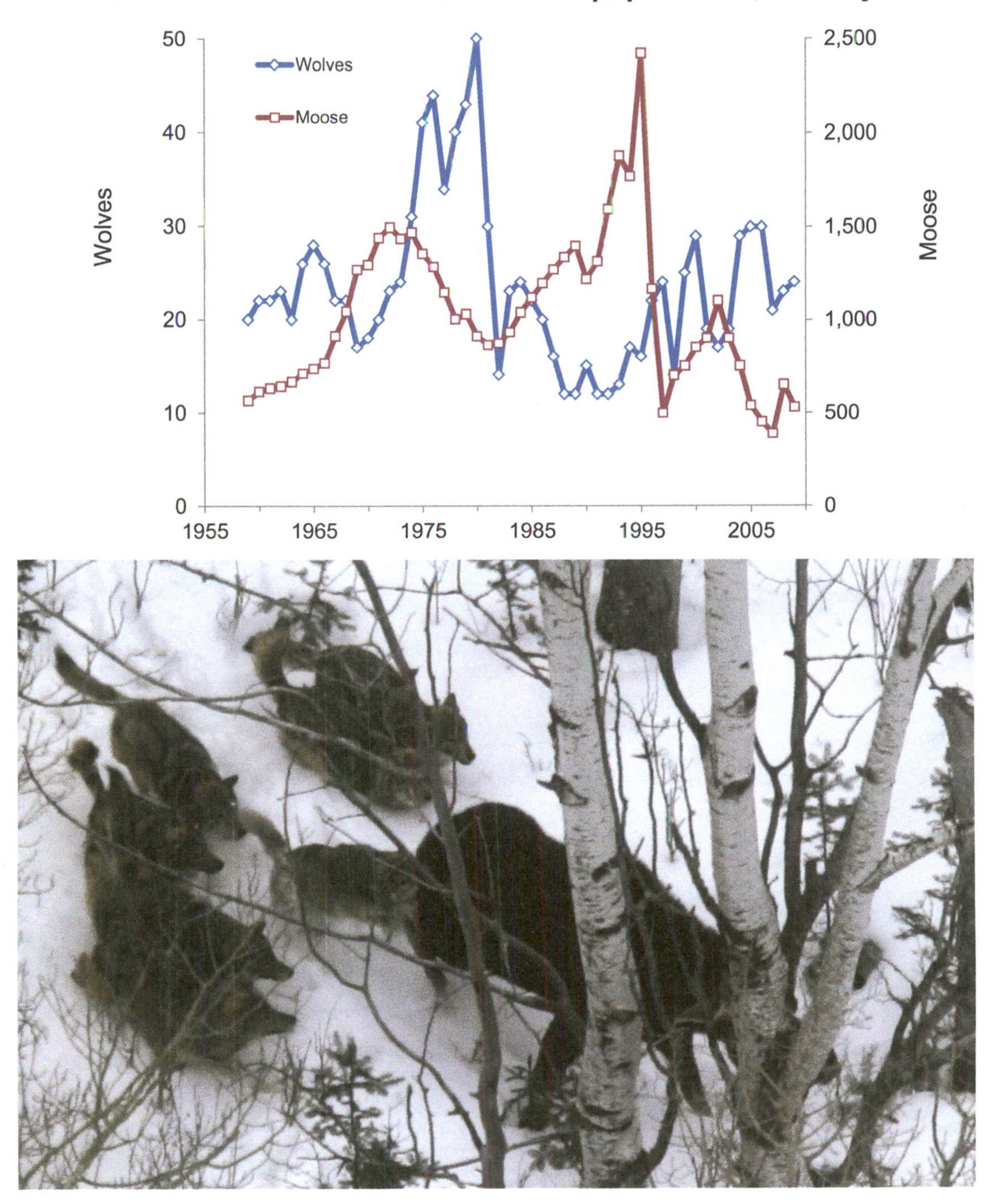

Figure 2.8 Interaction between moose and wolf populations on Isle Royale National Park, 1959–2009 (Source: http://www.isleroyale-wolf.org/). The photograph shows a rare observation of a common occurrence, wolves attacking and killing a bull moose (with permission, John and Rolf Vucetich).

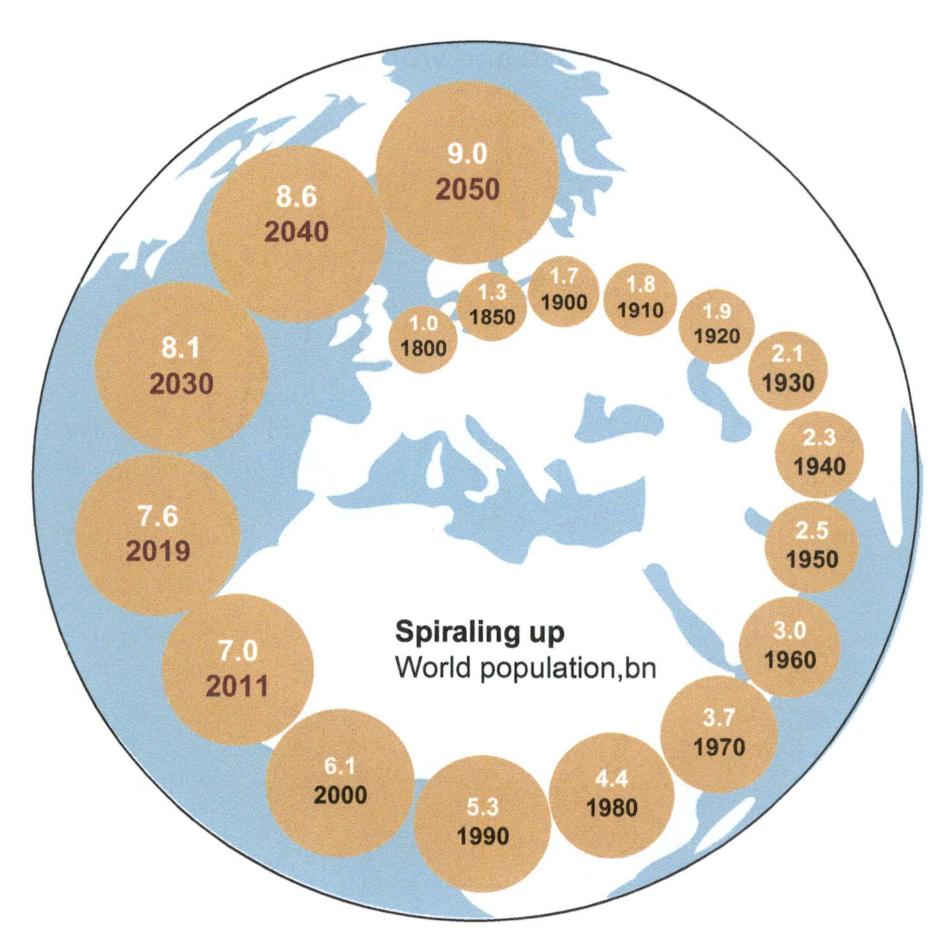

Figure 2.9 The human population spiral showing projected growth to nine billion-plus by 2050 (Source: UN and World Bank data).

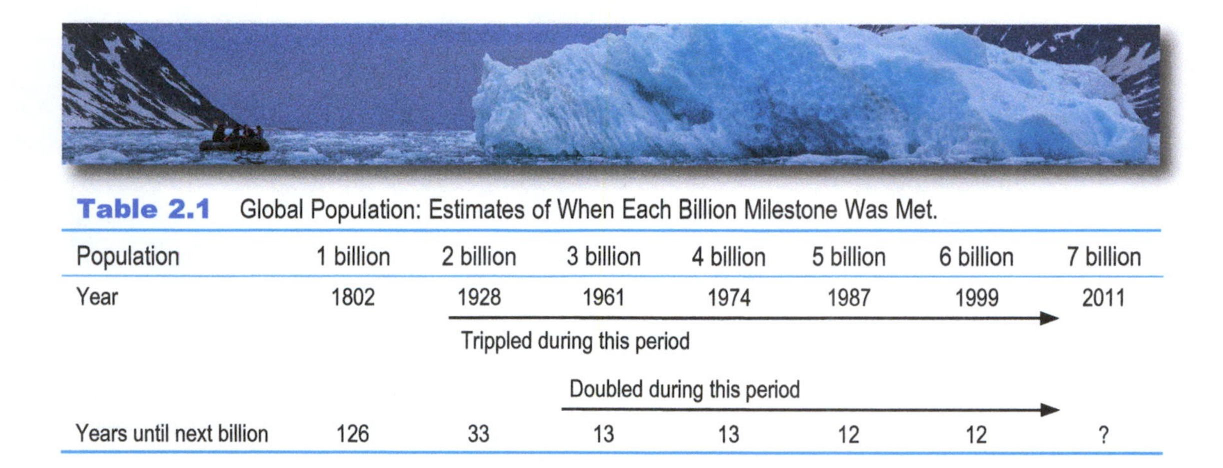

Table 2.1 Global Population: Estimates of When Each Billion Milestone Was Met.

Population	1 billion	2 billion	3 billion	4 billion	5 billion	6 billion	7 billion
Year	1802	1928	1961	1974	1987	1999	2011
		Trippled during this period (1928 → 2011)					
			Doubled during this period (1961 → 2011)				
Years until next billion	126	33	13	13	12	12	?

year 1999. The 20th century has clearly seen the biggest increase in global population in human history. An interesting way to think of this is as follows: anyone who died before approximately 1930 never lived through a doubling of the world's population; anyone who dies after 2050 almost certainly will not have either. At face value, these numbers seem alarming and give the impression of an exploding, out-of-control population. But is such growth likely to continue indefinitely?

The United Nations and The World Bank estimate that from 2008 to 2012, the world's population grew at the rate of 1.19% (or about 78–79 million people) per year. According to the rule of 70 that would mean a doubling of the global population in 59 years (i.e., we are marching toward 14+ billion people by the middle of 2072). That is an extraordinary projection to make but one that is not very realistic. A simple projection like this makes one enormous assumption: that the growth rate of 1.19% used in the calculation remains constant during the growth period, in this case, through to 2072. The reality is, the global growth rate (i.e., *rni*) has been steadily *decreasing* from its peak of 2.2% in 1963 to 1.09 in 2018 (see Figure 2.10), and will probably continue to do so. Remember, however, that any growth rate above zero means that a population will continue to grow, albeit more slowly as that growth rate approaches zero. It is also important to remember that different regions have vastly differing rates of population growth. In sub-Saharan Africa, growth rates remain well above 2% per year, with Malawi, Angola, and South Sudan all above 3%. These high regional growth rates are largely because birth rates have remained high. The **total fertility rate** (*tfr*), or the number of children per woman of child bearing age (defined as 15 to 44) in sub-Saharan Africa is currently 4.85 with countries like Chad, Mali, and Somalia all above 6, extraordinary by

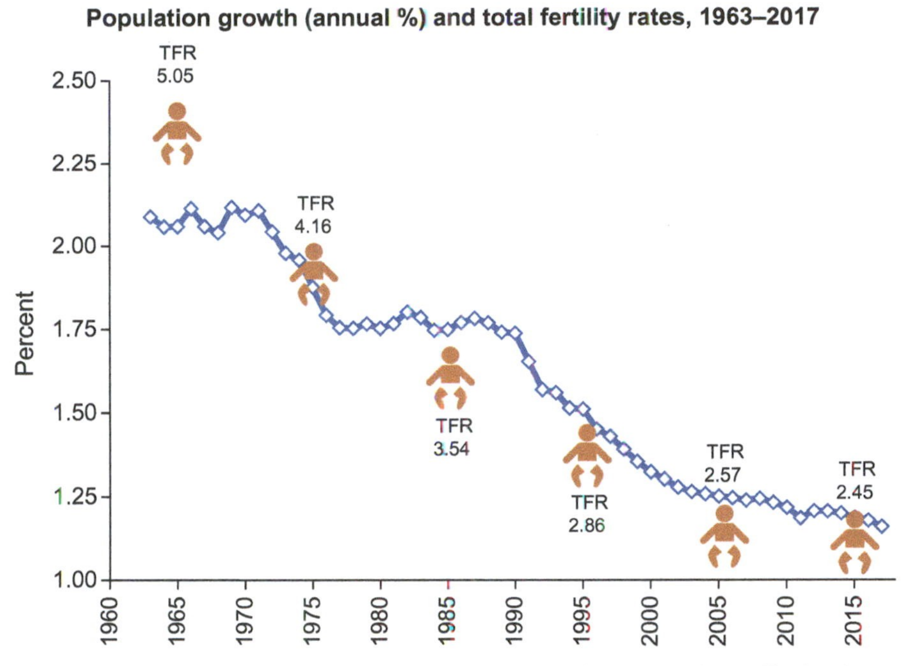

Figure 2.10 Global population growth and total fertility rates from 1963 to 2017 (Source: www.data.worldbank.org).

world standards.[7] Note that for a population to simply replace itself (i.e., neither grow nor shrink), *tfr* has to equal 2.1, which is also known as a population's **replacement fertility rate**. The 2.1 children per woman includes 2 children to replace the parents, with the extra one-tenth making up for the difference in sex ratio at birth, those who choose not to (or cannot) have children, as well as **infant mortality rate** (*imr*), defined as the number of number of children dying under one year of age divided by the number of live births that year. As shown in Figure 2.10, current global *tfr is* around 2.45, meaning women around the world are having 2.45 children on average.[8]

Some countries are currently experiencing negative population growth (i.e., a net decrease in population over time), especially in Europe where fertility rates remain low. Spain, Greece, Portugal, and Italy, for example, all have fertility rates of around 1.3. South Korea leads the way in terms of replacement fertility with a rate of just 1.05, exactly half the replacement rate[9] (as a frame of reference, the *tfr* in the United States in 2018 was around 1.75). Overall, below replacement fertility is expected in 75% of developing countries by the year 2050, according to the UN.

Forecasting World Population

The key question in all of this discussion is what is the global population likely to do over the next 50–100 years? For if we can get a better handle on how many people are likely to be on this planet by the middle of this century, then surely we will be better able to plan and manage resources more wisely, thereby minimizing the damage to the environment.

In the long run, the future population growth of the world is difficult to predict, although we do know that exponential growth of a population cannot continue indefinitely. Further, unfettered population growth toward 14 billion-plus people seems unlikely, given the quite dramatic declines in global fertility levels over the past decade, as noted above. The UN, which provides comprehensive reviews of past worldwide demographic trends and future projections in their publication *World Population Prospects*,[10] estimates that the world population will likely surpass 9 billion people by 2050 and possibly exceed 10 billion in 2100 (Figure 2.11). That seems fairly positive news, but we must remember that an increase in 2.5 billion people is equivalent to the size of the world's entire population in 1950! In addition, most of the added 2.5 or so billion people from now to 2100 will primarily be in developing countries, projected to rise to 8.0 billion in 2050 and to 8.8 billion in 2100. On the other hand, the population of the world's more developed regions is expected to change minimally by 2100 (Figure 2.11) and would have actually declined to around 1.1 billion were it not for the projected net migration from developing to developed countries.

As seen in Table 2.2, most of the global population growth will occur in the less developed regions such that, by 2100, over 80% of the world population will be in Africa and Asia alone. Europe, once the second most populous region in the world, will account for less than 7% of the world total by 2100. In fact, by 2050, just seven countries will account for 50% of the world's population

[7] Source: https://data.worldbank.org
[8] Source: https://www.cia.gov/library/publications/the-world-factbook/geos/xx.html
[9] There is an excellent BBC podcast on this at https://www.bbc.co.uk/programmes/w3cswf58
[10] http://esa.un.org/wpp/Documentation/publications.htm

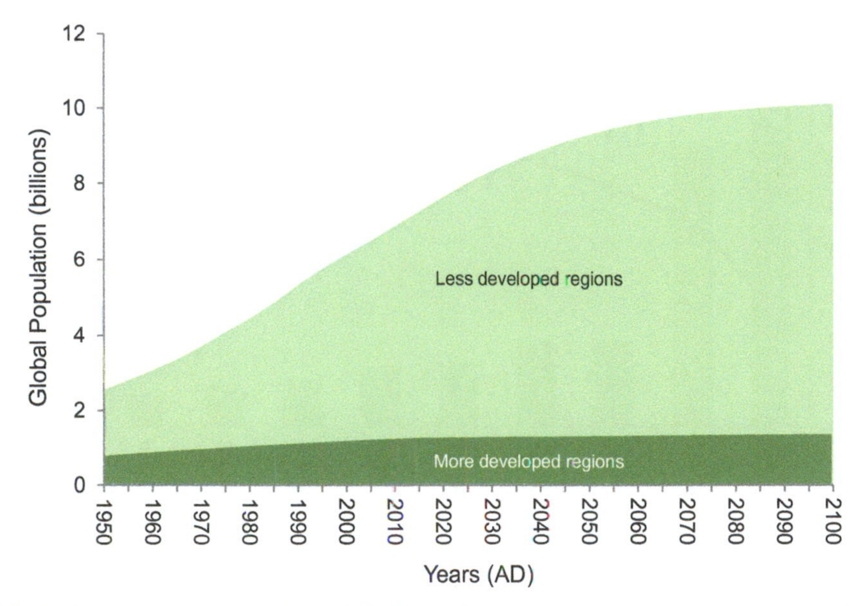

Figure 2.11 World population growth from 1950 to 2100, showing the scale of future growth in less developed countries (Source: http://esa.un.org/unpd/wpp/Excel-Data/population.htm).

Table 2.2 Distribution of the World Population by Development Group and Major Area, 1950, 2010, 2050, and 2100 (percentage)

	Population (%)			
Major area	1950	2010	2050	2100
More developed regions	32.0	17.8	14.1	13.2
Less developed regions	68.0	82.2	85.9	86.8
Africa	9.1	15.0	23.6	35.3
Asia	55.4	60.3	55.3	45.4
Europe	21.6	10.6	7.7	6.7
Latin America and the Caribbean	6.6	8.6	8.1	6.8
Northern America	6.8	5.0	4.8	5.2
Oceania	0.5	0.5	0.6	0.7

Source: World *Population Prospects: The 2010 Revision. Highlights.* New York: United Nations.

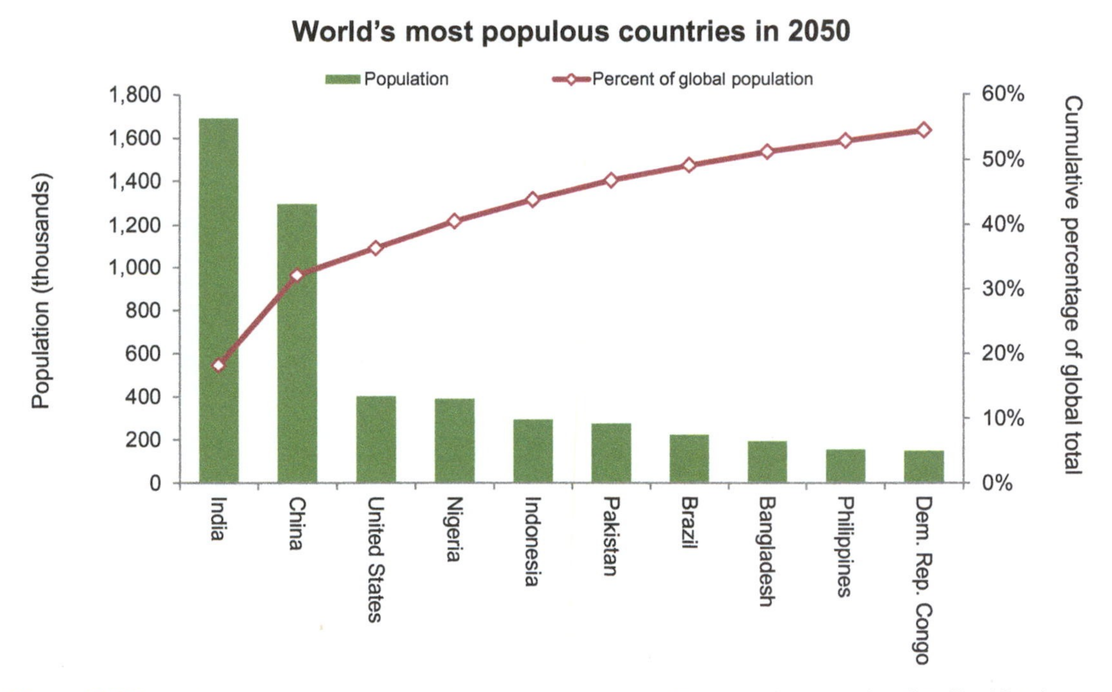

Figure 2.12 The top ten countries in terms of total and cumulative population (Source: http://esa.un.org/unpd/wpp/Excel-Data/population.htm).

(Figure 2.12), with India and China accounting for one-third (to get a sense of the scale of the population in East Asia, South Asia, and Southeast Asia, turn your attention to Figure 2.13). The average annual growth rate of the global population during this period is projected to be 0.76% (Table 2.3) with the least developed countries remaining high at 1.82%. Many in the group of least developed countries still have relatively youthful populations that are expected to age only

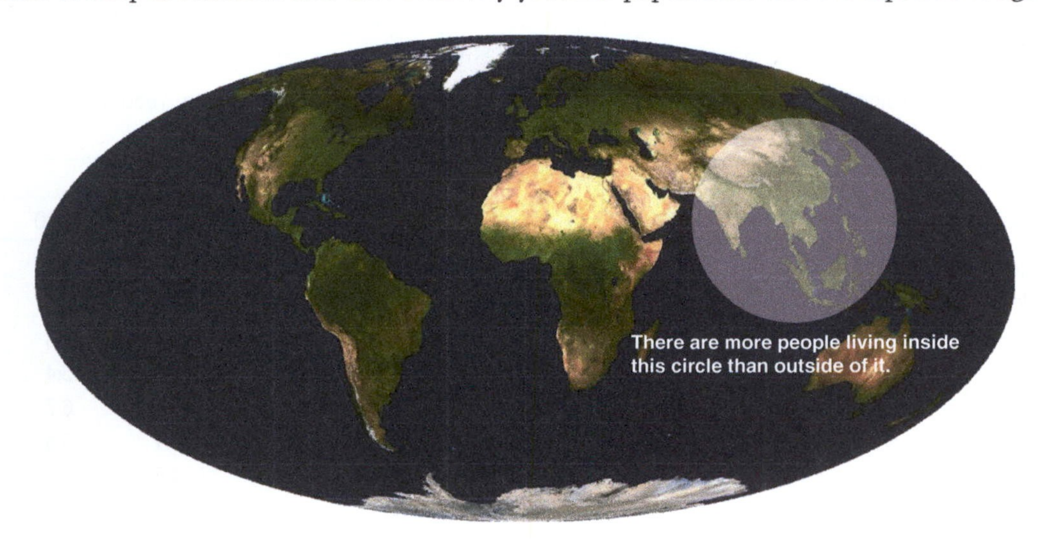

Figure 2.13 A map of the world showing where over half of the global population lives.

Table 2.3 Average Annual Growth Rates of the Total Population and the Population in Broad Age Groups by Major Area, 2005–2050 (percentage)

Major area	Total population (%)
World	0.76
More developed regions	0.05
Less developed regions	0.90
Africa	1.72
Asia	0.65
Europe	−0.21
Latin America and the Caribbean	0.71
Northern America	0.65
Oceania	0.84

Source: *World Population Prospects: The 2010 Revision. Highlights.* New York: United Nations.

moderately over the foreseeable future. Among the rest of the developing countries, rapid population aging is forecast, with the current youth choosing to have less children when they reach adulthood.

Note that in order to project population until 2050, the United Nations Population Division applies assumptions regarding future trends in fertility, mortality, and migration. Because future trends cannot be known with certainty, a number of projection variants are produced. For the population projection shown here, total fertility in all countries is assumed to converge eventually toward a level of 1.85 children per woman, although not all countries reach this level during the projection period, that is, by 2045–2050.

Future population growth will be highly dependent on the path that future fertility takes. In the United Nations projections described above, global fertility is assumed to decline from 2.5 children per woman (in 2010) to 2.17 children per woman in 2050 and 2.02 by 2100 (i.e., below replacement). In developed regions, *tfr*'s below 2.1 are now common, as noted earlier, and this decline is certainly expected to continue to 2050. In many developing countries, fertility has declined markedly since the late 1960s and is also expected to reach below replacement levels by 2050 in the majority of these countries. For example, although its population continues to grow, China is already below replacement fertility and has been for nearly 20 years, thanks in part to the highly controversial (though largely successful) one-child policy implemented in 1979. Chinese women, who were giving birth to an average of six children as recently as 1965, are now having around 1.5. In Iran, with the support of the Islamic regime, fertility has fallen more than 70% since the early 1980s. However, if fertility were to remain just half a child above the levels projected by the UN (i.e., 2.67 versus 2.17 in 2050), world population would reach 10.6 billion by 2050 and 15.8 billion by 2100, and the consequences for our planet would undoubtedly be disastrous.

A fertility path half a child below the medium would lead to a population of 8.1 billion by mid-century and 6.2 billion by the end of the century. This means that continued population growth until 2050 is inevitable even if the decline of fertility accelerates.

Finally, the age structure of a population is critical when trying to make projections of global population growth. In theory, when the *tfr* of a country reaches 2.1, **zero population growth** (*zpg*) occurs because the population is replacing itself. However, a fertility rate of 2.1 may not guarantee zpg. If at any time, a population has an unusually large number of children, they will, as they move into their childbearing years, increase the *r* of the population even if their *tfr* stays below two. Most childbearing is done by women between the ages of 15 and 49, so if a population has a large number of young people just entering their reproductive years, the rate of growth of that population is most likely to rise. Compare, for example, the so-called age structure diagrams (sometimes referred to as population pyramids) of the populations of the United States and India (Figure 2.14). These diagrams show the number of males and females in age clusters in 2010 and projected to 2050. In 2010, almost 30% of India's population were children (aged 15 years or less) who are yet to begin reproduction. When a large group like this begins reproducing, they greatly increase birth rates. Compare that to the United States where each group is about the size of the next until close to the top. Broad-based pyramids like India's are characteristic of populations with high birth rates, lower life expectancies, and reduced infant and child mortaility due to greater access to better healthcare. Nevertheless, the populations of many developing countries, like India, are poised to enter a period of rapid population aging, as is evident in the "bulge" in the 35–49-year-old age group by 2050.

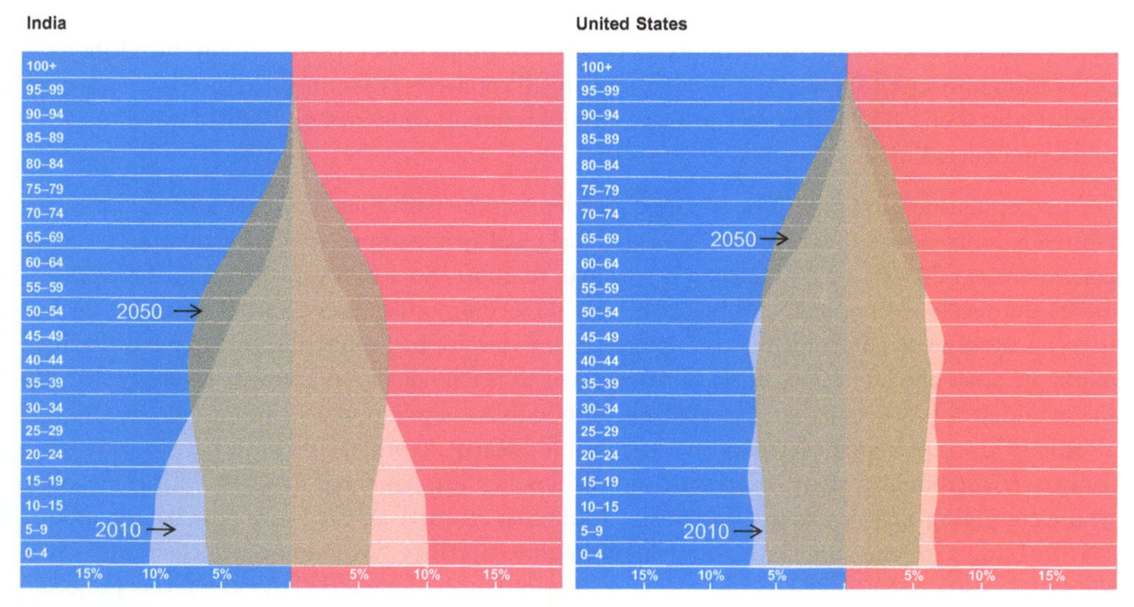

Figure 2.14 Population pyramids (or age structure diagrams) for India and the United States for 2010 and projected to 2050 (Source: http://esa.un.org/unpd/wpp/Excel-Data/population.htm).

THE CHALLENGE OF FEEDING 9 BILLION PEOPLE

From the preceding discussion, it seems the news is generally good when it comes to global population projections. Fertility rates are declining, not everywhere but in most geographic regions, a combination of many factors: falling teen pregnancy; women waiting longer to have children; children living longer (i.e., lower infant mortality); developing economies (with fewer children needed for work); access to family planning; women joining the workforce, and so on. However, a crisis still looms. According to both the World Bank and the Food and Agriculture Organization (FAO) of the United Nations, food production will need to *double* by 2050 to keep up with the projected demands of a growing population. On the plus side, we have seen a marked growth in food production over the past 50 years, dramatically reducing the number of the world's people that are hungry. Nevertheless, estimates suggest now that more than one in seven people still lack access to food or are chronically **malnourished**, stemming from continued **poverty**, and rising food prices (Figure 2.15). Even if food prices stabilize, it is likely that far more crop production will be required to guarantee future **food security** for all people. However, it is not simply a question of more mouths to feed. Slowing population growth is frequently correlated with increased wealth and with higher purchasing power comes higher consumption and a greater demand for processed food, dairy, meat, and so on. All of this

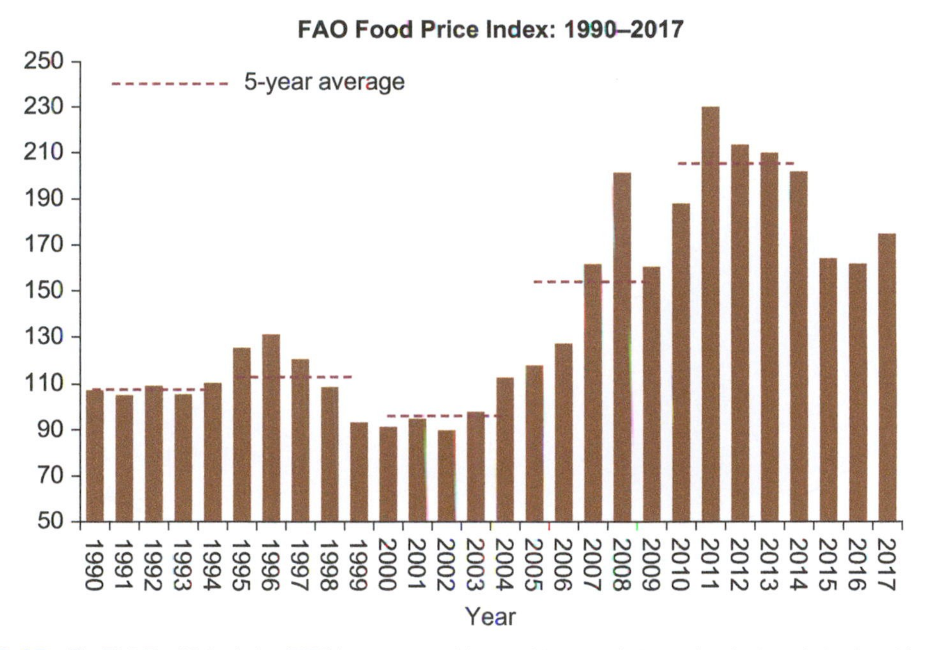

Figure 2.15 The FAO Food Price Index (FFPI) is a measure of the monthly change in international prices of a basket of food commodities. The FFPI averaged 228 points in 2011, 23% (42 points) more than in 2010, exceeding the previous high of 200 points in 2008 and the highest level since FAO started measuring international food prices beginning in 1990. Prices have eased since then (Source: http://www.fao.org/worldfoodsituation/wfs-home/foodpricesindex/en/).

adds pressure to the food supply system.[11] The problem is that agriculture must also address tremendous environmental concerns. Agriculture is now a major force behind many environmental threats, including climate change (Chapter 6), biodiversity loss (Chapters 7 and 8), and degradation of land and freshwater (Chapters 9 and 10). Thus, we face one of the greatest challenges of the 21st century: meeting society's growing food needs while simultaneously reducing agriculture's harm to the environment.

The State of Global Agriculture

Croplands currently cover about 1.5 billion hectares[12] (or 12%) of Earth's ice-free land, while pastures cover another 3.4 billion hectares (about 26%), as shown in Figure 2.16. In total, agriculture occupies almost 40% of Earth's terrestrial surface and is the largest use of land on the planet.[13] A new FAO assessment of land resources and global climates suggests that a further 2.8 billion hectares are possibly suitable for further crop production. This is almost twice as much as the current area and seems like an easy solution: plow more land under and into production! However, much of this potential land is, in effect, unavailable or already tied up in other valuable uses. For example,

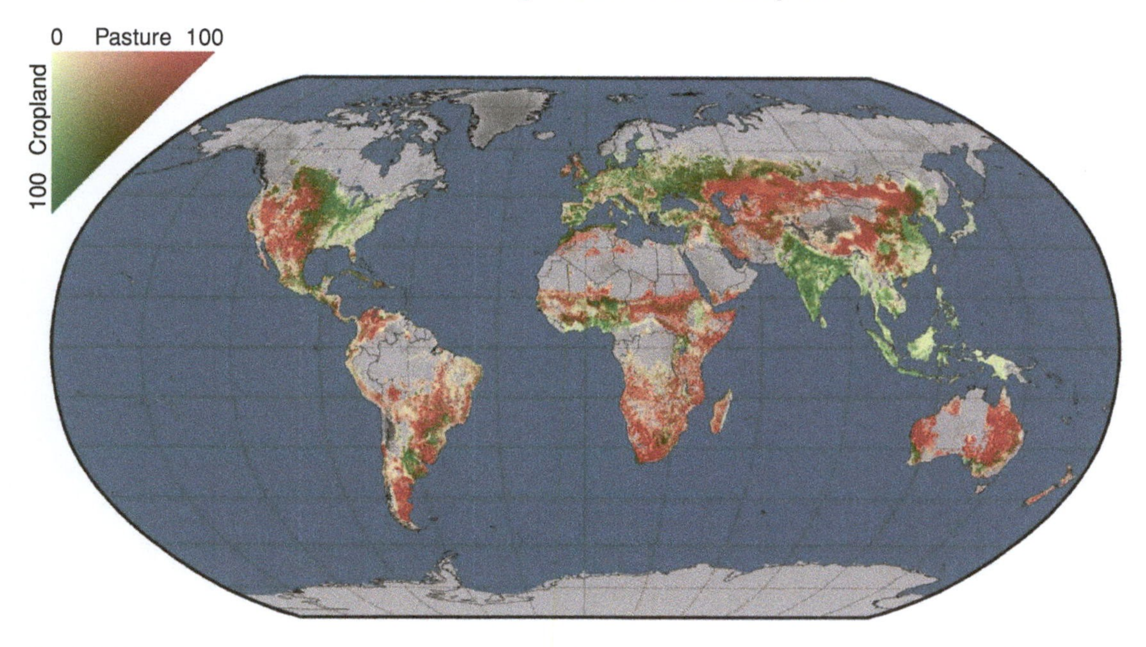

Figure 2.16 Global distribution of cropland and pasture (Source: Ramankutty et al. (2008), *Global Biogeochem. Cycles*, 22, GB1003, doi:10.1029/2007GB002952).

[11] See http://royalsociety.org/Reapingthebenefits and World Bank *World Development Report 2008: Agriculture for Development*

[12] A hectare is equal to 2.47 acres, about two football fields minus the end zones.

[13] Source: FAO (see http://faostat.fao.org/site/567/default.aspx#ancor)

approximately 45% is covered in forests, 12% sits in protected areas, and 3% is taken up by human settlements and infrastructure. More problematic is the fact that much of the potential land reserve has characteristics that make agriculture difficult, such as low soil fertility, soil toxicity, and difficult (often hilly) terrain.

Global food production has increased significantly in recent decades (Figure 2.17). Studies of widely used crops, such as cereals, suggest that crop production increased by almost 30% between 1985 and 2005.[14] This gain in production occurred as cropland area increased by just 3% (suggesting an increased yield greater than 25%). That's obviously great news in terms of food supply. But we allocate vast quantities of food to nonfood uses, such as animal feed, and this affects the amount of food actually available to the world. Globally, humans consume 62% of all crops produced (by mass), and 35% goes to animal feed. Of course, this does produce human food indirectly (through the consumption of meat and dairy products), but it does so much less efficiently. Indeed, the amount of land set aside for animal-based agriculture merits some very careful thinking, especially given the dual challenges of feeding a growing world while charting a more environmentally sustainable path for agriculture. This is illustrated through an info-graphic (see Figure 2.18). About 30% of our agricultural land (remember, 1.5 billion hectares) is under cropland; 70% (or 3.4 billion hectares) is used for grazing. We can then measure the crops and pasture grasses produced on agricultural land as the energy (originally from the sun) that is in the biomass that we (or our animals) grow or eat. So crops and pasture are used to produce

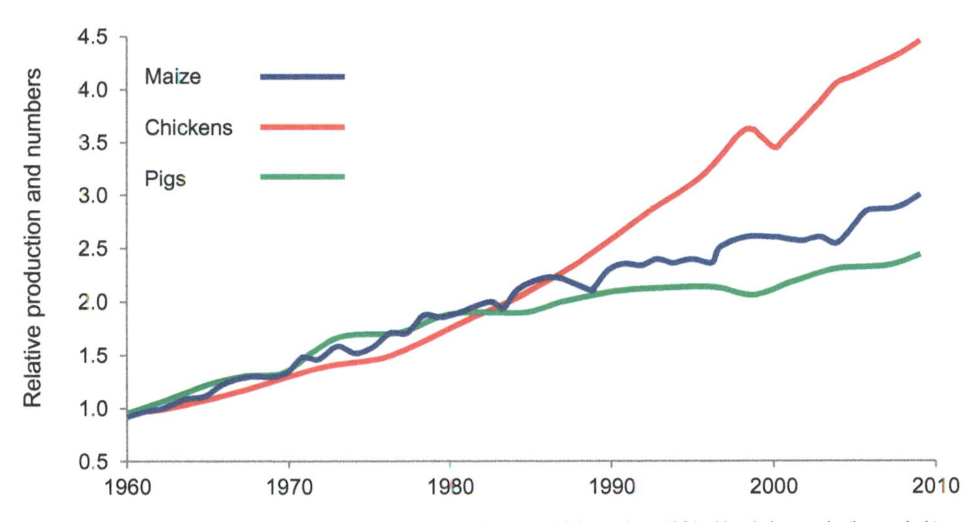

Figure 2.17 Changes in the relative global production of maize, pigs, and chickens since 1961 with relative production scaled to one equal unit in 1961 (Source: Godfray et al. (2010), *Science*, Vol. 327, p. 812–818).

[14] Foley et al. (2011), *Nature*, Vol. 478, p. 337–342.

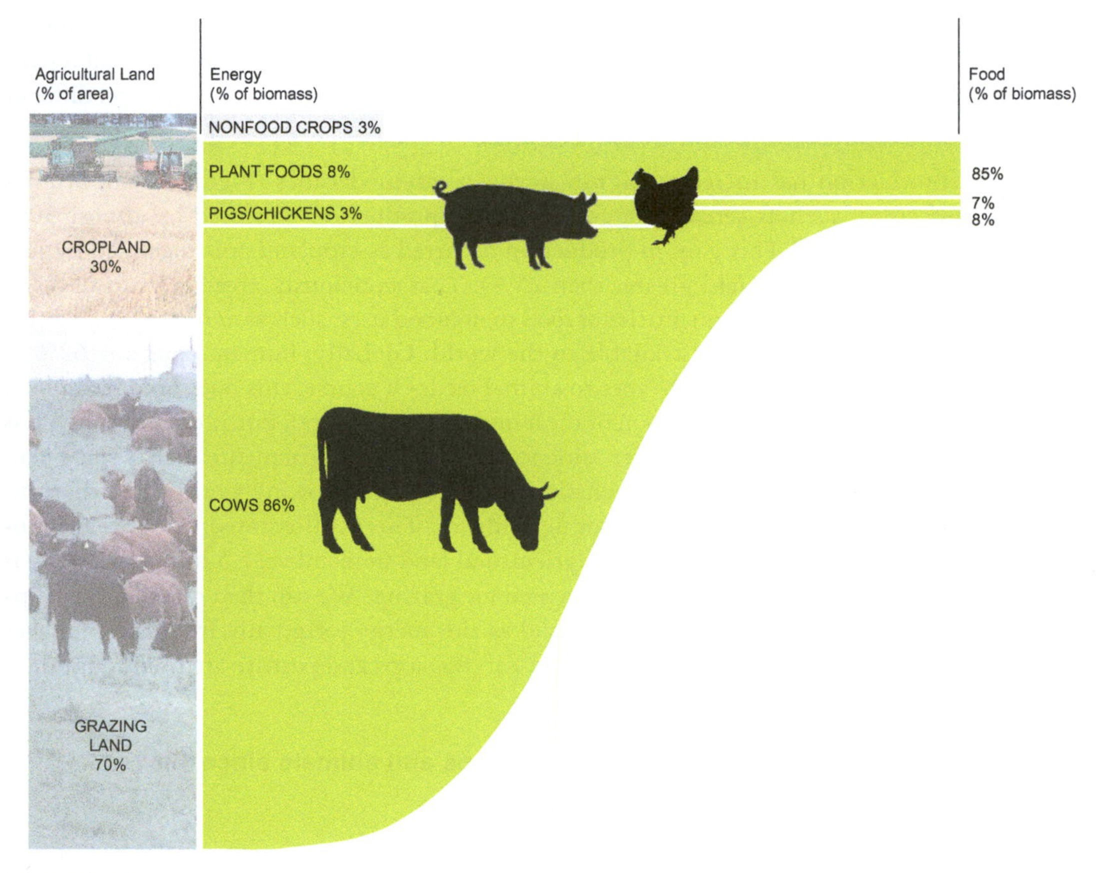

Figure 2.18 Cows use enormous amounts of biomass as well as the large majority of the planet's agricultural land, yet produce a very small amount of our food, about 8% of the total. This shows just how inefficient this part of the human diet is (Source of data: Union of Concerned Scientists and data from Smith et al. 2013; source of images: Shutterstock.com).

cattle, but crops like corn are also grown to feed people directly. We can then express all of the energy involved in these processes in billions of tons of biomass. What is the upshot of all of this? The vast majority of the energy from agricultural land is used for cattle (and other grazers), while only a small portion is used to produce plant-based food for direct human consumption or feed for pigs and chickens. Cows are the real hogs in the sense that they use enormous amounts of biomass as well as the majority of the planet's agricultural land yet produce very little food (just 8% of the total). In contrast, both plant based foods and chickens and pigs are much more efficient converters of biomass into things we can eat.[15] Using highly productive croplands to produce animal feed, no matter how efficiently we do it, represents a net drain on the world's potential food supply. Certainly, food for thought!

Currently, one of the major challenges to the global food system is the rapidly increasing demand for meat and dairy products. This has led to an estimated 1.5-fold increase in the global numbers

[15] Smith et al. (2013), *Global Change Biology*, Vol. 19, p. 2285–2302.

of cattle, sheep, and goats, with equivalent increases of roughly 2.5- and 4.5-fold for pigs and chickens, respectively[16] (look again at Figure 2.17). This is largely attributable to the increased wealth of consumers everywhere, most recently in countries such as China and India. This is not to say that all meat consumption is bad because, while large numbers of livestock are fed on grains that could feed humans, there remains a very substantial proportion that is grass fed. Nevertheless, reducing meat consumption and increasing the proportion of meat sourced from grass (or other feed not suitable for human consumption) offers an opportunity to feed more people.

Cereals (e.g., wheat, maize, and rice) are still the world's most important sources of food, both for direct human consumption and as feed for livestock. Many agronomists agree that what happens in the cereal market is crucial to world food supplies. Since the 1960s, we have managed to raise cereal production by almost a billion tons. It appears that over the next 25–30 years, we must repeat that feat. The task of increasing production that is currently facing world agriculture is massive. The problem is that, in developing countries, the demand for cereals has grown faster than production, leading to a dependence on imports. That dependence is only likely to increase in the years ahead, leaving these countries vulnerable to fluctuations in the price of foods.

Enhancing Food Production

So, how do we deliver enough food and nutrition to the world sustainably? As noted earlier, we will likely have to double food production globally to meet the projected demands of a growing population while, at the same time, minimizing the environmental impacts of agricultural expansion and intensification. Historically, the primary solution to food shortages has been to bring more land into agricultural production (and exploit new fish stocks). While some new land could be brought under the plow, the competition for land from other human activities makes this an increasingly costly and unlikely solution, particularly if protecting our biological resources is given higher priority.

The amount of land under cultivation has remained relatively stable for the past several decades (Figure 2.19). However, rapid population growth during the second half of the 20th century has resulted in a substantial decrease in the area of land available on a per capita basis. Now, with 7.6 billion people and 1.5 billion hectares under cultivation, it takes about 0.2 hectares to feed each person (down from 0.44 in 1961). What this means is that we face two scenarios in terms of food production: We either keep the area under cultivation constant through to 2050, which means the available arable land per person drops to approximately 0.16 hectares (Line A, Figure 2.19). Or, we keep the arable land per capita constant at 0.2 hectares, which would mean increasing the amount of land under production significantly to about 2.1 billion hectares (Line B, Figure 2.19). Confounding the issue (as we will see in Chapter 9) is the fact that large areas of once productive agricultural land have been (and continue to be) lost to soil erosion. Policy decisions to produce biofuels on good quality agricultural land have also increased pressure, as we will discuss in Chapter 3. Thus, the most likely scenario is that more food will need to be produced from the same amount of (or even less) land. We must also improve distribution and access to food, a daunting challenge by itself!

[16] Godfray et al. (2010), *Science*, Vol. 327, p. 812–818.

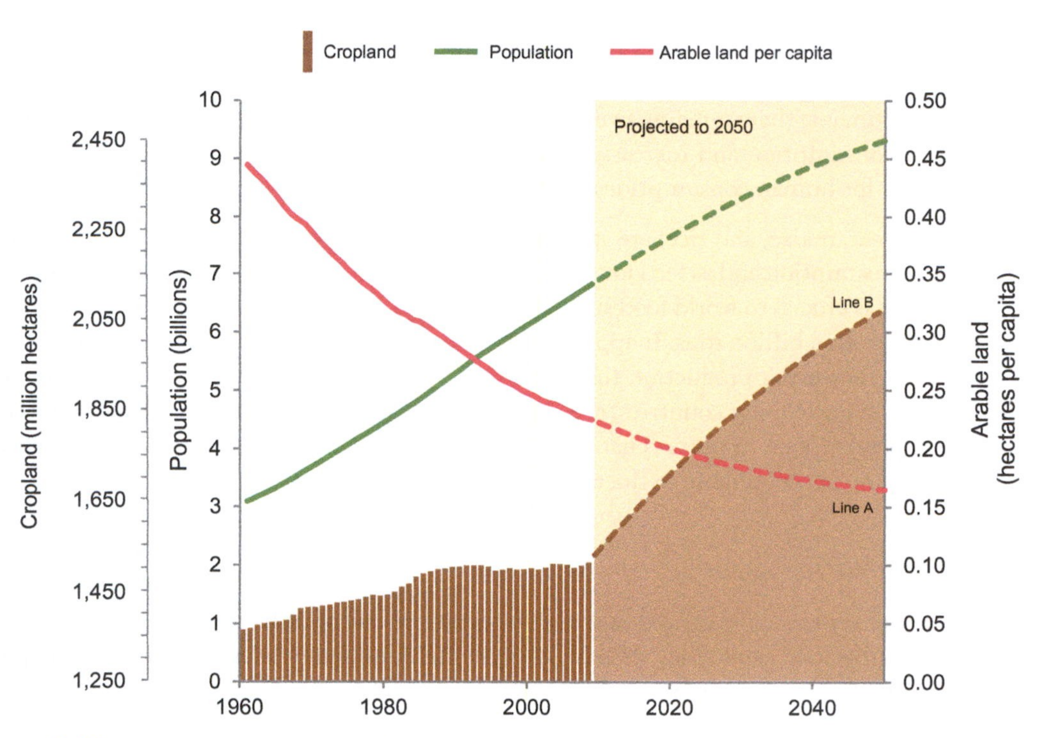

Figure 2.19 Human dependency on land. With the amount of land under cultivation staying essentially constant since the early 1990s, the growing population has resulted in the per capita land available dropping by 50% since 1961, to about 0.24 hectares per capita. The shaded area is the projected increase in arable land required by 2050 if we keep the per capita land constant (i.e., a 36% increase in land put into production).

An influential series of reports has suggested possible solutions to our interwoven food security and environmental challenges.[17] These include (1) a halt to the further expansion of agriculture, (2) the need to close yield gaps, (3) increasing agricultural resource efficiency, and (4) increasing food delivery.

The expansion of agriculture into sensitive ecosystems, such as tropical forests, has far-reaching effects on species and important environmental services, such as water-cycling and climate regulation. Interestingly, the food production benefits of tropical deforestation are often limited because many regions cleared for agriculture in the tropics have low yields compared with their temperate counterparts (see discussion in Chapter 9 on soil fertility in the tropics). Most scientists agree that expansion of agriculture into tropical forests must stop because the costs outweigh the benefits. This will be an important first step in moving agriculture forward onto a more sustainable path. Much of the world also experiences **yield gaps**, defined as the difference between the actual crop yield and the potential crop yield, given the agricultural practices present. For example, it is

[17] *Ibid.* and http://royalsociety.org/Reapingthebenefits

estimated that in parts of Southeast Asia, actual rice yields are only 60% of what they could be. Similar yield gaps have been observed in wheat in central Asia and cereals in Argentina and Brazil. However, research suggests that yields across many parts of Africa and Latin America (where population is expected to continue to grow) can be increased simply through improved management. Improvements in crop genetics will also likely be needed to increase potential yields during the 21st century. We also need to design more efficient irrigation systems, minimizing water loss through evaporation during both transport and storage, as well as become more efficient in our use and application of fertilizers.

One of the most shocking statistics to emerge from the FAO's database on food security relates to waste. Enormous quantities of food are never consumed but are simply discarded or degraded along the supply chain. Recent estimates suggest that between one-third to half of all our food grown is lost or wasted. This is a human catastrophe, given that almost one billion people suffer chronically from hunger. This simply must not continue.

CONCLUDING THOUGHTS

It is difficult to predict, with any degree of certainty, the precise path future population growth will take. Birth rates are declining but vary greatly between developed countries (where birth rates are frequently at or below replacement levels) and developing countries. There are also factors that are hard to predict, such as the potential impact of the worldwide HIV/AIDS **pandemic**. In Africa, birth rates are among the highest in the world, but if HIV/AIDS is controlled or even eradicated, world population could increase faster than predicted. But most projections of population growth indicate that the world's population will continue to grow until at least 2050, eventually cresting at approximately 9.2 to 9.5 billion people. Will such growth and the concomitant demand for resources eventually lead to a sudden population crash? Most demographers (i.e., scientists who study population) agree that such a scenario is unlikely. Perhaps the real question should be: Can Earth's resources sustain 9.5 billion, and, if not, how large of a human population will be able to live (with a decent quality of life) on this planet?

There is now little doubt that the majority of the world's population growth will occur in the less developed regions, with the most significant growth occurring in Africa and Southeast Asia. I think it is fair to say that many people think of developing countries in these regions as "them," with large families, low incomes, and shorter life expectancy, and more developed countries as "us," with small, wealthier families living longer lives. But seeing the world in this static way is far too simplistic. The reality is that the developing world is getting healthier and having fewer children as social services expand, and global demographic trends suggest that this will continue during the coming decades. In fact, since 1960, the number of children who die before age 5 has fallen by more than half. This is real progress. However, along the way, these countries are also getting richer and consuming more. We often hear that developing countries cannot grow to be like us with politicians echoing that they are catching up and that we (i.e., the developed world) must hurry to maintain our competitive advantage? Why should not they be able to live like us? Why should we "maintain our advantage?" Would not it be good if the people of Haiti or South Sudan had the same conditions of daily life as the United States? I believe that things are improving, and improving quickly, and this is, in many ways, a good thing. For example, in 2000, the country of Vietnam had a Gross National Income (GNI) per person of just $380; in 2017, that figure had jumped to $2,100. The issue is not whether these countries will continue to develop and improve the human condition as their populations grow but rather, whether the environment can absorb the pressure that will inevitably accompany such growth.

While there will undoubtedly be scientific and technological innovation in the food system, we face enormous challenges making food production more sustainable while, at the same time, ending world hunger. One recent study has suggested that food production may *only* need to increase by 25%–70% above current levels and this may be sufficient to meet 2050 crop demand.[18] That still represents a significant increase in global output. There is little doubt that agriculture's environmental footprint must shrink drastically to safeguard the ecosystems that humans rely on. I would also argue that we must not be tempted to sacrifice our planet's already hugely depleted biodiversity for the sake of easy gains in food production (as we shall see later on in Chapter 8). Biodiversity provides many of the public goods on which humankind relies. We also simply do not have the right to deprive future generations of the economic and cultural benefits of biodiversity. The challenges facing humanity today are unlike anything we have seen before, and they undoubtedly will require novel approaches to solving food production and other sustainability problems.

There is good news, of course, in that global fertility rates are falling and, in many parts of the world, falling rapidly. In sub-Saharan Africa, fertility is still high (approximately five children per woman), but in most of the world, family size has shrunk dramatically. The UN projects that the world will reach replacement fertility by about 2030. The bad news is that 2030 is just over a decade away with the largest generation of adolescents in history now entering their childbearing years. Even if each of those women has only two children, population will continue to grow under its own momentum for another quarter century.

[18] Hunter et al. (2017), *Bioscience*, Vol. 67, p. 387–391.

3

The Environmental Impact of Our Search for Energy

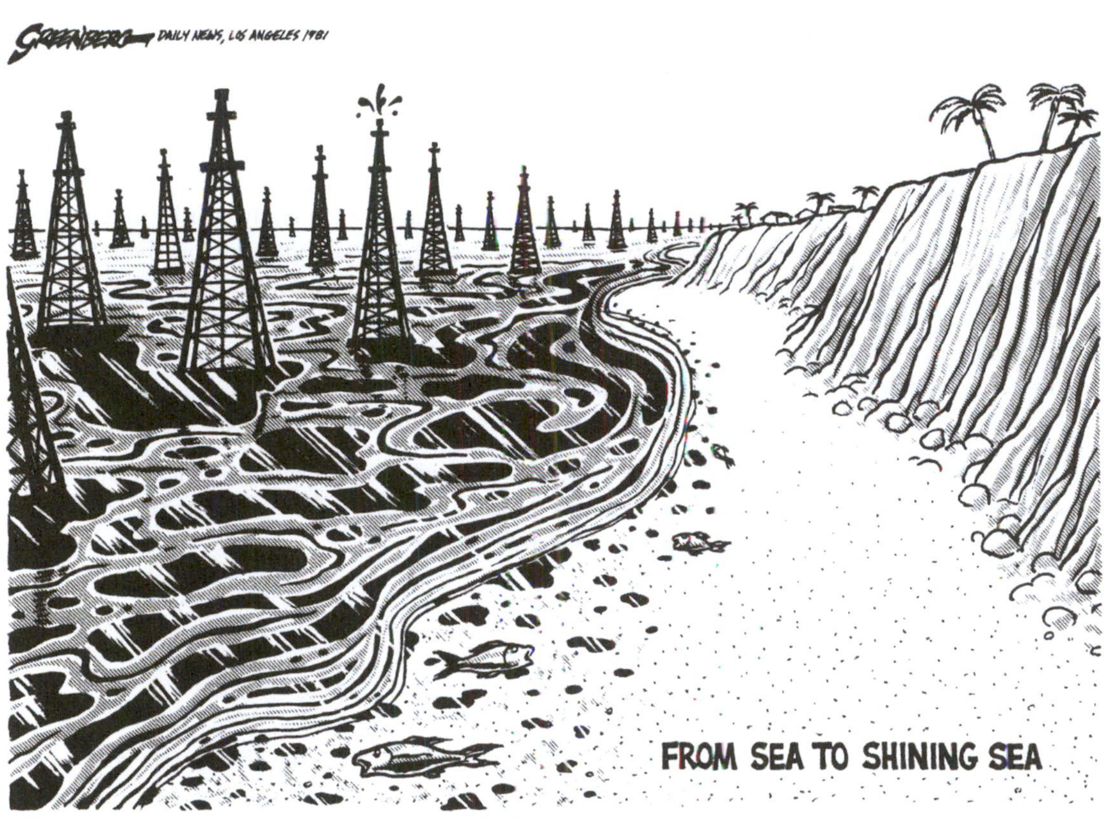

INTRODUCTION

We are an energy-based society. Indeed, almost everything we do depends on some form of energy. Think for a moment about how you spent the past 24 hours. If you turned on a light, drove a car, cooked a meal, took a flight somewhere, listened to music, powered up your computer—all of these required energy. But did you give any thought as to where that energy came from or what it took to get it from its source to your fingertips?

Energy is indeed a very hot topic. You have probably heard the statement that we need to reduce our dependence on foreign oil. President George W. Bush repeated it several times in a number of his State of the Union addresses. So too did President Obama. For example, in his 2013 State of the Union address, President Obama called on Congress to create an Energy Security Trust Fund[1], which focuses on shifting America's cars and trucks not just off foreign oil but off oil entirely. He even opened his 2014 State of the Union with the following:

> *Today in America . . . an autoworker fine-tuned some of the best, most fuel-efficient cars in the world and did his part to help America wean itself off foreign oil.*

Another phrase I see frequently banded about in the media these days, and one that goes hand-in-hand with reducing our dependence on foreign oil, is that of "energy independence." But no one really seems to know exactly what this means. Some people use the phrase energy independence as a synonym for producing as much oil as you consume, and that seems reasonable. However, we trade oil on a global market, and it is naive to suggest that America will really ever be independent from events overseas that affect our energy supply. Of course, if you do not use oil then, yes, you are considerably more energy independent than if you do, but herein lies one of the many "reality checks" that I will refer to in the coming pages: we currently use a lot of oil, and it takes a long time to reduce demand. We can increase oil production faster than we can cut oil use in cars and trucks because the typical car stays on the road for 15 years or longer. It takes a long time to turn the fleet over. It seems that oil is here to stay for that reason alone, at least for the coming decades.

In this chapter, we examine different types of energy sources, their advantages, their limitations, how they are used today, and how they could be used in the future. The scope of this chapter is squarely on the environmental impact of our search for energy.

POWERING OUR PLANET: CURRENT ENERGY SOURCES

You have likely heard the term **fossil fuels**. These are so-called **hydrocarbons** and refer to the remains of dead plants and animals that were exposed to heat and pressure within Earth's crust over hundreds of millions of years, eventually being converted into oil, coal, and natural gas. Currently,

[1] Source: www.whitehouse.gov

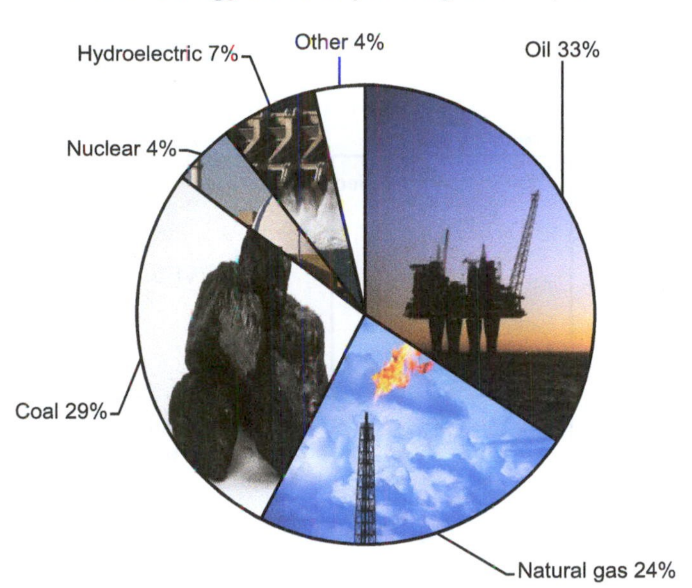

Figure 3.1 Worldwide commercial energy consumption (Source: www.eia.doe.gov/).

fossil fuels are our primary energy source, accounting for almost 90% of energy consumption worldwide (Figure 3.1).

We use oil primarily to produce transportation fuels, with an emphasis on gasoline and diesel. We consume natural gas in heating, cooking, and industrial applications, but its use for power generation is growing rapidly. We use coal for industrial applications, such as steel making, and for electric power generation. Although coal is no longer a significant heating fuel in industrialized nations, its use continues for residential and commercial heating where resources are abundant.

Along with fossil fuels, you also probably have heard about **alternative energy** technologies. These include nuclear power as well as renewable sources such as wind, hydroelectric, and solar. Nuclear energy, hydroelectric, and wind are used almost entirely for generating electricity, accounting for about 10% of global energy consumption (Figure 3.1).

Let us look a little closer at our fossil fuel resources given how important they are as an energy source. Globally, coal is the most abundant source of energy with **economically recoverable reserves** of nearly 1 trillion tons—estimated to be at least a 200-year supply of energy at current rates of production (Figure 3.2). Most of these reserves are located in North America, Latin America, South Africa, Australia, China, Indonesia, and India. Oil and natural gas reserves, which often are discovered close together, are also distributed throughout the globe. However, as Figure 3.3 illustrates, the distribution is very unbalanced with almost two-thirds of all known oil reserves situated in the Middle East alone. Of the estimated 1.48 trillion barrels of oil worldwide, Saudi Arabia, Iran, Iraq, Kuwait, and the United Arab Emirates (UAE) hold the lion's share of the reserves, about 50% (although the entire region is believed to have huge amounts of undiscovered oil and natural gas). In the United States,

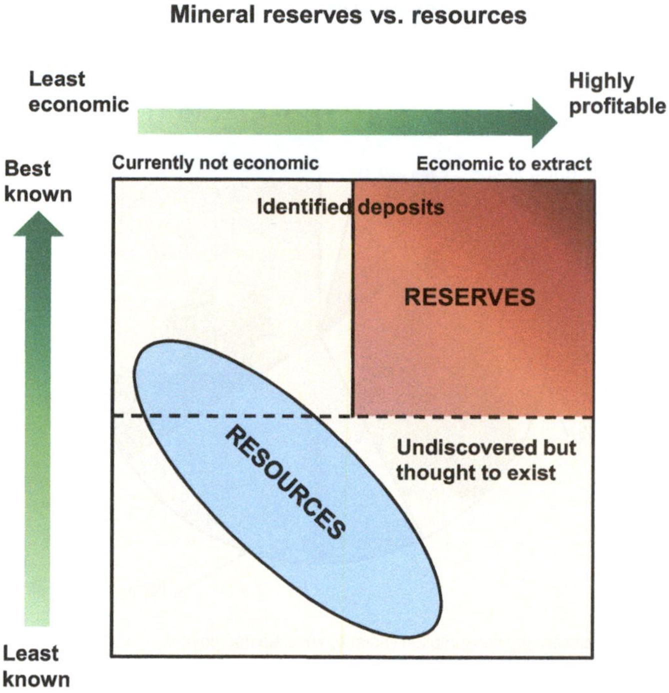

Figure 3.2 We refer to any mineral deposit that is currently economically profitable to extract as a reserve. Any other deposit, whether known but too expensive to mine or speculative in that we think they exist based on local geology, is defined as a resource.

which has less than 1.5% of the global reserves, almost 80% is found in Alaska, California, Texas, and the Gulf of Mexico. The reserves in Venezuela and Canada are both interesting and controversial in that they are **unconventional** oil reserves. The United States Geological Survey (USGS) estimates that Venezuela's **oil sands** deposits, known as the Orinoco Oil Belt, are approximately equal to the world's reserves of conventional oil but may be far larger. No one really knows how much of this heavy crude oil is economically recoverable. However, this is also very expensive oil, both economically and environmentally, as we shall discuss further in this chapter.

The United States and the countries of Europe, including those of the former Soviet Union, consume slightly more than half of all the crude oil traded in the world oil market. In 2017, the United States consumed a total of 7.26 billion barrels of petroleum products, an average of about 19.9 million barrels per day, or about 20% of the world total.[2] While we imported approximately 10.1 million barrels of oil per day (MMb/d) from about 84 countries, we also exported about 6.3 MMb/d around the globe (we will examine why the U.S. is both an importer and exporter of oil a bit later on). Importantly, from an energy independence point of view, imports have fallen dramatically since 2008 when we imported about two-thirds of our crude oil supply; we now import less than half.

[2] Source: BP Statistical Review of World Energy 2018.

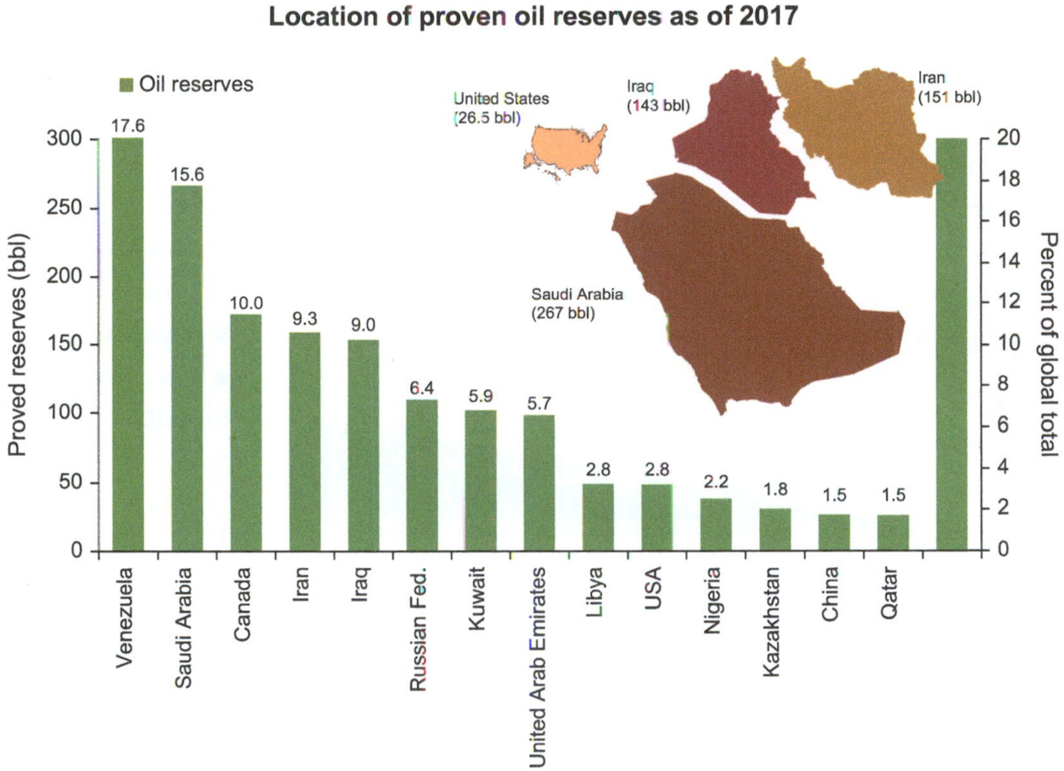

Location of proven oil reserves as of 2017

Figure 3.3 Proved reserves of crude oil in billion barrels as at 2017 (bbl). Proved reserves are those quantities of petroleum that, by analysis of geological and engineering data, are estimated with a high degree of confidence to be commercially recoverable from known reservoirs under current economic conditions, as explained in Figure 3.2. Saudi Arabia, for example, has 267 billion barrels of oil (or 17.6% of the world total). The inset shows Saudi Arabia, Iran, and Iraq scaled relative to the U.S. in terms of proven oil reserves. Note that Canada's proven reserves are due largely to the Athabasca oil sands deposit, discussed further in Box 3.2 (Source: www.eia.gov).

Where does this oil come from? In 2017, the United States imported 40% of its crude oil and petroleum products from Canada, 7% from Mexico, with Persian Gulf countries accounting for 17% (refer back to Figure 3.3). Many people think we get most of our oil from Saudi Arabia, but the Kingdom now supplies just 9% of our imported oil, which is still a significant number but far less than the 25% we imported from them in 2003.[3] China became a net importer of oil in 1993 and is now on a trajectory to compete with the United States for remaining reserves around the world. Never an exporter, India's oil appetite is increasing with an economy growing at over 7% in 2018, out-pacing China's growth rate.

In Chapter 2, we concluded that we are likely to add another 2 billion people to the planet by 2050 and that these people will seek food, water, and energy. Indeed, world energy consumption is projected to increase by about 35% by 2035 (Figure 3.4). Much of this growth is expected to be in developing nations, particularly in Asia, where energy demand is anticipated to double over the next

[3] Source: www.eia.gov

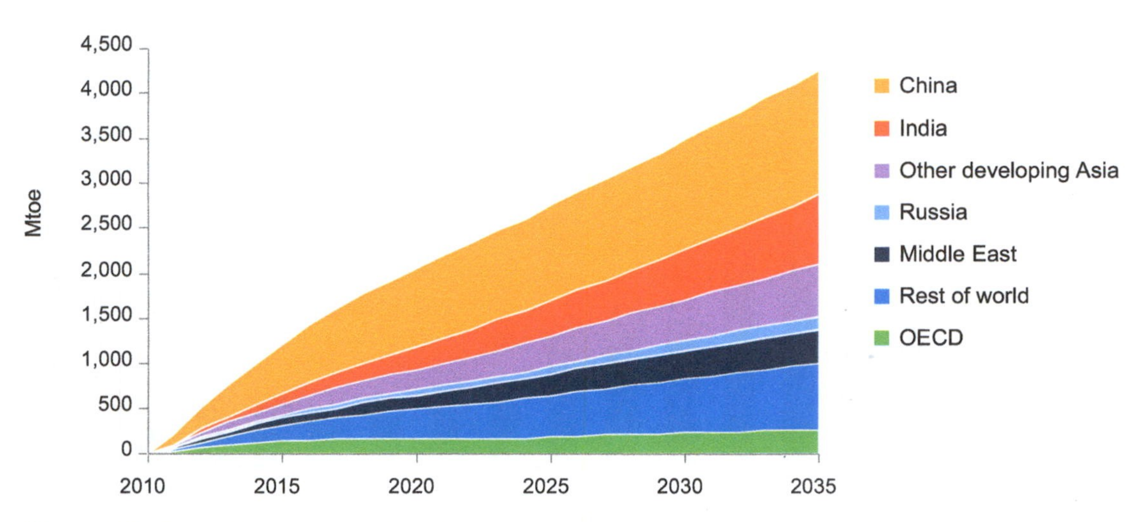

Figure 3.4 Total world energy consumption projected through to 2035. Consumption rises by about a third from 2013 to 2035, with China and India accounting for 50% of the growth. Note the vertical scale is in Mtoe, or million tons of oil equivalent, equal to the amount of energy released by burning one ton of crude oil (Source: BP Statistical Review of World Energy 2017).

two decades (even though countries like China are accelerating development of their own petroleum resources, but the speed of industrialization and development will likely outstrip their own energy reserves). As a consequence, China's oil imports have outstripped those of the U.S. to over 8 MMb/d. Huge amounts of oil will be needed to meet the growing demands for transportation fuels in the developing world alone, where per capita motorization is projected to more than double by 2020. A key question that emerges (one that will be common throughout this book) is: can we sustain such growth projections while maintaining healthy ecosystem functioning?

ENERGY AND THE ENVIRONMENT

In many ways, our choice of energy source reflects primarily economic values: what types of energy can we exploit that is affordable, secure, and will also allow industrialization and development to continue? What sources are easily available? How much does it cost to extract, purify, and/or construct different energy types relative to the profit of development? With our growing awareness of climate change and environmental degradation, we are realizing that we need to balance an equation that goes beyond economic considerations: how do we satisfy our energy demand while still protecting the environment?

In this next section, we will take a closer look at each of the major energy sources mentioned above. The information is intended to give you a broad overview of the pros and cons associated with each energy source, as well as give you a sense as to what our future energy options and policies may be.

FOSSIL FUELS

Coal

Coal-fired power plants account for nearly a third of the electricity generation in the United States (although, as Figure 3.5 shows, generation varies significantly by state). As a fuel source, coal is cheap and plentiful, but it is also dirty and dangerous work. Coal pollutes the environment when we mine, transport, store, and burn it. Once seen as a fuel of the past, "King Coal" has recovered

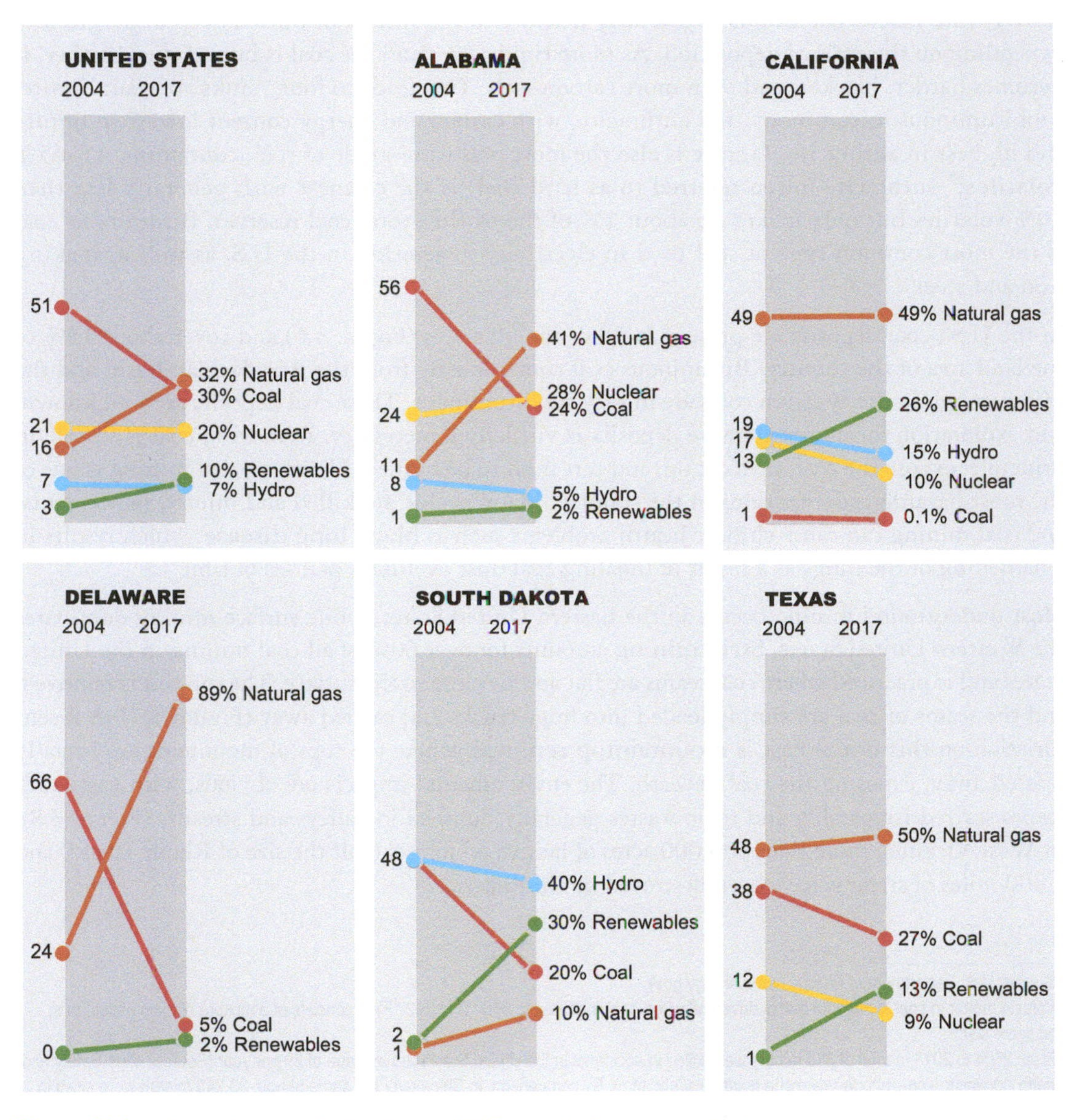

Figure 3.5 U.S. electricity generation from 2004 to 2017 (Source: U.S. Energy Information Administration).

and is now used in record amounts around the globe. Forecasts of future energy use give a prominent role to coal, with plans to build more than 100 new coal-fired power plants in the United States alone.[4] When President Trump took office in 2016, U.S. coal was in a state of decline, with production and consumption falling by more than a third between 2007 and 2016. He has promised to reverse that trend.

Coal is a sedimentary rock containing between 40% and 90% carbon formed from ancient plants accumulated in moist environments. As carbon-rich plants die, they form layers that become compressed by subsequent plant deposits and/or sediments. Eventually, the plant layers convert into black coal. Seams of coal may be close to the surface or buried deep underground, depending on the scale of deposition. As more time passes and the coal is buried even further, it becomes harder, blacker, and even more carbon-rich. This leads to four "ranks" of coal: lignite, subbituminous, bituminous, and anthracite, with carbon and energy content lowest in lignite and highest in anthracite. Lignite is also the most polluting grade of coal, containing 45–65% **volatiles**;[5] anthracite (often referred to as hard coal) is the cleanest with generally less than 10% volatiles but only makes up about 1% of the world's total coal reserves. Bituminous coal is the most common type of coal used in electricity generation in the U.S. as well as making iron and steel.

In the U.S., coal deposits are present in 38 of the 50 states (Figure 3.6) and cover about 13% of the land area of the country. Bituminous coal comes mostly from the Appalachian Basin and the Midwest, while the Western coals are mostly subbituminous. These coal deposits are well known, and exploration for more extensive deposits is virtually unnecessary. Removing coal is simple in principle: expose the coal, break it up, and cart it off to be burned. However, coal mining is one of the most hazardous occupations in the country. Many people are killed and injured in accidents, and coal mining can cause chronic health problems such as **black lung disease**, which results in a hardening of the lungs as a result of inhaling coal dust over long periods of time.[6]

Most underground mining occurs in the Eastern United States, while surface mining dominates the Western United States. **Strip mining** accounts for over 60% of all coal mining in the United States and is practiced where coal seams are flat and lie close to the surface. The top soil is removed, and the seams of coal are simply loaded into huge trucks and carried away (Figure 3.7). A recent variation on this in the East is **mountaintop removal**, where the tops of mountains are literally blasted away, exposing the coal beneath. The environmental impacts are obvious, with vast landscapes scarred irreparably and toxic wastes generally dumped in valleys and streams (Figure 3.8). In West Virginia, more than 300,000 acres of hardwood forests (half the size of Rhode Island) and 1,000 miles of streams have been destroyed by this practice.[7]

---

[4] Source: U.S. Department of Energy (www.energy.gov).

[5] Volatile matter in coal is material that is driven off when coal is heated to 950 °C (1,742 °F). It consists of a mixture of toxic gases, tars, and some sulfur.

[6] From 2004 to 2013, a total of 288 miners were killed in accidents in the United States, an average of 29 per year (Source: Mine Safety and Health Administration—http://www.msha.gov/fatals/fabc.htm). By comparison, in China over the same period, 33,240 miners were killed in mining related accidents (Source: www.usmra.com/).

[7] Source: West Virginia Department of Environmental Protection (www.wvdep.org/).

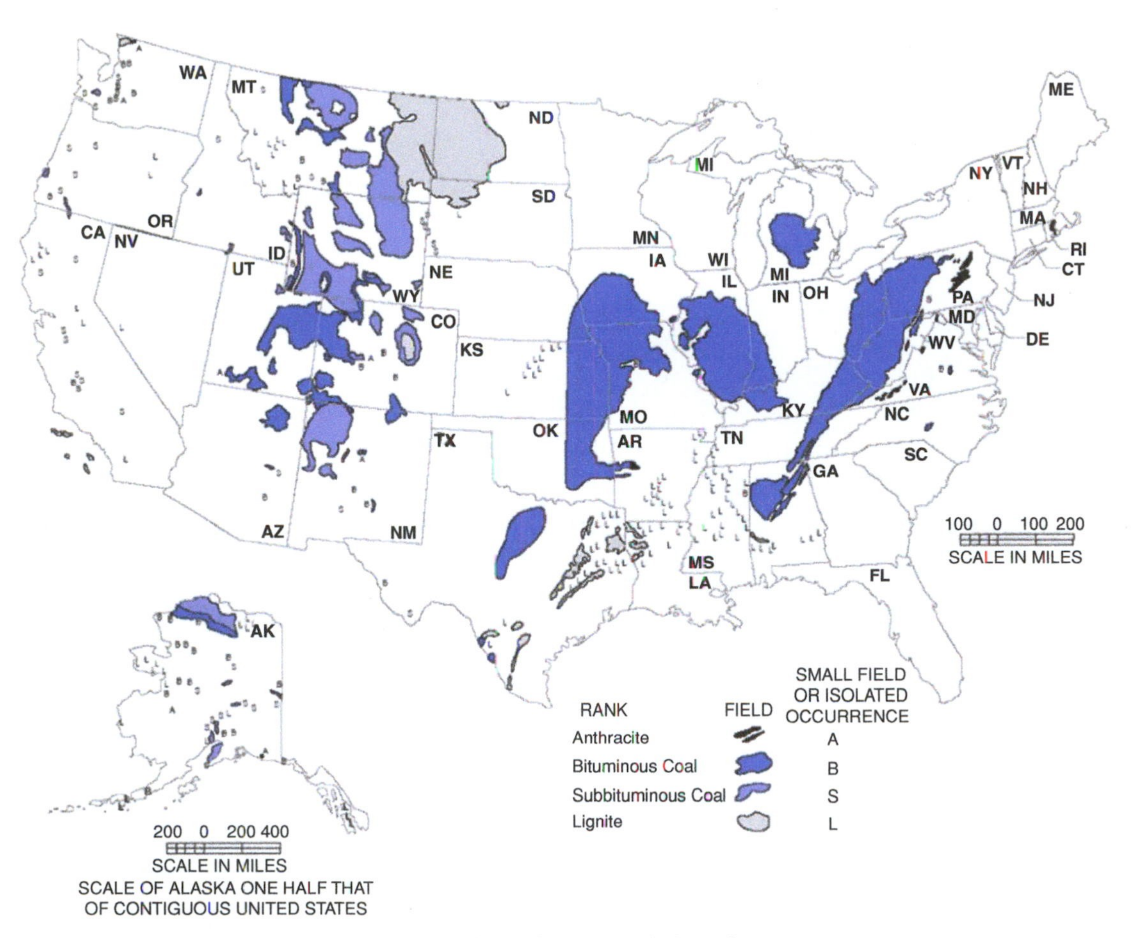

Figure 3.6 Map of coal-bearing areas of the United States (Source: www.eia.doe.gov).

The atmospheric impact of burning coal is enormous. It is a leading cause of smog, acid rain, global warming, mercury contamination, and air toxins, all topics that are covered in greater detail in upcoming chapters. In an average year, a typical 500 megawatt coal plant[8] burns about 1.4 million tons of coal, uses more than two billion gallons of water, and generates a variety of pollutants and toxins (Figure 3.9). Some of the particles, such as sulfur, can be partly removed with **scrubbers** or filters during combustion. However, the smallest particulates are less likely to be removed and pass out through the smokestack into the air. These tiny particulates are the major cause of concern in terms of global air quality, including the much-publicized smog in cities like Beijing, China, and New Delhi, India (which we discuss further in Chapter 4).

Coal-fired power plants also generate vast amounts of waste. A typical plant produces about 200,000 tons of sludge from the smokestack scrubbers each year, most of which is placed in unlined, unmonitored onsite landfills, and surface impoundments. Toxic substances in the waste,

[8] Such a plant would produce enough to power a city of about 140,000 people per year.

Figure 3.7 Open strip coalmine with overburden (i.e., topsoil) removed. Once the coal is mined, land is supposed to be levelled off and soil returned and then farmed as it was before mining, though many companies do not fulfill these requirements (Source: iStockphoto.com/sakakawea).

Figure 3.8 Largely hidden from most Americans, a highly destructive form of coal mining called mountaintop removal has devastated 1 million acres in the central and southern Appalachian Mountains. In many areas, Americans use electricity sourced as a result of mountaintop removal (Photo courtesy of iLoveMountains.org).

Figure 3.9 A typical 500 MW coal plant burns 1.4 million tons of coal each year. Coal plants are the leading cause of smog, acid rain, and air toxics (Source: Union of Concerned Scientists—www.ucsusa.org).

including arsenic, mercury, chromium, and cadmium, frequently leach out into local rivers, streams, and groundwater. Acid mine drainage occurs when exposed coal gets wet and toxic metals begin to dissolve. The resulting runoff is directly toxic to aquatic life and renders the water unfit for human use.

As mentioned above, coal-fired power plants also use large volumes of water each year from nearby lakes, rivers, or oceans to create steam for turning its turbines. The two billion gallons used by a typical plant described above is enough water to support a city of approximately 250,000 people. Once this water has cycled through the power plant, it is released back into the lake, river, or ocean. This water is hotter (by up to 20–25°F), causing thermal pollution in the receiving water body. Typically, power plants also add chlorine or other toxic chemicals to their cooling water to decrease algal growth. These chemicals are released into the environment with the discharged water.

Clean coal refers to technologies designed to enhance both the efficiency and the environmental acceptability of using coal, such as chemically washing out impurities and capturing carbon dioxide from the flue gas. In his 2007 State of the Union Address, President George W. Bush committed $2 billion over the following 10 years for development of clean coal technologies, citing it as one way to reduce the country's dependence on foreign oil. However, estimates suggest that it will be at least 2025 before any commercial scale clean coal power stations become economically viable and widely adopted. Environmental groups argue that clean coal is a myth. Indeed, everything to do with coal—from mining to processing to transportation to burning to waste disposal—adversely affects the environment, more so than any other energy source. And while some of these effects can be lessened with effort, others, such as carbon emissions, cannot as yet be removed from the power plant's exhaust and are an inevitable problem of coal use.

So, what is coal's future? As much as environmentalists lobby to reduce our dependence on coal and switch to cleaner, renewable alternatives, the stark reality is that coal is here to stay, probably for many decades to come. The physical supplies of coal are substantial and it is still the cheapest way to generate electricity. And to put coal's significance (and potential longevity) into a broader context, consider Figures 3.4 and 3.5 again. China now consumes four times as much coal as the United States in absolute terms with about 70% of its electricity generation sourced from this remarkable, yet highly problematic, sedimentary rock. And while there are alternatives to coal for power generation like wind and nuclear, for some industrial processes, like steel and cement, there simply are no alternatives. In China, 350 million people live on less than $2 a day, an entire United States of poverty. The country simply cannot provide enough energy to build homes and pave highways without coal.

Oil

Oil defined the 20th century. It shaped political boundaries, created and disrupted economies, and, for some, created enormous wealth. Simply put, we are addicted to oil. The United States now accounts for one-fifth of the world's oil consumption, about 19.9 million barrels of oil per day (China is a distant second, at 12 million barrels per day, see Figure 3.10), and we are producing and consuming more each year. In fact, the United States is now the largest *producer* of crude oil in the world, accounting for nearly 15% of global production. As noted earlier, we have dramatically reduced our oil imports to the point that we now produce more oil than we actually import.

Globally, we appear to be burning through our supplies of oil very quickly. More than 50% of our cumulative global production and consumption has occurred in the past 20 years. Estimates by the U.S. Energy Information Administration (among many others) suggest that worldwide reserves of oil will only last somewhere in the region of 40–60 years. By 2005, many analysts agreed that the global production of regular crude oil had hit a ceiling (or peaked) and had entered what some called an "inelastic" phase, meaning production is unable to respond to rising demand leading to wild price swings. This all seemed to support the concept of **peak oil**, or the notion that global production will reach a peak and then ultimately decline (Figure 3.11), an idea that had been around for decades. A noted British petroleum geologist, Colin J. Campbell, predicted that oil production would actually peak in 2007, as shown in Figure 3.11, and then begin a steady decline. However, a highly respected consulting company in the United States, IHS CERA, has shown quite convincingly that global oil productive capacity will actually grow through 2030 with no evidence of a peak in supply before that time (Figure 3.11). The latest global production figures of about 95 million barrels per day seem to support this projected growth.

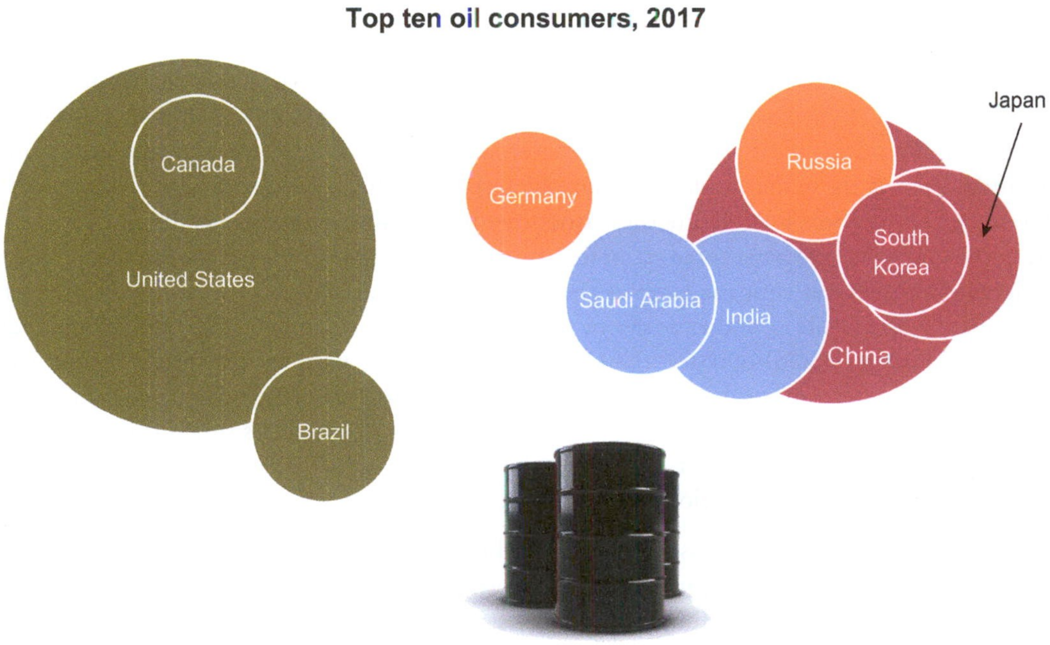

Top ten oil consumers, 2017

Canada
United States
Brazil
Germany
Saudi Arabia
India
Russia
South Korea
Japan
China

Figure 3.10 The world's top ten consumers of oil for year-end 2017. Each circle is scaled relative to total oil consumption (Source: www.eia.gov).

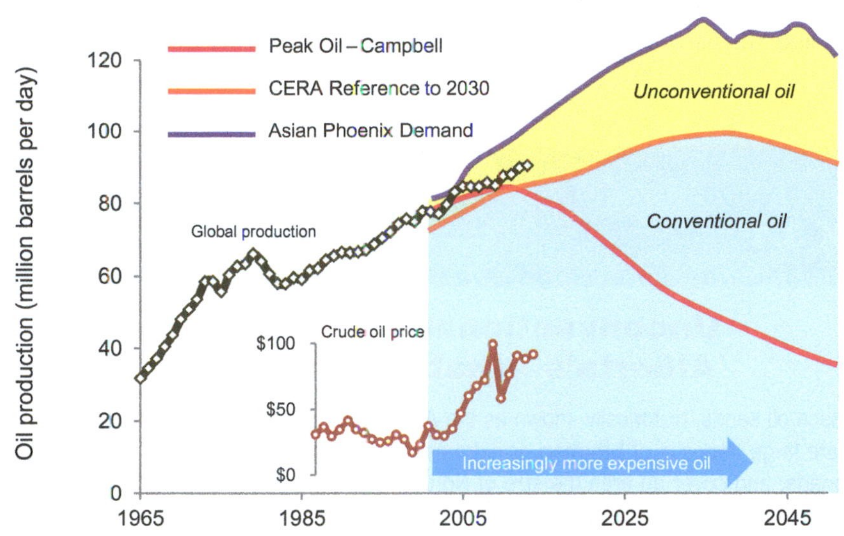

Figure 3.11 World oil production, price, and estimates of production decline based on the concept of peak oil. The red line is peak oil and subsequent decline according to British petroleum geologist Colin J. Campbell; the orange line is oil decline according to CERA; the purple line shows projected demand from Asia which will have to be satisfied by unconventional sources of oil (Source: IHS Cambridge Energy Research Associates—www.ihs.com).

While we may not be running out of oil in the very near future, it is a non-renewable resource. The EIA estimates that we will still require 22–25 MMB/d of *new* oil production by 2030 to keep up with global demand that appears to be increasingly driven by China (the so-called Asian Phoenix curve, Figure 3.11). Where will we get this oil? When oil prices hovered around $100/barrel, between 2000 and 2014, it suddenly became economically viable to tap the vast resources of **oil sands**, such as those that underlay Utah, the Orinoco Belt in Venezuela, and the province of Alberta in Canada (Box 3.1). While these **unconventional sources** are generally not counted as part of the global reserve base, some estimates suggest that the world's ultimate reserves of unconventional oil are several times as large as those of conventional oil and will be highly profitable for companies as a result of higher prices in the 21st century. In October 2009, the United States Geological Survey (USGS) updated the Orinoco tar sands (Venezuela) recoverable mean value to 513 billion barrels, more than twice that of Saudi Arabia's current conventional oil reserve. However, production of oil derived from Canada's tar sands—sometimes called the "oil junkie's last fix"—is expected to reach just 4.7 million barrels per day by 2035 with little prospect of a dramatic increase. The key question to consider is: should we go after this oil?

A key characteristic of the Athabasca deposit in Alberta, Canada, is that it is the only one shallow enough to be suitable for surface mining (the deposits are about 150 ft thick and are overlain by about 250 ft of **overburden**). However, extracting and refining the bitumen is a very labor and resource intensive process, requiring large volumes of water (which is heated to separate out the bitumen) and natural gas (for further refining the bitumen into synthetic crude). Thus, critics say that the oil sands industry is wasting a relatively clean fuel (i.e., natural gas) to make one of the dirtiest fuels, effectively turning "gold into lead."

Oil extracted from these sources also typically contains contaminants such as heavy metals, and the water, once used, is discharged into tailings ponds that now cover more than 50 square miles. The fine clay and silt particles in the wastewater take several years to settle, and when they do,

BOX 3.1

Unconventional Sources of Oil: Alberta's Athabasca Oil Sands

The Athabasca oil sands, historically known as the Athabasca tar sands due to perceived similarities with actual tar, are large deposits of **bitumen** or extremely heavy crude oil. They are located in northeastern Alberta, Canada, and cover an area the size of North Carolina (see map below). They lie under sparsely populated boreal forest and muskeg (peat bogs) and contain an estimated 2 trillion barrels of bitumen in-place (essentially 8 times more than Saudi Arabia). With modern unconventional oil production technology, at least 10% of these deposits, or about 170 billion barrels are considered to be economically recoverable, making Canada's total proven oil reserves the third largest in the world.

Canada's Athabasca Oil Sands and the Keystone Pipeline

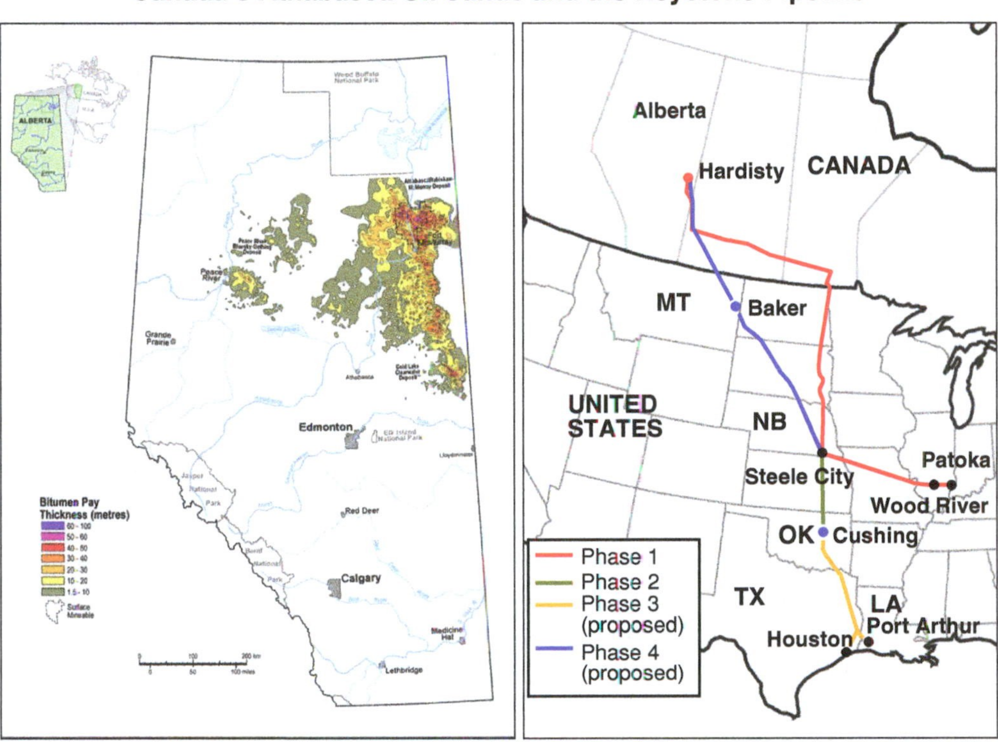

they produce a yogurt-like goop that is contaminated with toxic chemicals. Organizations like the Natural Resources Defence Council (NRDC) argue that the environmental effects of extracting unconventional sources are prohibitively high.[9] Alberta's boreal forest and wetlands are home to a diverse range of animals, including lynx, caribou, and grizzly bears and serve as critical breeding grounds for many North American songbirds and waterfowl. These habitats are being systematically destroyed by oil companies scraping thousands of acres to mine the oil sands. Nowhere on Earth is more earth being moved these days than in the Athabasca Valley, with the currently mined area several times larger than the District of Columbia (see Box 3.1).

Oil is transported from the Athabasca oil sands via the Keystone Pipeline System to multiple destinations in the United States, which include refineries in Illinois, the Cushing oil distribution hub in Oklahoma, and proposed connections to refineries along the Gulf Coast of Texas. The pipeline currently transports almost 600,000 barrels per day, with the proposed routes into Texas increasing capacity to 1.3 MMB/d. In November 2011, after 12,000 people encircled the White House in protest, President Obama delayed any decision on the pipeline extension (phases 3 and 4, see Box 3.1), a move seen as a victory for the environmental movement. Groups such as the NRDC argued that extending the pipeline would lock the United States into a dependence on hard-to-extract oil and generate a massive expansion of the destructive tar sands oil operations in Canada in coming decades. Executives from the Canadian oil companies argued that building the pipeline would further reduce the United States' reliance on Middle East oil, something that is already well underway. Whether or not this is accurate (remember, we import only 17% of our oil from Persian Gulf countries), the oil sands do raise questions about whether we are going to get serious about alternative energy and a clean energy future or proceed further down the unconventional oil track. While President Trump fully supports completion of the Keystone project, significant hurdles remain that continue to cast doubts on its prospects, most notably securing landowner easements along the proposed route and lawsuits brought by environmental and landowner groups.

One development that has had a dramatic effect on the future of the oil sands has been the American shale-drilling boom that has produced an abundance of crude, enough to make the United States a major exporter. In the Permian Basin in West Texas, a historic oil boom is underway, and U.S. production is at record levels. In May 2018, oil production in the Permian soared to 3.2 MMB/d (more than six times what the Keystone project can potentially deliver). This pushes total U.S. production above 10 MMb/d, which is the highest level since the federal government began keeping records in 1920! Oil executives are now focused squarely on the Permian, with the oil sands of Canada drifting farther and farther out of view.

Oil has taken a heavy toll on the environment, primarily through oil spills and air pollution. Spills are the most graphic type of impact. The International Tanker Owners Pollution Federation Limited (ITOPF),[10] a nonprofit organization funded by the world's ship owners, maintains a database of oil spills from tankers that carry, on average, 524 billion gallons (or about 12.5 billion barrels) of oil across our oceans each year. The vast majority of spills are small (i.e., less than 50 barrels) and

[9] See http://www.nrdc.org/energy/dirtyfuels_tar.asp
[10] Source: www.itopf.com

result from routine operations, such as loading and unloading. These spills normally occur in ports or at oil terminals. Not surprisingly, more attention is paid to large spills that result from groundings, ship structural damage, fires, and explosions. Figure 3.12 lists data for spills greater than 5,000 barrels from the years 1970–2017. Clearly, the number of large spills has decreased significantly during the past 30 years: by the 1990s, the average number of large spills had decreased by 66%, largely the result of improved tanker technology such as building double hulls. A few very large spills are, however, responsible for a high percentage of the oil spilled into our oceans. For example, the three largest record spills (the *Atlantic Empress* in 1979 spilling 2 million barrels, the *Castilloe de Bellver* in 1983 spilling 1.7 million barrels, and the *ABT Summer* in 1991 spilling 1.8 million barrels) account for almost 15% of all oil spilled during the 40-year reporting period. It is notable that all three spills, despite their large size, caused relatively little environmental damage as the oil did not impact coastlines. The *Exxon Valdez* spill in 1989, however, attracted an enormous amount of media attention even though it was well down the scale in world terms, spilling "just" (my quotations) 259,000 barrels of crude into Prince William Sound, Alaska. Despite the utilization of a massive number of vessels, booms, and skimmers, less than 10% of the original spill volume was recovered from the sea surface, and oil subsequently affected a variety of shores, mainly rock and cobble, to varying degrees over an estimated 1,800 km of the sound.

In terms of the total volume released, the *Valdez* spill was the largest ever in U.S. waters until the 2010 *Deepwater Horizon* oil spill (also referred to as the *BP* spill or the *Gulf of Mexico* spill). This spill resulted from an explosion on the off-shore oil well, *Deepwater Horizon*, killing 11 men working on the platform and injuring 17 others (Figure 3.13). This explosion caused a sea-floor oil gusher that spewed unabated into the Gulf of Mexico for three months (April 20, 2010 to July 15, 2010). After it had released about 4.9 million barrels of crude oil, *Deepwater Horizon*'s leak was

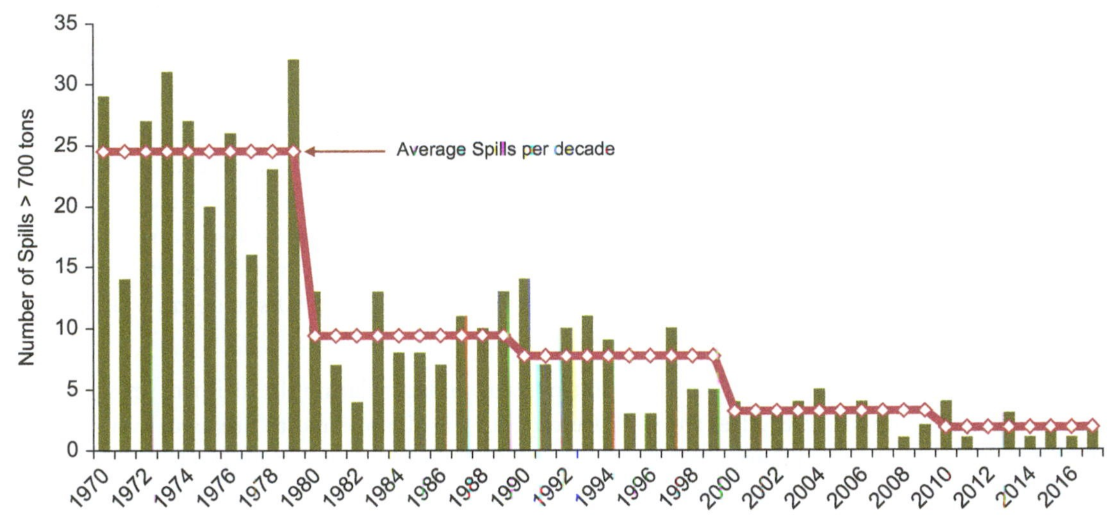

Figure 3.12 Number of oil spills worldwide over 700 tons or 5,000 barrels, 1970–2017 (Source: International Tanker Owners Pollution Federation Limited—www.itopf.com).

Deepwater Horizon Oil Spill

Figure 3.13 The BP Deepwater Horizon fire on 22 April, 2010. Photo courtesy of the U.S. Coast Guard.

finally stopped by capping the gushing wellhead (Figure 3.14). An estimated 53,000 barrels per day escaped from the well just before it was capped. The spill caused extensive damage to marine and coastal habitats, and the Gulf's fishing and tourism industries were also hit with severe losses. In an attempt to protect hundreds of miles of beaches, wetlands, and estuaries from the spreading oil, BP tried skimmer ships, floating containment booms, anchored barriers, sand-filled barricades along shorelines, and even chemical dispersants. Despite these efforts, scientists still reported immense underwater plumes of dissolved oil not visible at the surface as well as an 80-square-mile "kill zone" surrounding the blown well.

Assessing the scale of the BP spill has proven controversial, and calculating the ultimate impact of the spill will likely be impossible. After the well was capped, the oil appeared to dissipate more rapidly than expected due to a combination of factors: the natural capacity of the region to break down oil, winds from storms, and the cleanup response by BP and the government. Some scientists have suggested that as much as 40% of the oil may have simply evaporated at the ocean surface; others argue that around 50–75% of the material that came out of the well remains in the water or on the sea

Deepwater Horizon Oil Spill—Cumulative Oil Slick Footprint, April 25 – July 16, 2010

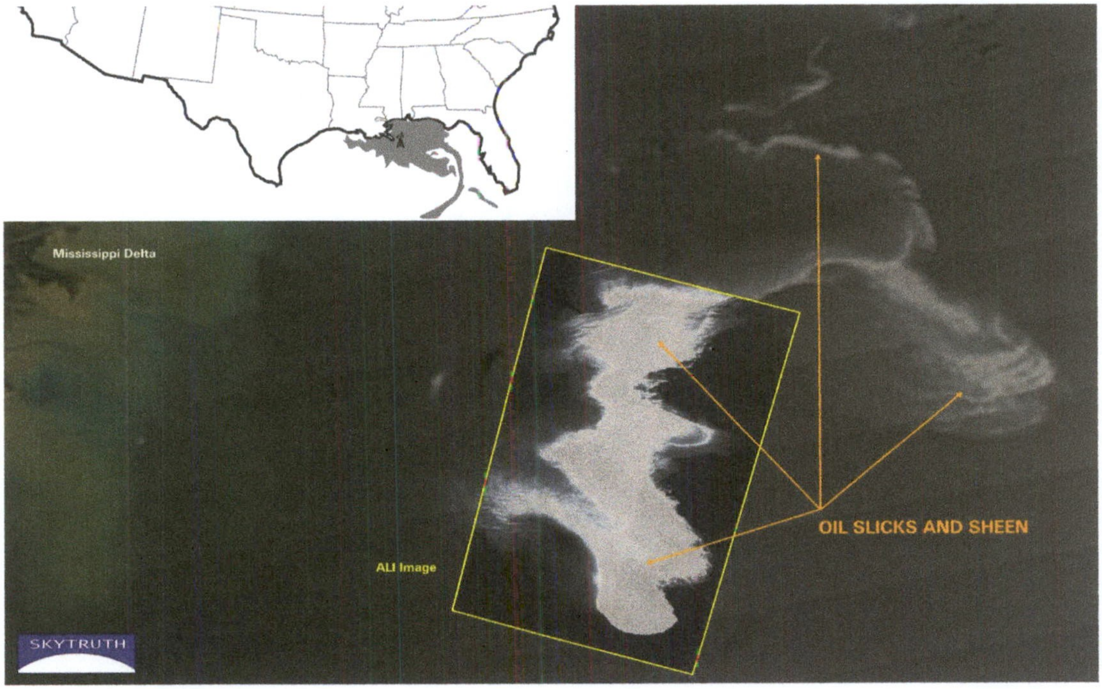

Figure 3.14 Graphic showing the cumulative oil slick footprint of the BP Deepwater Horizon oil spill in the northeast Gulf of Mexico (top left). I created the map by overlaying all of the oil slicks mapped by Sky Truth on satellite images taken between April 25 and July 16, 2010. The satellite image shown here is one of those used in creating the slick map. Cumulatively, surface oil slicks and sheen observed on satellite images directly affected 68,000 square miles of ocean, an area the size of Oklahoma. Images courtesy of Sky Truth.

floor. It left behind a disrupted coastal economy and a devastated ecosystem: the deaths of as many as 105,400 sea birds, 7,600 adult and 160,000 juvenile sea turtles, up to a 51% decrease in dolphins in Louisiana's Barataria Bay, and as many as 8.3 billion oysters lost.[11] BP admitted that they made mistakes which led to the spill (the main cause of the blowout was a defective cement job around the well) and settled for $20.8 billion, the largest environmental damage settlement in U.S. history.

As bad as marine oil pollution can be, air pollution resulting from the use of petroleum products is arguably even worse. While it is easy to point fingers at the large oil companies, the reality is that they are not really the culprits—it is each and every one of us who owns and drives a car. We use two-thirds of the oil in the United States for transportation, and we have been driving increasingly larger cars greater distances every year (Figure 3.15). Transportation accounts for half of nitrogen oxide emissions in the United States, a third of carbon dioxide emissions, and a host of other air emissions, including carbon monoxide, ozone, sulfur oxides, particulates, and toxic metals (we will discuss these in much greater detail in Chapter 4). These emissions contribute to

[11] Source: www.noaa.gov

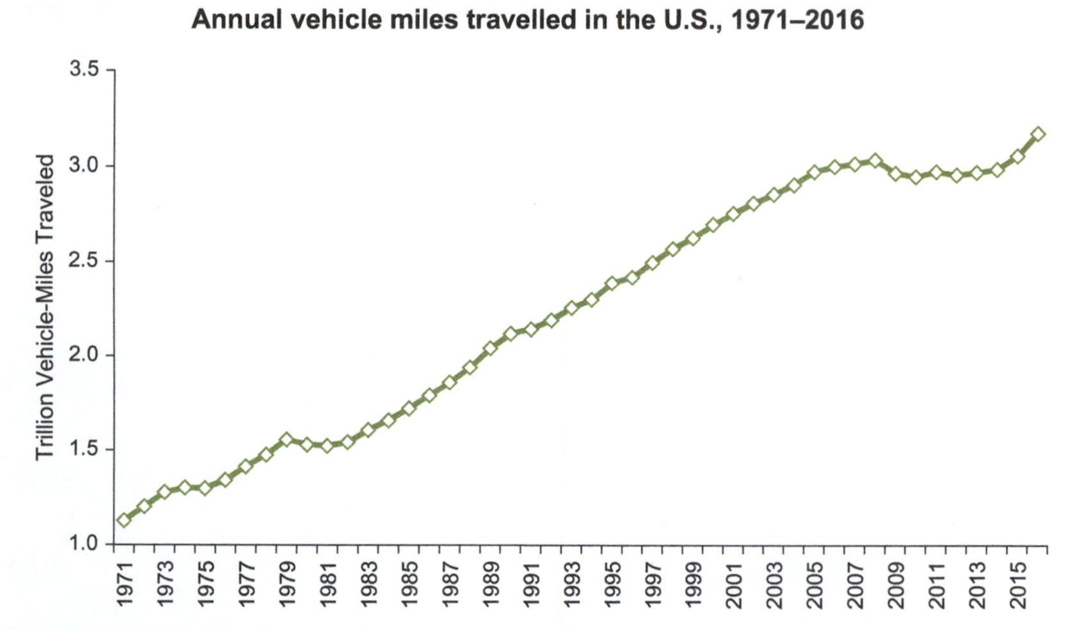

Figure 3.15 Historical vehicle miles travelled in the U.S. (Source: U.S. Department of Transportation—www.fhwa.dot.gov/policyinformation/travel/tvt/history). Note the impact of the global financial crisis in 2008.

urban smog, acid rain, global warming, health problems in humans and animals, damage to crops, forests, and buildings, degradation of habitat . . . the list seems endless. And while car makers and gas producers certainly bear some responsibility in continuing our dependence on polluting oil, we must also take some personal responsibility by choosing vehicles with higher fuel efficiency and by minimizing the amount we drive.

Natural Gas

Natural gas has been called "the prince of hydrocarbons" by some, and it is becoming an increasingly important fuel source in the world energy system. The main ingredient in natural gas is methane, a gas (or compound) composed of one carbon atom and four hydrogen atoms, and it is possible to burn it. Chemically, this process consists of a reaction between methane and oxygen. When this reaction takes place, the result is CO_2, water (H_2O), and a great deal of energy! The gas itself formed millions of years ago from the remains of oceanic plankton decaying and building up in thick layers on the ocean floor. Over time, they were covered by layers of sand and silt which, in turn, changed to rock, trapping the organic material beneath. Pressure and heat changed some of this organic material into coal, some into oil (petroleum), and some into natural gas—tiny bubbles of odorless gas.

Domestically abundant (Box 3.2), natural gas appears to offer a number of environmental benefits over other sources of energy, particularly other fossil fuels. Emissions from natural gas are much less than coal or oil, the latter being composed of much more complex molecules, with higher carbon, nitrogen, and sulfur contents. This means that when combusted, coal and oil release

BOX 3.2

Are We Living in the "Golden Age" of Natural Gas?

Natural gas has a long history as a reliable fuel source for home heating, industrial manufacturing, and electrical generation. However, securing long-term supplies has always been tied to the discovery and development of conventional oil and gas reservoirs that over time became more difficult and expensive to find and often occur in environmentally or politically sensitive areas throughout the world.

Geologists have long known that the source for our known oil and gas deposits was actually deeper plankton-rich mud layers that, over geologic time, hardened into black shales. These shale source rocks are subjected to heat and pressure that can transform the original organic matter into oil and natural gas that migrates upward into overlying geologic traps to form major targets for drilling worldwide. For over 100 years, conventional thinking has always been that these shales, which often still contain about 80% of the original hydrocarbons, were too impermeable to ever produce commercial supplies of either oil or gas. All that changed in 2002 near Fort Worth, Texas, when two small independent producers (Mitchell Energy and Devon Energy) decided to drill horizontally and fracture ("frac") the gas rich but nonproductive Barnett Shale source rock. Their engineers pumped millions of gallons of water mixed with sand, under very high pressures (over 5,000 psi), down the drill hole. The "water-sand frac" hit the tight, brittle Barnett Shale like a hydraulic sledgehammer, freeing up tremendous amounts of stored natural gas. Continuously underlying over 5,000 square miles in North Texas, the Barnett is now the largest producing gas field in the United States with estimated reserves measured in trillions of cubic feet, but it is not the only organic shale in the country.

This unconventional technique of fracturing shale source rocks has changed the entire outlook for domestic supplies of natural gas in this country. As the figure below shows, there are many other gas-rich shales throughout the country covering more than 26 states. Taken together, there is a tremendous potential for domestic production of cleaner burning shale gas and perhaps reducing dependence on imported oil until the next generation of fuels are developed. However, it should be pointed out that developing all this shale gas has also led to concerns about water use, disposal, and/or treatment of well flow-back fluids, as well as issues related to urban drilling sites and pipeline infrastructure development throughout the country.

In addition to the discovery of gas-rich shales throughout our country, large organic shale deposits are also known to exist throughout Europe, Asia, and South America. With so much potential for developing vast supplies of unconventional shale gas, many believe that we may be entering the long anticipated Golden Age of Natural Gas. If true then natural gas (methane, CH_4) will likely play a major role in our energy future as a source of hydrogen. Freeing up the hydrogen from methane already accounts for 95% of all hydrogen produced in the United States using a process called steam methane reforming (SMR). As a matter of fact, hydrogen produced from SMR is used to help lift the Space Shuttle off the launch pad. Technologies are already being developed to fully capture the carbon (known as carbon sequestration) that is produced as CO_2 in the process.

Contribution by Dr. Ken Morgan

(Continued)

(Continued)

higher levels of harmful emissions, including higher carbon emissions[12], nitrogen oxides (NOx), and sulfur dioxide (Table 3.1). Compared to coal, natural gas produces somewhere in the region of 40% fewer carbon emissions for each unit of energy produced and about 25% less than oil. Gas also produces no solid waste, unlike the massive amounts of ash from a coal plant, and very little sulfur dioxide and particulate emissions. It is easy to transport, easy to use, and seems to be a vast improvement over coal and oil.

So why do not we simply use more natural gas and less coal, particularly for generating electricity? Well, the simple answer is, we are! Look again at Figure 3.5. The main reason for this ongoing switch to natural gas is technology, specifically our ability to extract tightly held gas trapped in **shale** formations deep underground. In shale reservoirs, engineers must split the rock to release the gas by pumping pressurized water, sand, and chemicals underground to open fissures and improve the flow of gas to the surface, a process called **hydraulic fracturing** or **fracking** (Figure 3.16). This

[12] Some people use carbon and carbon dioxide interchangeably. The fraction of carbon in carbon dioxide is the ratio of their weights. The atomic weight of carbon is 12 atomic mass units, while the weight of carbon dioxide is 44 because it includes two oxygen atoms that each weigh 16. So, to switch from one to the other, use the formula: One ton of carbon equals 44/12 = 11/3 = 3.67 tons of carbon dioxide. Thus, 11 tons of carbon dioxide equals 3 tons of carbon.

Table 3.1 Fossil Fuel Emission Levels: Pounds per Billion Btu[13] of Energy Input.

Pollutant	Natural Gas	Oil	Coal
Carbon Dioxide	117,000	164,000	208,000
Carbon Monoxide	40	33	208
Nitrogen Oxides	92	448	457
Sulfur Dioxide	1	1,122	2,591
Particulates	7	84	2,744
Mercury	0.000	0.007	0.016

Source: EIA—Natural Gas Issues and Trends

has proven controversial on several fronts. First, fracking uses large volumes of water, about 1 million gallons per well. There are also questions surrounding the impact of drilling, specifically, that it causes earthquakes and can pollute groundwater aquifers. In December 2011, the U.S. EPA announced that compounds likely associated with fracking chemicals had been detected in the groundwater beneath a small community in central Wyoming, where residents say their well water reeked of chemicals. However, the gas industry (and many in the scientific community) contends that fracking is indeed safe and that, because wells are located in such fine-grained rock with exceptionally low porosities so far below the water table, the likelihood of groundwater contamination is extremely remote. Data from many thousands of hydraulic fracturing jobs support this assertion. In fact, research shows that hydraulic-fracture heights are relatively well contained within the shale formation itself and that any groundwater contamination, while possible, is most likely the result of poor sealing around the well casing near the surface, rather than the migration of chemicals from the fissures in the shale itself.

New technologies have greatly reduced the number and size of areas disturbed by drilling, sometimes called "footprints". For example, advanced seismic technologies are making it possible to discover natural gas reserves while drilling fewer wells. The use of **horizontal drilling** and **directional drilling** now makes it possible for a single well to produce gas from much bigger areas. This is especially important when drilling occurs in urban areas (Box 3.2), where opposition to shale gas is especially strong. Residents cite having to endure noise and associated movement of vehicles in and around the pad site in addition to the potential of groundwater contamination.

While carbon emissions from natural gas are lower than those of coal or oil, methane is itself a very potent greenhouse gas. In fact, methane is much more effective than carbon dioxide at "trapping heat" in the atmosphere on a pound-for-pound basis. According to the EIA, although methane emissions account for only 1.1% of total U.S. greenhouse gas emissions, they account for 8.5%

[13] A BTU, short for British Thermal Unit, is a basic measure of thermal (heat) energy. One BTU is the amount of energy needed to heat one pound of water one degree Fahrenheit.

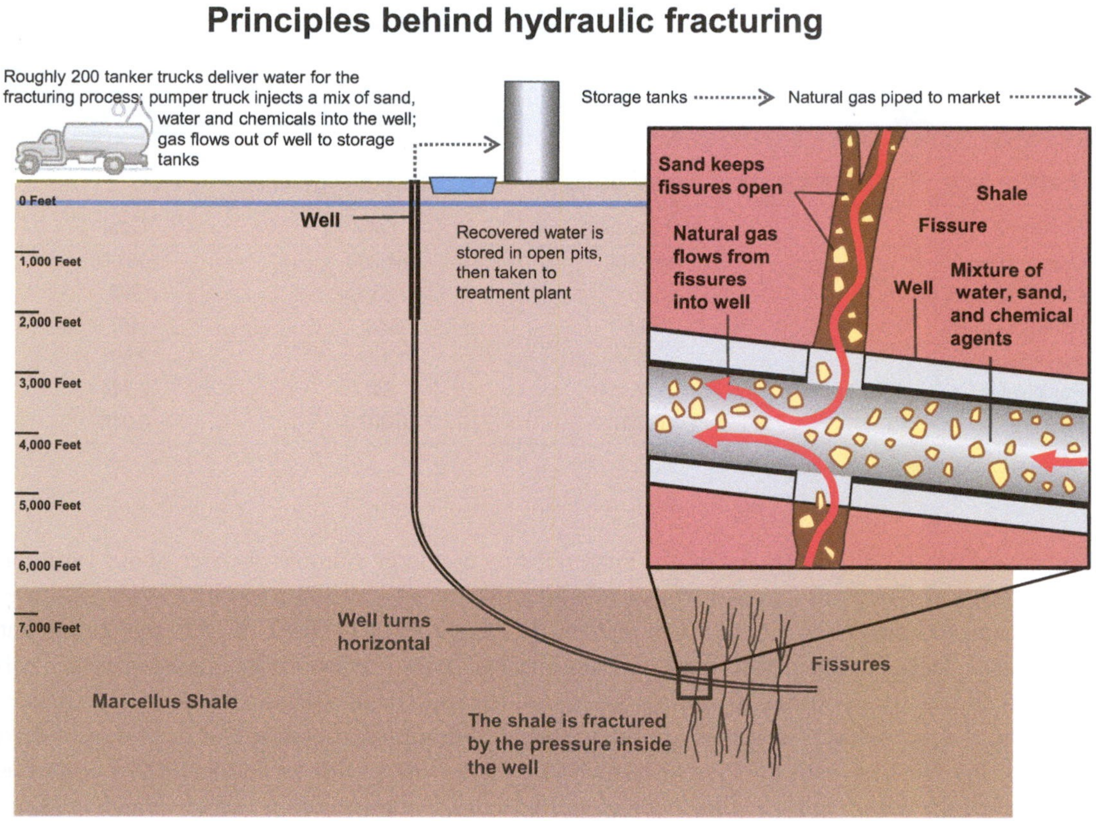

Principles behind hydraulic fracturing

Figure 3.16 Hydraulic fracturing, or "fracking", involves the injection of more than a million gallons of water, sand, and chemicals at high pressure down and across horizontally drilled wells several thousand feet below the surface. The pressurized mixture causes the rock layer, in this example the Marcellus shale in Pennsylvania, to crack. These fissures are held open by the sand particles so that the natural gas trapped in the shale can flow up the well.

of the greenhouse gas emissions based on *global warming potential* (we discuss global warming in much greater detail in Chapter 6). We tend not to hear about methane in the debate over global warming because the sheer volume of carbon dioxide emissions into the atmosphere is so high. However, a paper by scientists from Cornell University found that between 3.6% and 7.9% of the methane from shale-gas production escapes to the atmosphere in venting and leaks over the lifetime of a well.[14] When this is accounted for, scientists argue, carbon emissions associated with shale gas are no better—or may even be worse—than those from coal. So while the jury on this issue is very much still out, on balance, I think the reduction in emissions from increased natural gas use appears to outweigh the potential detrimental effects of increased methane emissions. Thus, the increased use of natural gas in the place of other, dirtier fossil fuels could serve to lessen the emission of greenhouse gases in the United States and elsewhere.

[14] www.sustainablefuture.cornell.edu/news/attachments/Howarth— EtAl— 2011.pdf

NUCLEAR ENERGY

Nuclear power is an extremely important source of energy worldwide. As of September 2018, 38 countries worldwide are operating 451 nuclear reactors providing about 12.5% of the world's electricity production, with 56 new nuclear plants under construction.[15] In total, 13 countries rely on nuclear energy to supply at least one-quarter of their total electricity, with France (72%), Ukraine (55%) and Slovakia (54%) generating the largest percentage of their electricity from nuclear. In the United States, 99 nuclear power reactors provide about 20% of the country's electricity, making it the third-largest electricity source as a percentage after coal and natural gas. (Figure 3.17). In terms of total generation, the United States leads the way with 805.3 billion kWh produced in 2017, more than double that of France (Figure 3.18).

The **nuclear fuel cycle** uses the element **uranium** (U) that, like the fossil fuels, is processed to produce efficient fuel for generating electricity. Figure 3.19 shows this cycle, which is rather complex. To prepare uranium for use in a nuclear reactor, it is mined, enriched, and fabricated into small pellets (about the size of pencil eraser heads) of uranium dioxide (UO_2) powder. These pellets are then inserted into thin, 12-foot long tubes of stainless steel to form fuel rods. Once the rods are sealed, they are assembled in clusters of about 100 rods to form fuel assemblies. Several hundred fuel assemblies are then placed in the core of the nuclear reactor. These steps make up what is termed the "front end" of the nuclear fuel cycle.

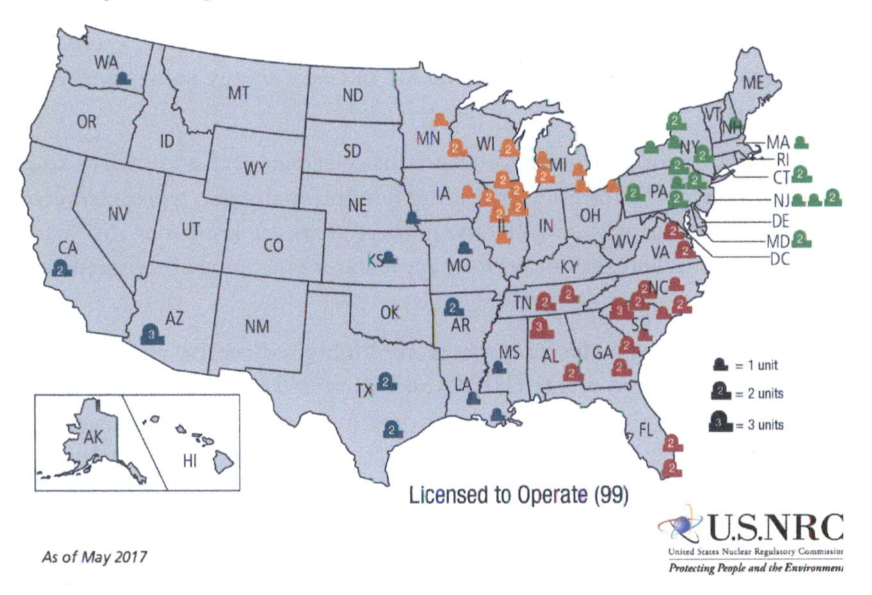

Figure 3.17 Map showing the 99 operating nuclear power reactors in the U.S. (Source: U.S. Nuclear Regulatory Commission).

[15] Source: Nuclear Energy Institute (www.nei.org).

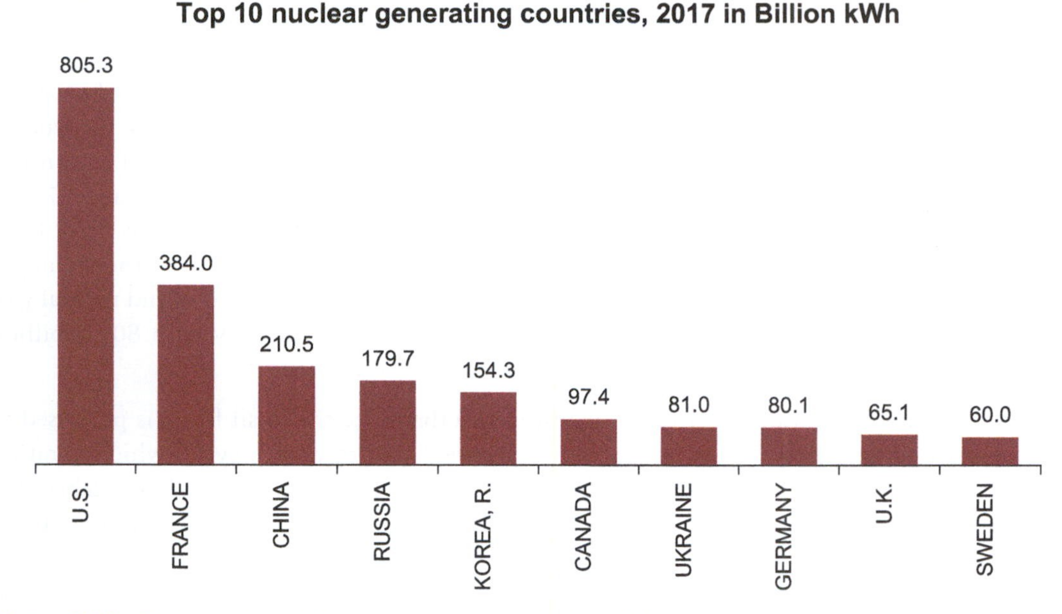

Figure 3.18 The ten countries with the greatest nuclear generation in 2017 (Source: www.iaea.org).

In the reactor core (see Figure 3.20), the uranium isotope (in the pellets) is bombarded with neutrons, causing it to fission, or split. This produces an enormous amount of heat in a continuous process called a **chain reaction**. The process is fully controlled and not, as some think, a nuclear explosion. The heat is used to produce steam to drive a turbine and an electric generator. The process is highly efficient (one uranium pellet equals one ton of coal's energy equivalent) and produces no atmospheric emissions.

After the uranium has been used in a reactor to produce electricity, it is known as **spent fuel.** This fuel may undergo a further series of steps, including temporary storage, reprocessing, and recycling, before its eventual disposal as waste. Collectively these steps are known as the "back end" of the fuel cycle. To maintain efficient reactor performance, about one-third of the spent fuel is removed every 12 to 18 months.

Spent fuel assemblies taken from the reactor core are highly radioactive and give off a lot of heat. They are **high-level radioactive waste** (HLW) and are stored in special ponds usually located at the reactor site. The water in the ponds serves the dual purpose of acting as a radiation barrier and cooling the spent fuel. Spent fuel can be stored safely in these ponds for relatively long periods. However, it is intended only as an interim step before the spent fuel is either reprocessed or sent to final disposal. The longer it is stored, the more **radioactive decay** occurs and the easier it is to handle.

This is as far as the nuclear fuel cycle goes at present. The final disposal of non-reprocessed spent fuel has not yet taken place. This is worth emphasizing: *Currently, no permanent storage site of nuclear waste exists anywhere in the world.* It is envisaged that spent fuel rods will be encapsulated in corrosion-resistant metals such as copper or stainless steel and buried in stable rock structures deep

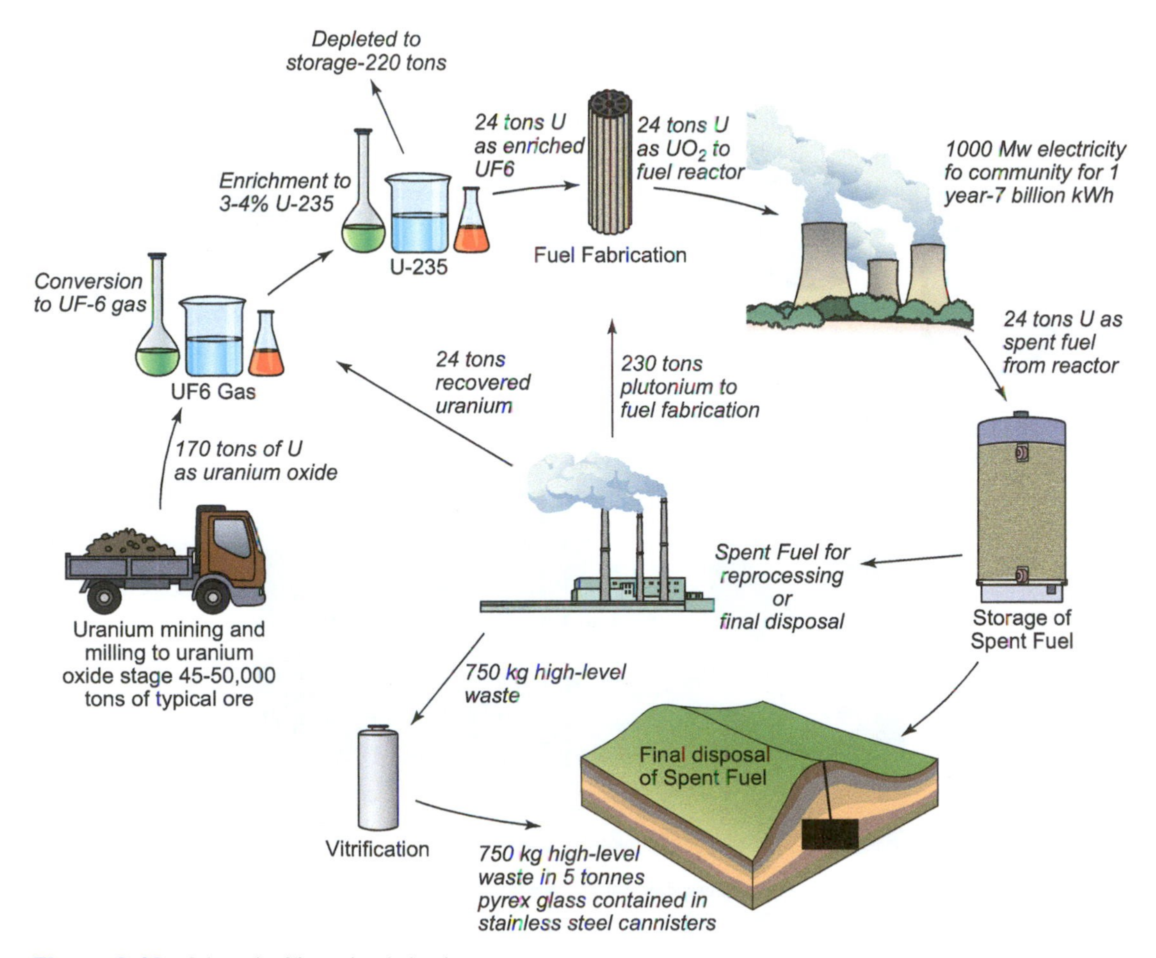

Figure 3.19 Schematic of the nuclear fuel cycle.

underground. The first permanent disposal was expected to occur in 1998 at Yucca Mountain, a site about 100 miles northwest of Las Vegas in the Nevada desert. However, the project faced a series of delays due to legal challenges, concerns over how to transport nuclear waste to the facility, and political pressures resulting in underfunding of the construction. In March 2010, the Obama Administration officially cancelled the project, and Yucca Mountain has now come to be known as the "$20 billion hole in the ground."

So then, is nuclear energy a viable alternative to fossil fuels? From an atmospheric perspective, the answer seems to be a resounding yes! First, nothing is burned during the nuclear process, which means that atmospheric emissions are, effectively, zero. This is an arguing point for proponents of nuclear energy, who contend that such facilities help meet state, national, and international clean-air goals. For example, in 2017, U.S. nuclear power plants avoided the emission of almost half a million tons of nitrogen oxide (the same amount emitted by 25 million passenger cars in a year), 400,000 tons of sulfur dioxide, and about 540 million tons of carbon dioxide from entering Earth's

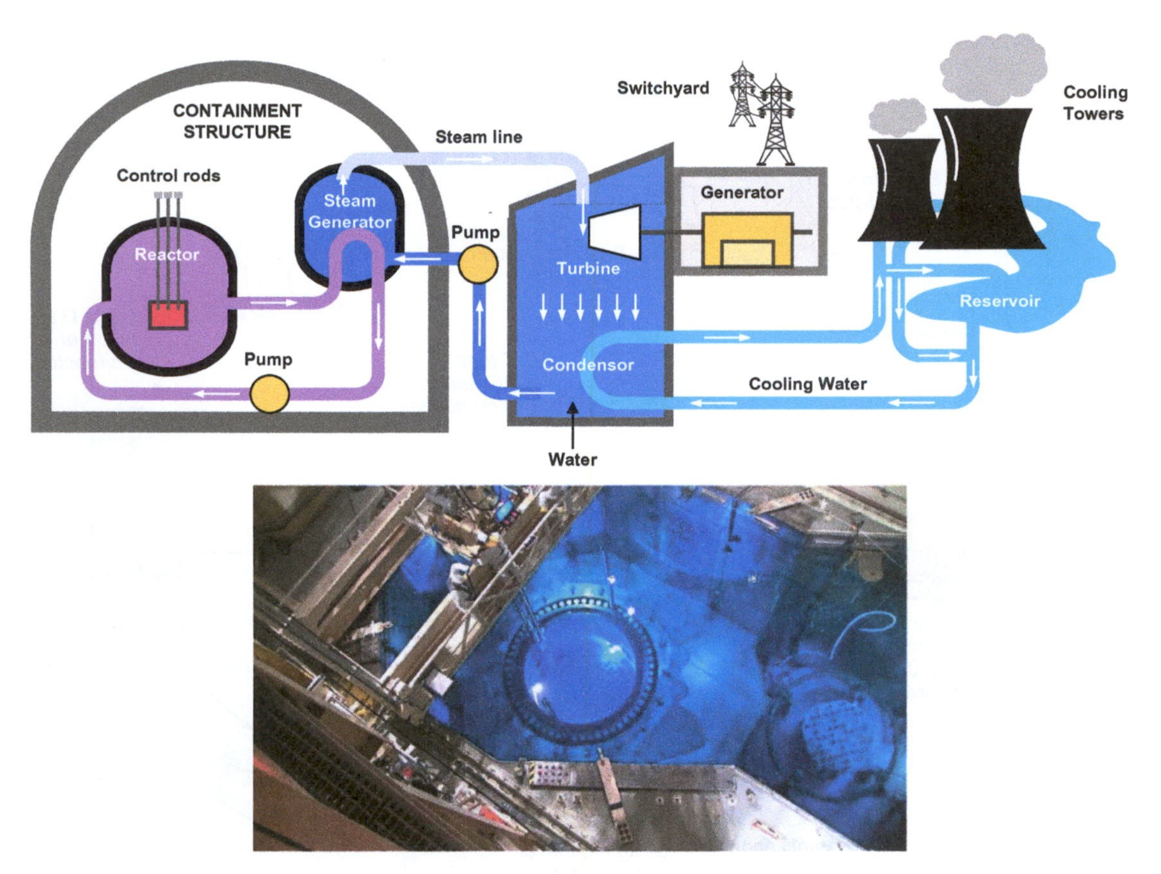

Figure 3.20 Schematic diagram of a typical pressurized water nuclear reactor. The photograph shows the reactor core (Photo credit: © Southern Nuclear. Reprinted by permission).

atmosphere.[16] That last number is especially significant: it represents as much carbon dioxide as is released from 113 million cars, about as many automobiles as there are currently on U.S. roads In fact, worldwide, nuclear energy avoids on average the emission of about 2.5 billion metric tons of carbon dioxide per year or about 7.5% of the global total.[17]

The benefits of nuclear energy extend beyond atmospheric emissions. Cooling water discharged from a nuclear plant contains no harmful pollutants, and it meets federal Clean Water Act requirements and state standards for temperature and mineral content. In fact, nuclear power plants often provide excellent habitats for wildlife and plants. Some companies have developed extensive wetlands, providing better nesting areas for waterfowl and other birds, more habitat for fish, and sanctuaries for other wildlife, flowers, and grasses. These environmental activities have been recognized by the nation's best-known environmental organizations, including the Audubon Society, Ducks Unlimited, and the U.S. Fish and Wildlife Service.

[16] Source: Nuclear Energy Institute (www.nei.org).

[17] In 2009, worldwide, emissions of CO_2 totaled 30.4 billion metric tons (Source: www.eia.gov).

The second Bush administration and much of Congress pushed hard during the 2000s to revive and expand the nuclear industry. President Bush endorsed nuclear as an environmentally friendly energy source. In 2006, the administration's budget increased nuclear power funding by 5%, and Congress followed suit: it gave the nuclear industry $7 billion in research-and-development subsidies and $7.3 billion in tax breaks. In February 2010, President Obama announced over $8 billion in loan guarantees from the Recovery Act to help fund several new nuclear reactors, the first new plants in our country in three decades. As this book went to press, there were just two new nuclear reactors actively under construction in Georgia with 18 applications for new reactors at varying stages of review. It appears that the three-decade hiatus of nuclear plant construction in the United States may well be ending.[18]

Given all of its advantages, why do we not simply build more nuclear reactors as quickly as possible? Unfortunately, that solution is not as simple as it seems. As with almost anything, there are numerous (and sometimes hidden) costs associated with nuclear energy that are not widely publicized. As they say, the devil is in the details.

The number one drawback of nuclear energy is security and safety concerns. Many environmentalists center their critique of nuclear energy on the potential of nuclear reactor meltdowns or malfunctions and the lack of permanent waste facilities. In 1979, the United States had a serious nuclear accident at Three Mile Island, a plant located within 100 miles of Philadelphia, Baltimore, and Washington, DC. Although no radiation was released during the accident, the partial meltdown of the reactor caused widespread panic and dealt a significant blow to the nation's nuclear power industry, particularly in so far as public confidence.

Just one month after the British magazine, *The Economist*, declared in its focus article that nuclear technology was "as safe as a chocolate factory" (1986), a catastrophic nuclear accident at Chernobyl in the former Soviet Union threatened the lives of 130,000 people within a 20-mile radius of the plant and potentially exposed 300–400 million people in 15 nations to radiation. All those living around the plant had to be permanently evacuated. Forecasts of cancer deaths attributable to the Chernobyl accident range from at least 5,000 to 75,000.

The issue of nuclear safety became front-and-center of media reporting in March 2011 following the Japanese tsunami, when the Fukushima Daiichi nuclear power plant suffered a series of equipment failures, **nuclear meltdowns**,[19] and releases of radioactive materials. This happened because the tsunami broke the reactors' connection to the power grid, leading the reactors to begin to overheat, with flooding hindering external assistance. The Japanese government estimates that the total amount of radioactivity released into the atmosphere was approximately one-tenth as much as was released during the Chernobyl disaster, although significant amounts of radioactive material were released into groundwater and oceanic waters. By December 2011, Japanese authorities declared the plant to be stable.

[18] As of February 2012, 20 proposed nuclear facilities were under review for licensing by the U.S. Nuclear Regulatory Commission.

[19] Nuclear meltdown is an informal term for a severe nuclear reactor accident that results in core damage from overheating. It has been defined to mean the accidental melting of the core of a nuclear reactor.

Scientists are divided on the scale of the Fukushima disaster. Some say that Fukushima is worse than the 1986 Chernobyl accident (which it shares a maximum level-7 rating on the sliding scale of nuclear disasters) with many suggesting that there will be "horrors to come" in Fukushima.[20] On the other side of the nuclear fence are more industry-friendly scientists who insist that the crisis is under control, and radiation levels are mostly safe. Notwithstanding, two things appear certain. First, there will always be risks of a major accident with nuclear power, potentially releasing large quantities of radioactivity into the environment, even though the risks today are relatively low, given improved security and new technologies. Second, whether the Fukushima plant is stable or not, it will most likely take decades to decontaminate the surrounding areas and to decommission the plant altogether, not to mention the health problems caused by radiation that may take decades to present symptoms. The cost of Fukushima, in human terms, may never really be knowable.

Permanent storage of high-level waste remains a key issue, given that spent nuclear fuel is initially thermally hot, highly radioactive, and potentially very harmful. Radioactive waste can only dissipate its energy with time, meaning we must find ways to store spent fuel that provides adequate protection for the public permanently. Meanwhile, in almost all countries, nuclear waste is stored in bunkers (both above and below ground) that are expensive to construct and require strict security measures. It is good to know that the amount of high-level waste requiring permanent storage is actually quite small. The United States currently produces approximately 2,000 tons of spent fuel rods per year. If these fuel rods were stacked together end-to-end and side-by-side, they would fill a football field to a depth of only seven yards. However, the cost of storing nuclear waste is high, and those costs are mounting. The Department of Energy estimates that the government will have to pay $23 billion in damages to nuclear power utilities, which for the past 30 years have paid a fee to DOE on the promise that the feds would begin collecting their waste in 1998. They have not, and industry argues that the damages are actually closer to $50 billion.

Another disadvantage to nuclear energy is trying to gauge the costs of plant construction and maintenance. Each plant costs a lot of money to build, around $5 billion on average, according to the International Atomic Energy Agency, with a mean construction time of 7.5 years.[21] This means that companies have to finance construction costs through time. The permit alone can cost anywhere from $45 million to $70 million.[22] The economics of nuclear power plants are further complicated by the fact that gas-fired power plants are far faster and cheaper to build. This, coupled with the revolution in drilling techniques known as directional drilling and fracking, has produced a boom in cheap natural gas, as discussed earlier.

Most people do not realize that nuclear reactors have a finite life and have to be decommissioned after 40 years.[23] Decommissioning, including restoration of the land surrounding a closed plant, costs between $300 and $500 million per plant, according to the Nuclear Regulatory Commission. This means that somewhere in the region of $32 to $52 billion must be set aside to decommission

[20] Helen Caldicott, a physician and former professor at Harvard medical school, in an Op-Ed piece in the New York Times in December 2011, wrote that the people of Fukushima may face a medical catastrophe beyond all proportions in the wake of the disaster.
[21] Source: https://pris.iaea.org/PRIS/home.aspx
[22] Source: https://www.nrc.gov/about-nrc/regulatory/licensing/general-fee-questions.pdf
[23] In the U.S., nuclear plants are licensed for 40 years although all plants have either been granted or are expected to be given a 40-year extension.

all U.S. plants by the middle of this century, which is an astounding sum of money. Since 1988, the Nuclear Regulatory Commission required all nuclear power plant owners to set money aside in trust funds (the so-called Nuclear Decommissioning Trust, or NDT) in order to finance the considerable costs of decommissioning nuclear power plant sites after the reactors shut down. However, there are concerns of significant shortfalls in the trust fund while projected costs continue to rise. The most recent study conducted on decommissioning found a deficit of $27 billion in the trust between the current balance and the estimated $91 billion required to decommission all U.S. nuclear reactors.[24]

Arguably, nuclear power remains the most controversial form of energy. No one doubts that a severe nuclear accident has the potential to do catastrophic harm to people and the environment. A combination of human and mechanical error, or even a natural disaster, could result in the accidental killing of several thousand people, injuring several hundred thousand others, contaminating large areas of land, and costing billions of dollars. Still, we have had more than 25 years of advanced technology since Chernobyl, and proponents argue that the industry has an excellent safety record bar this one major accident. However, the Fukushima disaster could dramatically alter the landscape for further development of nuclear power.

RENEWABLES

Hydroelectric Power

Hydroelectric power is currently the world's largest renewable source of electricity, accounting for 15% of the world's electricity. In the United States, hydropower accounts for about 7% of the total electricity generation, but over half of this hydroelectric capacity is concentrated in just three states: Washington, California, and Oregon. In the Pacific Northwest, hydropower provides about two-thirds of the region's electricity supply.

Hydroelectric power plants convert energy contained in flowing water into electricity by forcing the water, often held at a dam, through a hydraulic turbine that is connected to a generator. The water exits the turbine and is returned to a stream or riverbed below the dam. Using dams to impound water can also store water during rainy periods and release it during dry periods, which results in consistent and reliable electricity production.

Traditionally thought of as a cheap and clean source of electricity, most large hydroelectric schemes planned today are facing increased opposition from environmental groups. The construction of hydropower plants can alter sizable portions of land mainly by flooding areas that may have once served as wildlife habitat, farmland, or scenic retreats. The size of reservoirs (i.e., the lake formed by a dam) created can be extremely large. For example, the Three Gorges Dam in China is the largest hydropower station and dam in the world, with a 370-mile-long reservoir (Figure 3.21). The project is also the largest power station in the world: the 32 turbines have an electric generating

[24] Source: https://www.callan.com/wp-content/uploads/2017/09/Callan-2017-NDT-Survey.pdf

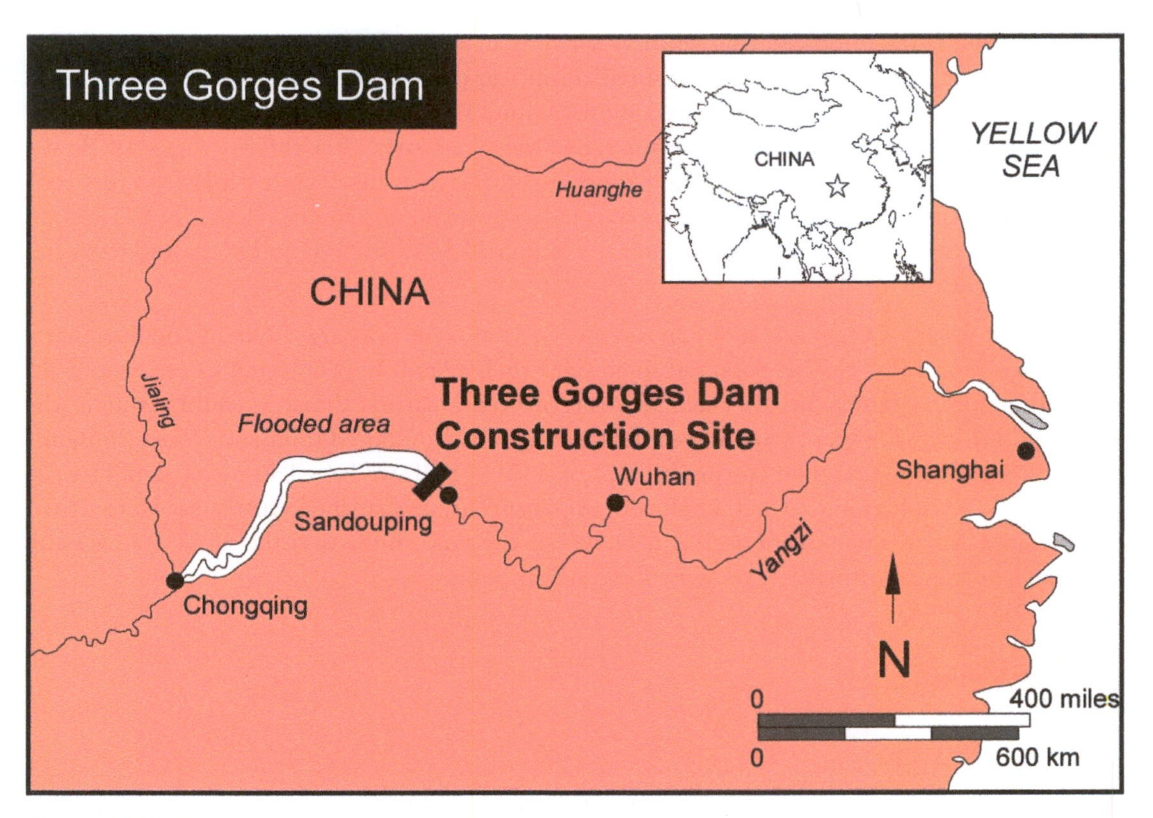

Figure 3.21 The Three Gorges dam project in central China.

capacity of 22,500 MW, equivalent to about 25 nuclear power reactors! However, an estimated 31,000 hectares (approximately 76,500 acres) of farmland has been submerged in a country already suffering from a severe shortage of arable land.[25] The reservoir has also caused massive resettlement: two cities and 140 towns were inundated, and more than 1.1 million people had to be moved. The reservoir has also inundated some 1,300 archaeological sites.

Construction and operation of hydropower dams can also significantly affect natural river systems. Damming a river can alter the volume and quality of water downstream of the dam, as well as prevent fish from migrating upstream to spawn and reproduce. Sediment normally carried downstream is trapped by a dam and deposited on the bed of the reservoir. This sediment can slowly fill up a reservoir, decreasing the amount of water that can be stored and used for electrical generation. Furthermore, the river downstream of the dam is deprived of sediment that helps maintain aquatic habitats and soil functioning.

In North America and Europe, a large percentage of hydropower potential has already been developed. Public opposition to large hydro schemes will probably result in very little new development of big dams and reservoirs. Much of the remaining hydro potential in the world exists in the

[25] Arable land (from the Latin word *arare*, meaning to plow) is an agricultural term, meaning land that can be used for growing crops.

developing countries of Africa, Asia, and South America. Because hydroelectric facilities generally have very high construction costs, harnessing this resource requires billions of dollars. In the past, the World Bank has spent billions of foreign aid dollars on huge hydroelectric projects in developing countries. Opposition to hydropower from environmentalists and native people, as well as new environmental assessments at the World Bank, will most likely restrict the amount of money spent on hydroelectric power construction in the developing countries.

The environmental impacts of dams, though significant, must be weighed against the environmental impacts of alternative sources of electricity. Hydroelectric power plants do not emit any of the standard atmospheric pollutants such as carbon dioxide or sulfur dioxide. In this respect, hydropower is far better than fossil fuels to produce electricity because it does not contribute to climate change, acid rain, or other consequences previously discussed (Figure 3.22). Similarly, hydroelectric power plants do not result in the risks of radioactive contamination associated with nuclear power plants. Hydroelectric power has always been an important part of the world's electricity supply providing reliable, cost-effective electricity and will continue to do so for decades to come. The future of hydroelectric power will depend on future demand for electricity as well as how societies value the environmental impacts of hydroelectric power compared to the impacts of other sources of electricity.

Wind Power

Wind power is the world's fastest growing electricity generation technology. In the United States, wind power continues to increasing at a rate of about 20% per year, becoming a mainstream option to meet growing electricity demand. As of August 2018, the United States had over 91,000 MW

Figure 3.22 Lake Arenal in Costa Rica. Renewable sources like hydroelectric, geothermal, and wind energy supply more than 95% of the energy output for the country. Note the wind turbines on the horizon toward the right of the photo (Photo: Mike Slattery).

of installed wind capacity (Figure 3.23), second only to China. Total U.S. wind power capacity is now capable of powering almost 24 million American households (see Boxes 3.3 and 3.4 for an explanation of the difference between power and energy as well as how much wind power turbines actually generate). While wind energy accounts for just 6.3% of the electricity generation in the United States, many states rely heavily on this renewable resource. In Texas, for example, wind energy's share of the state's electricity generation is 15%. Iowa and Kansas get almost 36% of their electricity demand supplied through wind power.

Wind turbines use two or three long blades to collect the energy in the wind and convert it to electricity. To create enough electricity for a town or city, several wind turbine towers need to be placed together in groups or rows to create a wind farm (Figure 3.24). Generally, wind is consistent and strong enough in many parts of the United States, especially over the Great Plains states and over various mountain ranges, to generate electricity using wind turbines (Figure 3.25). In fact, America's wind resource is vast. The Rocky Mountain and Great Plains states, sometimes referred to as the "Saudi Arabia of Wind," have sufficient wind resources to meet the electric power requirements of the United States several times over! The difficulty, of course, is being able to capture that wind efficiently and then transport the power to where the population is—namely, the west and east coast. In terms of total installed capacity, Texas is the leader in wind power development with over 23,000 MW installed as of mid-2018 (Figure 3.26).

Wind power is arguably the most environmentally benign source of energy. It causes no emissions of harmful pollutants including the greenhouse gas carbon dioxide. In addition, wind does not

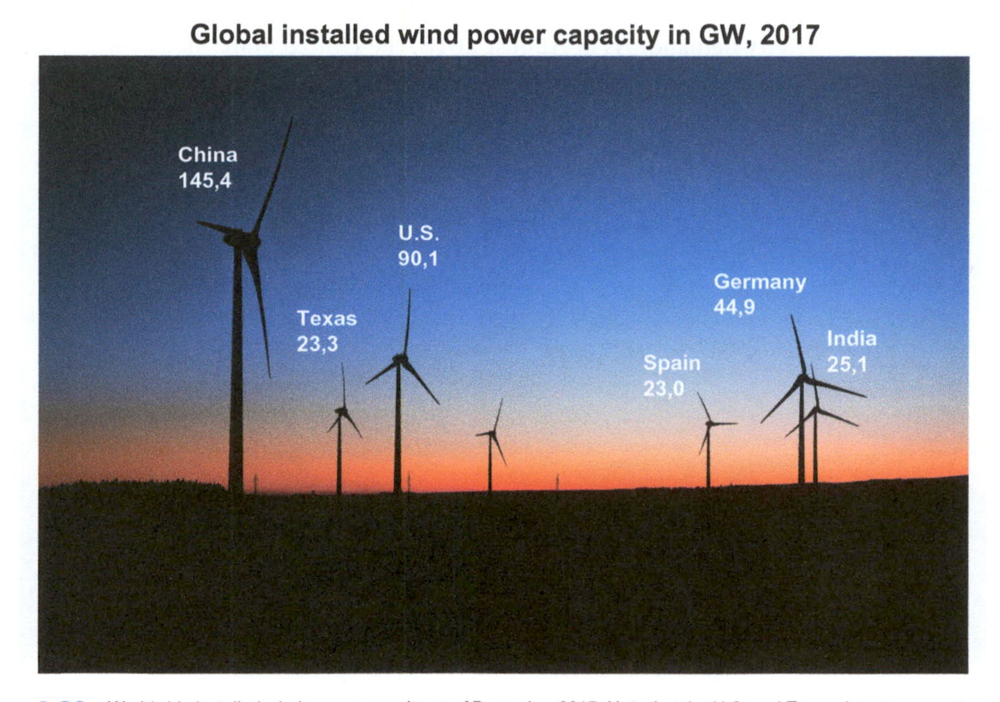

Figure 3.23 Worldwide installed wind power capacity as of December 2017. Note that the U.S. and Texas data are current up to June 30, 2018 (Sources: www.gwec.net and www.awea.org, Shutterstock.com).

BOX 3.3

Power vs. Energy—Understanding the Difference

Many people assume that power and energy are the same thing. They are not. *Power* is the **rate** at which energy is consumed, expressed in watts or kilowatts (or Joules per second). *Energy* is the actual **amount** of power consumed, usually expressed in watt-hours or kilowatt-hours (kWh). To really understand energy use, and consequently those dreaded monthly utility bills, we must factor in the amount of *power* our devices and appliances use and how long we use them.

Let us use the example of a typical light fixture with a 60-watt light bulb. Sixty watts represents the amount of *power* the lamp consumes, or the rate at which the lamp actually uses energy. Now, if you run a 60-watt light bulb from sunSunup to sunSundown for 12 h, you will consume 720 watt-hours of *energy* (or 0.72 kWh). In Texas, we currently pay around 8 cents per kilowatt-hour for electricity, so that light bulb would cost 5.76 cents per night, or $21.02 annually. If you look at your utility bill, you will see that you are charged for the number of kilowatt-hours that you consume. So to reduce the *energy* you use, you must either reduce the amount of *power* you use (i.e., reduce wattage) or reduce the amount of time you use that *power* (or, preferably, both!). So, while *power* and *energy* are closely connected, they are not the same.

How Much Power Do Wind Turbines Really Supply?

Every wind turbine is rated to produce a certain amount of power, just like the light bulbs you buy from the store (Box 3.3). As an example, a 1.5 MW wind turbine will generate its full rated power when the wind reaches, or exceeds, a certain speed, approximately 11 meters per second (m/s). If the wind blows at this speed every hour for an entire year (i.e., 8,760 h), the turbine would generate 13,140 MWh of energy (i.e., $1.5 \times 8{,}760$). Because wind is intermittent, and sometimes does not blow at all, wind turbines produce less energy than their full power rating, typically around 30%, which is known as the *capacity factor*. This means that our 1.5 MW turbine would generate about 3,900 MW of energy each year. Given that the average annual energy consumption of U.S. households is about 11.5 MWh, each 1.5 MW turbine supplies enough energy to power about 340 households (i.e., 3,900/11.5). Energy here is couched in terms of "households equivalent." It does not mean you can put up a single 1.5 MW turbine in one location and assume that it will provide enough electricity consistently for 340 homes. But expressing the energy produced by wind turbines in terms of households equivalent allows for comparison across generation technologies and is more easily understood by the general public.

Figure 3.24 Several wind turbines grouped together as part of a wind farm near Sterling City, Texas (Photo: Mike Slattery).

require mining or drilling for fuel, does not cause radioactive or hazardous wastes, and does not use water for steam generation or cooling. Wind farms can spread out over large areas and farmers can continue to work the land around the turbines. In fact, most land uses remain the same when a wind farm is installed. In addition, farmers and ranchers can earn income from the wind turbines, essentially reaping a "second-crop" year-round. The wide, open landscape where once-ubiquitous windmills helped homesteaders and ranchers pump water for their cattle now plays host to a new generation of wind turbines that generate clean, inexhaustible power.

Critics of wind power cite three categories of environmental impacts: (1) visual impacts, (2) noise pollution, and (3) impacts to wildlife. These impacts can vary tremendously from site to site. Because wind farms are composed of large numbers of turbines, each mounted atop tall towers most frequently in rural areas, they can often be seen for long distances. Some find wind turbines to be enduring symbols of self-sufficiency; others see them as stark intrusions in the natural landscape. Wind turbines, particularly older designs, emit noise that can be heard in the vicinity of the wind farms. However, in reality, the level of noise produced by new technology wind turbines is equivalent to that of a washing machine, which seems a small price to pay for the delivery of clean energy.

Arguably, the most controversial environmental impact of wind turbines is the impact on bird and bat populations. We do know that building a large wind farm causes some habitat loss and that the towers can displace certain species. However, it is also true that birds face daily threats far more lethal than wind turbines. According to a 2014 study of anthropogenic (human-caused) bird mortality in the United States, wind turbines account for a tiny fraction of bird deaths, about two to three

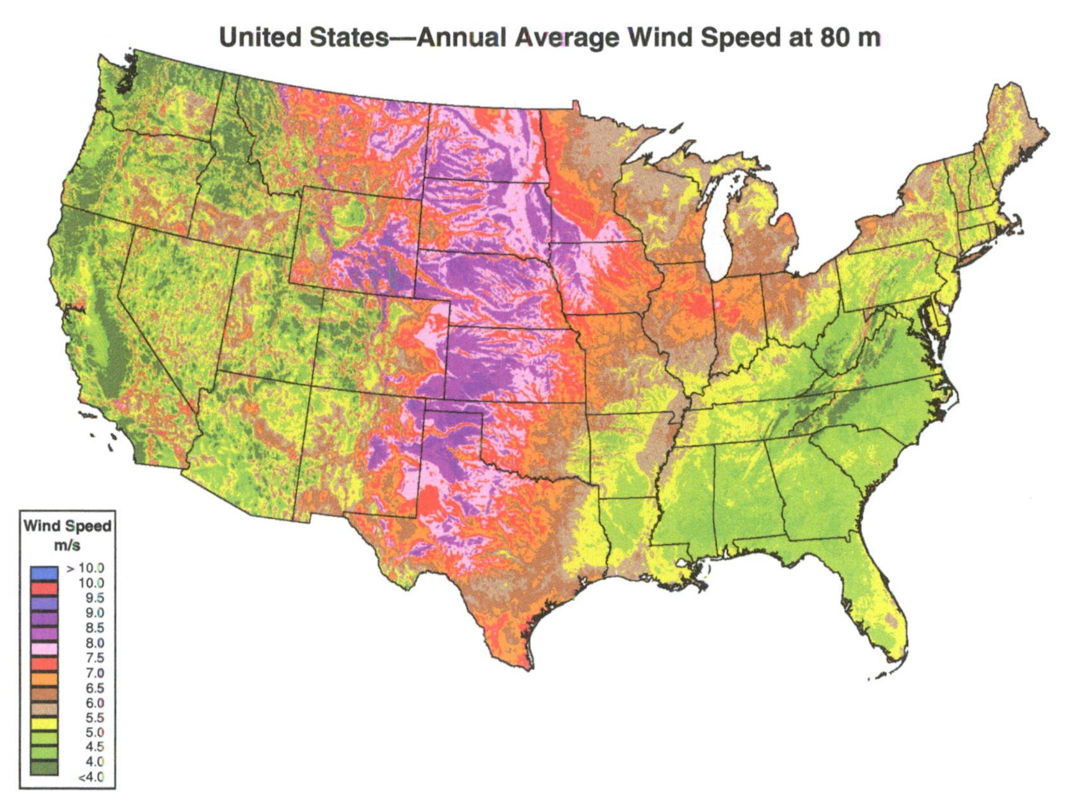

Figure 3.25 Wind resource map of the U.S. The wind speed is measured at 80 meters, as this is generally the height of the nacelle that houses the blades. Speeds of > 6.5 meters per second are considered economically viable winds (Source: National Renewable Energy Laboratory—www.nrel.gov).

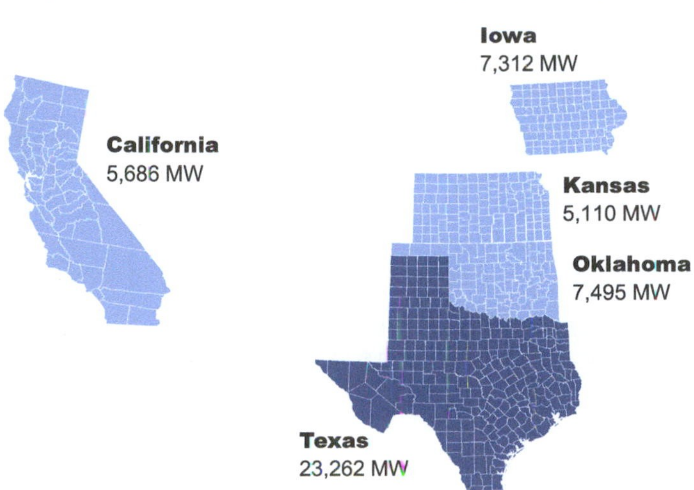

Figure 3.26 The U.S. states with the most installed wind capacity, as of July 2018 (Source: www.awea.org).

birds/MW of installed capacity. Other human-related sources of bird deaths (e.g., communication towers, buildings (including windows), and domestic cats) have been estimated to kill millions to billions of birds each year.[26] There is concern regarding the impacts on bats, especially those that migrate seasonally down through the high-wind resource area shown in Figure 3.25. However, improvements to wind turbine technologies and turbine position have helped mitigate bat mortality, and power companies have shown willingness to curtail (i.e., shut down) turbines during active migration periods. Current wind turbine technology offers solid tubular towers to prevent birds from perching on them, and turbine blades now rotate more slowly than earlier designs, reducing the potential for bird collision.

In a survey of over 1,500 residents throughout the Midwest, we found that less than one quarter of all people strongly agreed that wind turbines either create noise, are unreliable, or are a danger to wildlife (Figure 3.27). One third felt that turbines were unattractive. However, three quarters of all respondents to the survey viewed wind farms very favorably, citing job creation and economic benefits to their counties and state.[27]

Wind power is ultimately clean, cost-effective, inexhaustible, and readily available. It looks to become an essential element of the solution to both climate change and America's increasing demand for electricity. Many industry experts suggest that wind power could supply 20% or more of the electricity used in the United States.[28] However, in order to achieve this, there will have to be far more aggressive limits set on carbon emissions (carbon dioxide is currently an unregulated gas) in order to create incentive to exploit wind and a concerted effort to build transmission lines from where

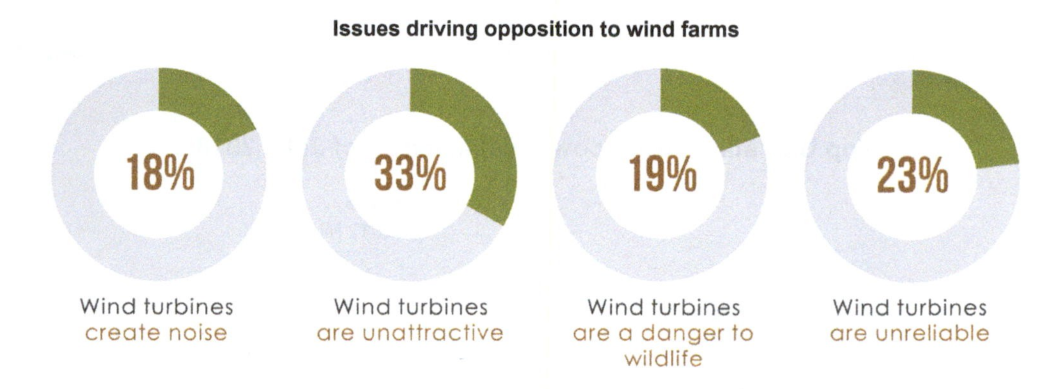

Figure 3.27 Issues driving opposition to wind farms in the Midwest of the United States, based on a survey of 1,500 people. Dark green and percentages correspond to those who agreed and strongly agreed with the question. (Source: Slattery et al. (2012), *Renewable and Sustainable Energy Reviews*, Vol. 16, p. 3690–3701).

---

[26] Erickson, W.P. et al., 2014, PLoSOne, September 15, DOI: 10.1371/journal.pone.0107491.

[27] Slattery et al. (2012), *Renewable and Sustainable Energy Reviews*, Vol. 16, p. 3690–3701.

[28] See www.20%wind.org.

the wind blows to major urban centers. Because wind is intermittent and varies over hours, days, and even seasons, storage of the electricity generated by wind farms must be developed. In West Virginia, the world's largest lithium-ion battery has been installed alongside 98 MW of wind (that is 61, 1.6 MW turbines) at the Laurel Mountain wind farm. The battery is able to store 32 MW of electricity, which will then be fed into the grid during periods of high demand.[29]

Solar Power

The largest potential source of energy is, of course, the sun. Some estimates suggest that solar power has the potential to provide over 1,000 times total world energy consumption. It currently provides less than 0.02% of that total. Like wind, solar power is experiencing rapid growth, doubling in capacity every two to three years. If this growth track continues, it may well become the dominant source of energy this century.

With solar power, sunlight is converted directly into electricity using a solar or **photovoltaic** (PV) cell or indirectly with **concentrating solar power** (CSP) systems. CSP systems use lenses or mirrors to focus a large area of sunlight into a small beam directed at a receiver (Figure 3.28). The

Figure 3.28 An array of parabolic mirrors at the world's largest solar energy generating center in the Mojave Desert, California (Photo courtesy of NextEra Energy Resources).

[29] The energy storage industry is still in its infancy. Over 99% of the energy storage installed globally is made up of pumped hydro, whereby surplus power is used to pump water uphill and then the water flows down, turning turbines, when the spare power is needed.

receiver is a tube positioned right above the middle of the mirror and is filled with fluid, normally synthetic oil, which heats to over 400 °C (750 °F). The reflected light focused at the central tube is 70 times more intense than the ordinary sunlight! The synthetic oil then transfers its heat to water, which boils and drives a steam turbine, thereby generating electricity. The mirrors are made to follow the sun during the daylight hours, repositioning to absorb the maximum amount of sunlight.

Photovoltaics were initially used to power small- and medium-sized applications, from the calculator powered by a single solar cell to off-grid homes powered by a PV array, but larger multi-megawatt photovoltaic plants are being built. The largest solar plants are CSP facilities, such as the Solar Energy Generating System (SEGS) plant in southern California, which has 354 MW of installed capacity on the ground (as shown in Figure 3.28). Here, incoming solar radiation from the sun is among the best available in the United States (see Figure 3.29). However, a frequent criticism of solar power is that the sun only shines brightly for part of the day and that many of the times when there is significant electrical demand (e.g., when people get home from work in winter), the sunlight will be weak. Although the SEGS plant has a large installed capacity, average gross solar output for the site is only around 75 MW, a **capacity**

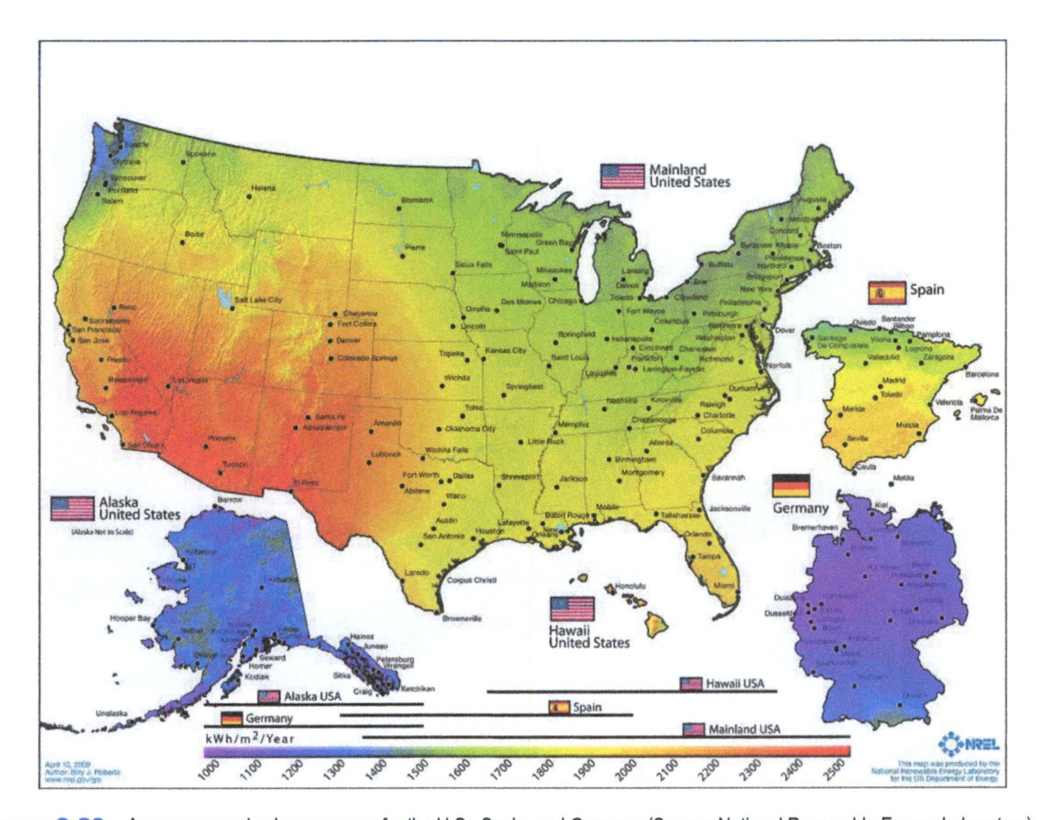

Figure 3.29 Average annual solar resources for the U.S., Spain, and Germany (Source: National Renewable Energy Laboratory).

factor of just 21%.[30] This is one of the reasons that electricity storage has become such an important area of research. Notwithstanding, solar plants such as SEGS power 232,500 homes and displace 3,800 tons of pollution per year that would have been produced if the electricity had been provided by fossil fuels, such as oil.[31]

Terrestrial solar power, much like wind, is a predictably intermittent energy source, meaning that while solar power is not available at all times, we can predict with a very good degree of accuracy when it will and will not be available. Arguably, the biggest disadvantage of solar facilities is their large geographic footprint. A general rule of thumb is that approximately 1 km^2 is needed for every 20–60 MW generated. The SEGS plant discussed above has a total of 936,384 mirrors that cover more than 1,600 acres (6.5 km^2). Lined up, the parabolic mirrors would extend over 229 miles (370 km).

The large amount of land required for utility-scale solar power plants poses problems, especially where wildlife protection is a concern. For example, in the eastern Mojave Desert, power companies are seeking permission to erect over 400,000 mirrors over more than six square miles. The site is also prime habitat for an endangered species of desert tortoise, and this habitat would be permanently lost if the solar plants get built. The Sierra Club and other environmentalists like the project and support the growth of alternative energy, but they also want it relocated to preserve the ecosystem. Federal and state biologists reviewing the plan have proposed that the power company move the tortoises and preserve 12,000 acres elsewhere, which will cost an estimated $25 million. The dispute is likely to echo for years as more companies seek to develop solar and wind plants on sensitive land. The Bureau of Land Management has received more than 150 applications for large-scale solar projects on 1.8 million acres of federal land in California, Nevada, Arizona, New Mexico, Colorado, and Utah. As California pushes to generate one-third of its electricity from renewable sources by 2020, the question that will have to be answered is: what habitat is worth preserving and at what cost?

Biofuel

Biofuel, generally defined as liquid or gas transportation fuel derived from biological material, is being widely touted as a viable alternative to fossil fuels, specifically oil. In the United States, attention is focused primarily on ethanol for use in automotive transport. Ethanol, which is the most common biofuel worldwide, is produced by fermentation of sugars from corn and other crops, such as wheat and sugar cane. You may have heard of E-85, an alcohol mixture that typically contains a mixture of up to 85% ethanol fuel and 15% gasoline. E-85 is used in engines that have been modified to accept higher concentrations of ethanol, and vehicles designed to run on it are known as flex-fuel vehicles (FFVs). Today, the United States has more than 6 million FFVs on the road. These vehicles are available in a range of models, including sedans, pickup trucks, and

[30] The net capacity factor of a power plant is the ratio of the actual output of a power plant over a period of time and its output if it had operated at full capacity the entire time. To calculate the capacity factor, total the energy produced by the plant during a period of time and divide by the energy the plant would have produced at full capacity (see Box 3.4 as a worked example for wind power).

[31] Source: www.nexteraenergyresources.com

minivans, and several auto manufacturers have announced plans to greatly expand the number of FFV models they will offer.

The concerns behind the renewed interest for biofuels include rising oil prices, concerns over the potential oil peak (which seems to be decades away now), greenhouse gas emissions, and instability in the Middle East. Indeed, President George W. Bush said in his 2006 State of the Union speech that the United States should replace 75% of imported oil with biofuel by 2025. Is this feasible and, if so, what (if any) are the potential costs to the environment?

At first glance, ethanol, like other biofuels such as biodiesel, appears to be a no-brainer when it comes to finding alternatives to fossil fuels. Indeed, much of the increased interest in ethanol as a vehicle fuel is due to its potential to replace gasoline produced from imported oil. The United States is currently the world's largest ethanol producer, and most of the ethanol we use is produced domestically from corn grown by American farmers. Statistics from the Renewable Fuels Association[32] show that ethanol production more than tripled between 2005 and 2010, from 3.9 to 15.8 billion gallons (Figure 3.30), allowing the United States to surpass Brazil as the major producer of ethanol.

In 2007, the U.S. Congress gave the agribusiness industry a mandate it simply could not refuse via the Energy Independence and Security Act: corn ethanol production must rise to 15 billion gallons by 2015.[33] This proved a boon to the iconic American farmer: ethanol increased demand

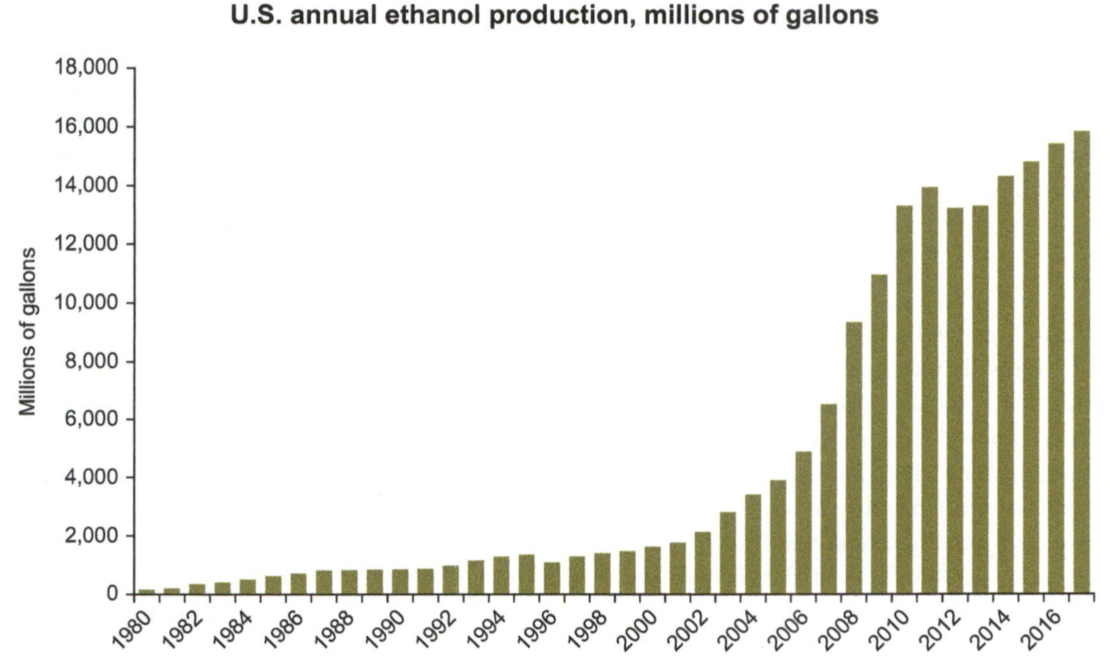

Figure 3.30 U.S. ethanol production in billion gallons from 1980 to 2017 (Source: www.ethanolrfa.org).

[32] Soucre: www.ethanolrfa.org/

[33] Source: www.ethanolrfa.org/resource/standard/

for crops such as corn, increased the prices farmers received for these crops, and brought economic development opportunities to the rural areas where the ethanol is made. However, in early 2009, the industry experienced financial stress due to the effects of the economic crisis of 2008. Motorists drove less (Figure 3.15), gasoline prices dropped sharply (Figure 3.11), and less financing was available. The upshot of all of this is that the number of plants under construction declined dramatically and production of ethanol leveled off at about 13 billion gallons annually (Figure 3.30).

The story of biofuel expansion is being repeated in many countries around the world: from Thailand to Malawi, crops as diverse as oil palms, soybeans, and coconuts are being grown for fuel. This begs the question: Are we seeing the inevitable transition from petroleum to next-generation fuels right in front of our very eyes? The answer is, *possibly*, but no one expects oil to disappear overnight or even in the next two to three decades given the increase in domestic production and the tapping of unconventional resources. The fact is farm-grown biofuels like ethanol and biodiesel still account for only a small fraction of fuel use, as do other renewables such as wind and solar power. Also, as of January 2018, less than 2% of our nation fueling stations offered E-85, and they are often few and far between.[34]

Serious questions remain as to whether we can successfully scal-up biofuels to take on oil. Principally, is there enough land on which to grow energy crops without putting the squeeze on food production? Estimates vary tremendously, depending on factors like how efficiently we become in converting plant biomass to biofuel at any given yield of biomass per hectare. One study, published in *Newsweek* magazine, suggested that for either the United States or Europe to replace just 10% of transport fuel using today's crops and technology would require around 40% of cropland.[35] No one doubts that we are going to need a lot more acreage and big yield improvements if corn production is going to keep up with increased demand for biofuels. And if we go this route, other key questions arise, including: what will this do to the price of corn used for hog, cattle, and chicken feed and, by extension, the price of meat, poultry, eggs, etc?[36] How will this affect aid agencies being able to get affordable food to the hungry? How will this affect people's ability to buy maize for food? Will biofuels really be able to take hold without tax credits and subsidies, especially if oil prices head downward?[37] How will the United States be able to compete with super-efficient countries like Brazil, with year-round growing seasons and cheap farm labor, who now sells ethanol at less than half the cost of U.S. ethanol?

The ramifications of significantly increasing biofuel production, especially in terms of food production and on the environment, have yet to be determined. Given our discussion about population projections in Chapter 2, considerably more land will likely be needed to feed a further 2.5 billion people, and (as we shall see in Chapter 9) we are essentially out of virgin land that could be brought into long-term production. We could turn to the tropical forests and subtropical grasslands, but (as will be discussed in Chapter 7) the ecological cost of bringing such land into

[34] Source: www.e85refueling.com/

[35] Source: *Newsweek*, 8/8/2005.

[36] Since 2000, the price of beef is up 31%, eggs up 50%, and corn sweeteners up 33%, all of which makes food more expensive exacerbating world hunger.

[37] The current ethanol subsidy is a flat 51 cents per gallon of ethanol paid to the agent (usually an oil company) that blends ethanol with gasoline.

production may be prohibitive. With the most suitable land already under cultivation, agricultural expansion into marginal areas to expand biofuels may not be a viable long-term strategy. But that is precisely what is happening in southern countries growing sugarcane, where farmers can get up to five times as much biofuel from each acre of land compared to northern farmers in colder climates. Biofuel experts at the International Energy Agency suggest that, without too much effort, producing ethanol from sugarcane in developing countries like Brazil and India could replace 10% of global gasoline fuel,[38] but at what cost to the environment? One thing we do know is that use of large-scale mono cropping almost always leads to significant biodiversity loss and soil erosion.

Proponents of biofuels argue that we can "grow" our way out of our dependency on oil. Many environmentalists hail the new fuel as cleaner than oil in the long run. Willie Nelson, the legendary country singer, markets his vegetable oil-based BioWillie® to truck stops to "Help eliminate America's dependence on foreign oil, put the American family farmer back to work, and clean up the environment."[39] Opponents from the environmental side are quick to warn that higher corn production is bad for the environment: it requires more fertilizer and produces more chemical runoff into water sources (a link we will make explicitly in Chapter 10). They also point to the fact that ethanol is more inefficient, and hence more costly, than gasoline, which is true. Some scientists have suggested that turning plants such as corn into fuel actually uses much more energy than the resulting ethanol generates.[40] In a report published by the UN, scientists noted that "Unless new policies are enacted to protect threatened lands, secure socially acceptable land use, and steer bioenergy development in a sustainable direction overall, the environmental and social damage could in some cases outweigh the benefits."[41] Scientists have even shown, quite convincingly, that clearing land to produce biofuels will actually do more to exacerbate climate change than using gasoline or other fossil fuels.[42] Converting rainforests, peatlands, savannas, or grasslands in Southeast Asia and Latin America to produce biofuels will remove vast areas of land that have been absorbing carbon for decades. With biofuel, as with many environmental issues, the devil is in the details. Producing 15 billion gallons of ethanol sounds like a lot, yet the United States consumed 143 billion gallons of gasoline in 2017. Simply stated, ramping up biofuel production will do very little, if anything, to reduce our dependence on foreign oil.

Demand for ethanol may be destroyed by the development of a cheaper biofuel. One alternative receiving both attention and research dollars is cellulosic ethanol, the so-called second-generation biofuels, made from plant-based materials like wood and grass. Production of ethanol from lignocellulose (a structural material that comprises much of the mass of plants) has the advantage of abundant and diverse raw material compared to sources like corn and cane sugars but requires a greater amount of processing to make the sugar monomers available to the microorganisms that are typically used to produce ethanol by fermentation. Like all sources of energy, cellulosic ethanol is

[38] Source: www.iea.org/
[39] Source: www.biowillieusa.com
[40] Pimental, D. and Patzek, T.W., 2005, Natural Resources Research, Vol. 14, pp. 65–76.
[41] http://esa.un.org/un-energy/pdf/susdev.Biofuels.FAO.pdf
[42] Searchinger, T. et al. (2008), *Science*, Vol. 319: p. 1238–1240.

controversial. In his State of the Union address in 2006, President George W. Bush said that "We will fund additional research in cutting-edge methods of producing ethanol, not just from corn but also from wood chips and stalks or switch grass. Our goal is to make this new kind of ethanol practical and competitive within six years." The result has been, well, disappointing, to say the least. Under President Bush, and then President Obama, the federal government pumped at least $1.5 billion of grants and loan subsidies to fledgling cellulosic producers. Congress passed, and President Bush signed into law, a mandate to produce 250 million gallons of cellulosic ethanol in 2011, rising to 10.5 billion gallons by the end of this decade. Despite the taxpayer enticements, cellulosic fuel production will not come anywhere near these targets. In fact, the EPA, which has the authority to revise the biofuel mandates, quietly reduced the 2011 requirement by 243.4 million gallons to a mere 6.6 million gallons. In 2013, the agency further reduced the 2013 figures to less than 1 million gallons.

An October 2011 report on biofuels by the National Academy of Sciences[43] concluded that the mandates may not only be an ineffective way to reduce global greenhouse gas emissions but also, because production is so low, advanced cellulosic fuels may do very little to reduce our dependence on foreign oil. The report notes "currently, no commercially viable bio-refineries exist for converting cellulosic biomass to fuel." This is largely due to the high cost of producing cellulosic biofuels compared with petroleum-based fuels. The future of cellulosic fuels remains uncertain, to say the least.

Finally, algae, or third-generation biofuel, has begun to receive serious attention. In principle, biofuels can be made from just about any plant material, and it has some obvious advantages. It would not compete for arable land as it is grown, in water, and it grows like, well, a weed, allowing for significant yields. The two avenues of third-generation development being considered so far are microalgae (pond scum, etc.) and macroalgae (seaweed). Research is ongoing into both harvesting algae from its natural environment and creating artificial growing environments (Figure 3.31). The process is simple and elegant. Algae gather energy from the Sun through the process of photosynthesis. Oil is the by-product of this process, which can then be utilized to create biofuel. The algae itself can be transformed into ethanol through the process of fermentation. But this is still very much a developmental fuel, and no one expects commercialization for several years, as costs will need to come down and production successfully scaled up. Still, some research looks promising. For example, scientists from Utah State University have shown that algae yields about 2,500 gallons of biofuel per acre per year, in contrast to soybeans (approximately 48 gallons) and corn (about 18 gallons). A major advantage is that microalgae can be produced in non-arable areas unsuitable for agriculture. The researchers estimate untillable land in Brazil, Canada, China, and the United States could be used to produce enough algal biofuel to supplement more than 30% of those countries' fuel consumption.[44]

With all of its apparent advantages over other biofuels, algal fuel seems destined for a role in our energy future. However, the environmental backlash against first- and second-generation biofuels is now being replicated with the focus on algae. If the energy needs and greenhouse gas emissions of

[43] National Academies of Sciences (NAS) study, entitled "Renewable Fuel Standard: Potential Economic and Environmental Effects of U.S. Biofuel Policy", available from the NAS (ISBN—10: 0-309-18751-6).
[44] Moody, J.W. et al. (2014), *Proceedings of the National Academy of Sciences*, www.pnas.org/cgi/doi/10.1073/pnas.1321652111.

Figure 3.31 The Solix BioSystems pilot plant facility located outside Durango, CO. The facility has 0.8 acres of photobioreactor cultivation capacity (Photo credit: Jason Quinn, Utah State University).

algal fuel production cannot be lowered, algal fuels may be limited to areas in which they can feed off of wastewater and power plant effluents. Nevertheless, significant venture capital is beginning to flow into the sector, and the economics of algal fuels seems likely to improve. When the world decides that it is time to move on from oil, algae very well may be the best option.

WHERE TO NOW?

The central theme of the debate surrounding our energy future is this: there is no silver bullet and no single quick fix. Wind and other renewables, such as hydropower and solar, should continue to be exploited as viable energy sources, but their expansion and role will depend on how climate change will shape future policy. Based on the current evidence of global warming (see Chapter 6), it seems clear that we must begin to decarbonize energy production, *now*. Several studies have shown that energy efficiency and renewable energy technologies can reduce carbon emissions enough to significantly slow global warming. But significantly ramping up renewables will cost money, and we must be prepared to invest and, possibly, pay more to speed up the integration of these technologies into our daily lives (see Figure 3.32).[45]

[45] Slattery et al. (2012), *Renewable and Sustainable Energy Reviews*, Vol. 16, p. 3690–3701.

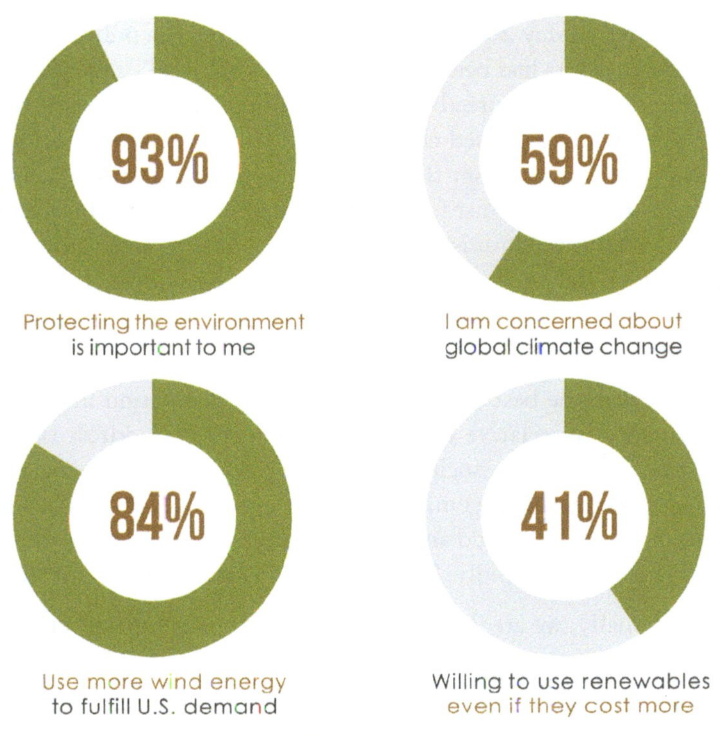

Figure 3.32 Results from a survey of 1,500 residents of Texas, Oklahoma, and Iowa conducted by the author on environmental attitudes and renewable energy use. Note that 84% of respondents agreed or strongly agreed (green) that we should use more wind energy in the U.S., but that level of support dropped to 41% if doing so would cost more (Source: Slattery et al. (2012), *Renewable and Sustainable Energy Reviews*, Vol. 16, p. 3690–3701).

The nuclear industry is hoping that concerns over climate change will result in growing support for nuclear power, but high costs, long construction times, high environmental risk, and problems resulting from waste management suggest that perhaps it is not a sustainable solution to climate change. There are some clear signs in this regard: Germany will phase out nuclear power by 2022. Italy and Switzerland have decided against it, and anti-nuclear advocates in Japan have gained traction. China remains cautious on nuclear power. Yet there is growing enthusiasm for more nuclear power in the United States, Britain, Russia, and Canada. Despite broad international concern that the safety risks of nuclear power are unacceptable even after half a century of widespread use, proponents have argued that a new generation of reactors and strong U.S. regulations justify making it part of the mix to meet the nation's energy needs. And in November 2013, four top environmental scientists released an open letter calling on world leaders to support development of safer nuclear energy systems.[46] The scientists believe embracing nuclear is the only way to reverse the looming threat of climate change which they blame on fossil fuels.

The coal industry argues that rising oil prices, our dependence on foreign sources, uncertainties surrounding the storage of nuclear waste, and the intermittency of wind and solar will justify

[46] A full copy of the letter can be accessed at https://plus.google.com/104173268819779064135/posts/Vs6Csiv1xYr

building more coal plants. In May 2007, the Department of Energy published a list showing that 151 new coal-fired power plants had been proposed, indicating coal's resurgence in electric power generation. The list received widespread publicity with opponents arguing that, if built, these plants would simply bury any proposed carbon targets, unless enormous resources are spent on trying to capture and store carbon from the plants, an approach that is currently uneconomical. The federal government now counts just four new coal projects on the list of planned power plants nationwide, even as the Trump administration has vowed to revive the ailing industry. There is now just one coal facility under construction: a tiny plant being built by the University of Alaska, Fairbanks. Despite all the political rhetoric, the industry is slowly contracting as plants retire and utilities replace them with natural gas and renewables.

Natural gas appears to offer the best solution for electricity generation in the near-term, at least from an emissions perspective relative to coal, but that does not address the issue of increased demand for transportation and its associated emissions. And while bioenergy represents a real opportunity to reduce greenhouse gas emissions, rapid growth in biofuels production will make substantial demands on the world's land and water resources at a time when demand for both food and forest products is also rising rapidly. This complex situation has no easy answers.

One thing is certain: globally, we are in the midst of a power-plant construction boom, and China is the dominant player. China now uses more coal than the United States, Europe, and Japan combined, making it the world's largest emitter of gases that are warming the planet. Chinese corporations are building or planning to build more than 700 new coal plants at home and around the world, some in countries that today burn little or no coal. The Shanghai Electric Group, one of the country's largest electrical equipment makers, has announced plans to build coal fired power plants in Egypt, Pakistan, and Iran. Overall, 1,600 coal plants are planned or under construction in 62 countries, according to the Global Coal Plant Tracker portal, which would expand the world's coal-fired power capacity by almost 50%.[47]

Largely missing in the discussion on emissions from power plants is this: China has also emerged as the world's leading builder of more efficient, less polluting coal power plants, mastering the technology and driving down the cost. While the United States is still debating whether to build a more efficient kind of coal-fired power plant that uses extremely hot steam, China has begun building such plants at a rate of one a month. Indeed, this is good news, but only half of China's coal-fired power plants have the emissions control equipment to remove sulfur compounds that cause acid rain, and even power plants with that technology do not always use it. China has not even begun regulating some of the emissions that lead to heavy smog in big cities, but by continuing to rely heavily on coal, which supplies 70% of its electricity, China ensures that it will keep emitting a lot of carbon dioxide. Even an efficient coal-fired power plant emits twice the carbon dioxide of a natural gas-fired plant. However, coal remains the cheapest energy source in China by a wide margin, and the country has the world's third-largest coal reserves, after the United States and Russia. No matter how much renewable or nuclear energy is in the mix, coal will remain a dominant power source.

[47] https://endcoal.org/tracker/

World energy consumption is undoubtedly going to increase. There is also little doubt that liquids (primarily oil and other petroleum products) are expected to continue providing energy over the next three decades. Liquids remain the most important fuels for transportation, largely because there are few alternatives to replace them. The use of renewable energy sources, such as wind, is expected to continue to expand over the coming years, so long as government policies and programs continue to support renewable energy. The reality with our energy supply, then, is that, for the near future, it is going to be an all-of-the-above approach, as shown in Figure 3.33.

Here in the United States, we are implementing a broad strategy that will move us from an economy dependent on foreign oil to one that relies on homegrown fuels and cleaner energy. Sources such as natural gas and nuclear power, though not perfect from an environmental perspective, are going to be part of that mix. There is no doubt that we need to responsibly expand conventional energy development and exploration here at home, which should strengthen our energy security and create jobs, but we also need to aggressively (and responsibly) develop renewable sources of energy. In this regard, I would hope to see many more wind farms and other renewable technologies built in parts of the United States where the geography is appropriate.

However, government and industry cannot do this alone. We all can (and must!) be part of the solution by taking a few simple, yet essential, steps toward reducing our dependence on fossil fuels. Because transportation accounts for nearly 30% of U.S. annual CO_2 emissions, raising fuel economy is one of the most important things we can do. For each gallon of gas you burn, 20 pounds of heat-trapping

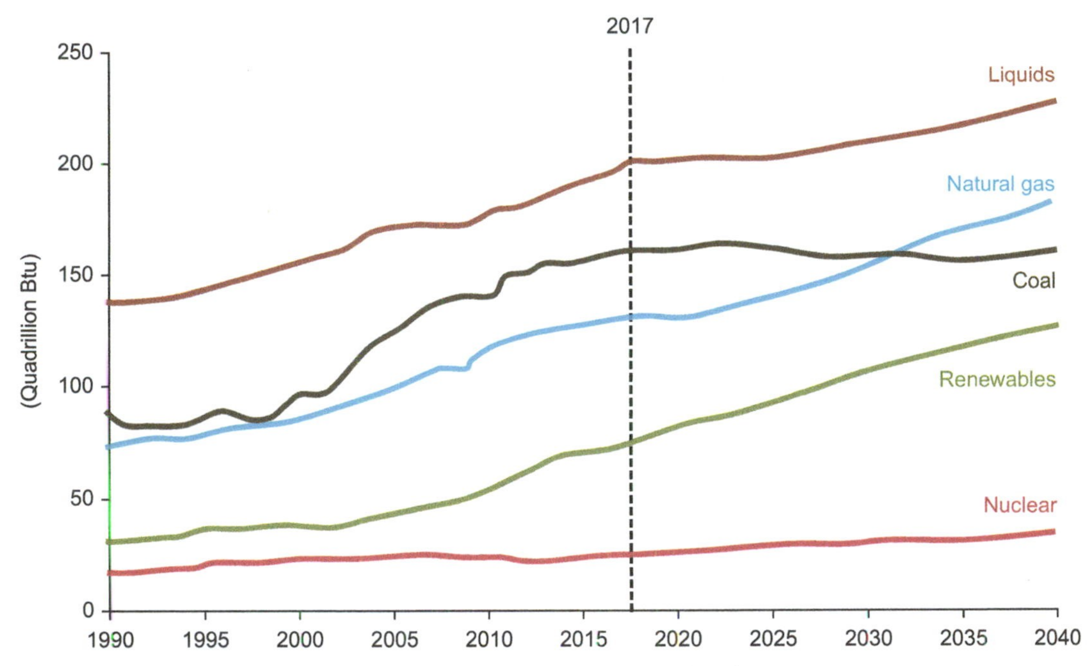

Figure 3.33 World energy consumption by fuel type, 1990–2040 (Source: www.eia.gov).

CO_2 is released into the atmosphere. Better gas mileage not only reduces global warming but also will save you thousands of dollars at the pump over the life of the vehicle. Next time you are in the market for a new car, check the fuel economy sticker on the car you are considering buying, or, better yet, buy a hybrid vehicle. We have the technology to build cars and SUVs that are just as powerful and safe as vehicles on the road today but get 40 miles per gallon (mpg) or more. According to the Union of Concerned Scientists, doing this alone would be equivalent to taking 44 million cars off the road.[48]

We must also become much more energy efficient. Household energy savings really can make a difference, not only in terms of saving you money but also in reducing our impact on the environment. Simple solutions like replacing existing appliances with the most efficient models available, unplugging that extra refrigerator or freezer you rarely use, or turning off lights you are not using all add up to real energy savings and reduced carbon emissions. One of the easiest things to do is replace your incandescent light bulbs with more efficient compact fluorescent lights, which now come in all shapes and sizes. Some of these steps may cost a bit more initially, but the energy savings will pay back the extra investment in no time. Many utilities offer free home energy audits to help you increase energy efficiency. Take advantage of this service. In some states, you can even switch to electricity companies that provide 50–100% renewable energy. In other states, utilities offer "green power" choices. Do some research and make a commitment to implement such changes. They *will* make a difference. Only by moving away from fossil fuels can we both ensure a more robust economic outlook and address the challenges of climate change and other environmental impacts. This will be a decades-long transformation that needs to start immediately.

[48] Source: www.ucsusa.org

4

Air Pollution and Atmospheric Deposition

INTRODUCTION

I live in the Dallas/Fort Worth (DFW) Metroplex, an urban metropolis that sprawls across 12 counties in North Texas. DFW is home to almost 6 million people, the world's third busiest airport, and several Fortune 500 corporate headquarters. It is also an area experiencing tremendous growth, with the population expected to double to 12 million by 2040. More people in cities and their surrounding counties mean more cars, trucks, industrial and commercial operations, and, generally, more pollution. It is no surprise then that the EPA has designated 9 of the 12 counties in the Metroplex as **non-attainment** areas (due to ozone, in this case), meaning that the air quality does not meet the national ozone standard. A similar situation is found in many urban centers around the United States (Figure 4.1). In Southern California, the situation is particularly acute with several counties failing air quality criteria for multiple regulated pollutants (i.e., not just ozone).

Millions of people live in areas where urban smog, very small particles, and toxic pollutants pose serious health concerns. Worryingly, a recent analysis by the World Health Organization (WHO) found that 90% of the global urban population (based on 1,600 cities) are exposed to unsafe air pollution. New Delhi, India, which we will discuss in a bit more detail later on, has particulates in the air at concentrations more than 1,500% the recommended level,[1] but the problems of air pollution are not

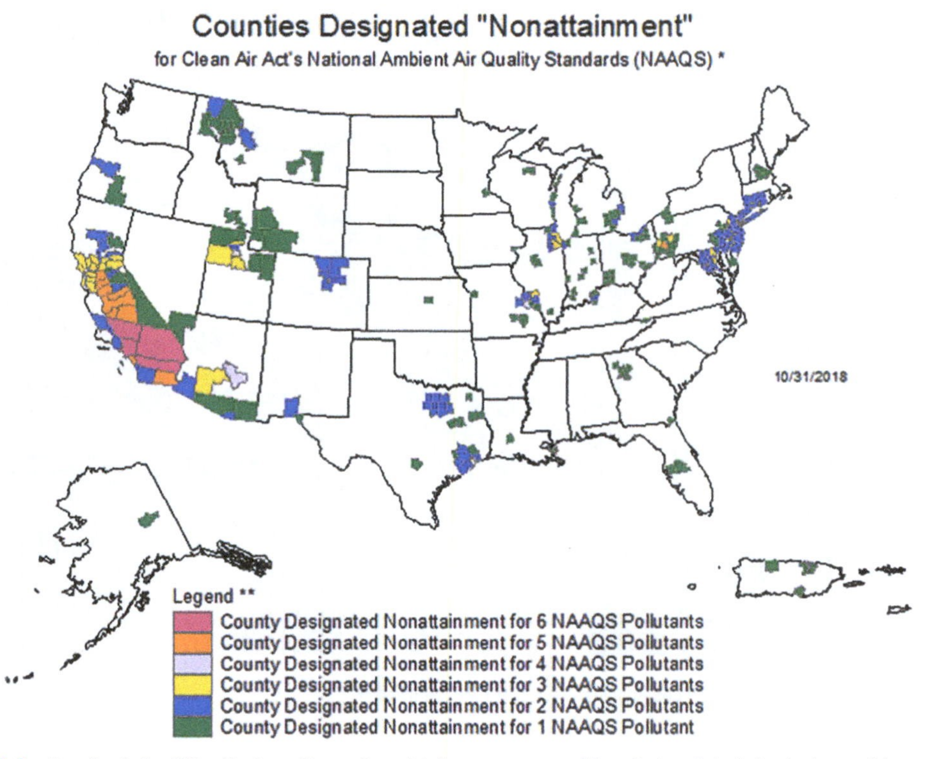

Figure 4.1 Counties in the U.S. with air quality monitors violating one or more of the criteria pollutant standards, as of January 2018 (Source: www.epa.gov).

[1] http://www.who.int/phe/health_topics/outdoorair/databases/cities/en/

limited to urban areas. Many air pollutants remain in the environment for long periods of time and are carried by the winds hundreds of miles from their origin. For example, researchers have found that coal combustion in Florida contributes significant amounts of mercury to Lake Superior, over 1,000 miles away! Long-term exposure to air pollution has been linked to cancer and damage to the immune, neurological, reproductive, and respiratory systems. In extreme cases, it can even cause death.

Whenever I lecture on air pollution, I poll my students on their perception of air quality. The results are always the same: most believe that the air quality where they live has gotten worse and that industrial development is to blame. The most frequently cited reason for the decline in air quality is population growth and an increase in the number of cars on the road. Invariably, someone will mention public health and make the link between pollution and increasing cases of asthma. Medical statistics show that the prevalence of asthma has risen in recent years: 1 in 12 people (about 25.5 million or 8.1% of the population) had asthma in 2012, compared with 1 in 14 (about 20 million or 7%) in 2001.[2] Asthma is now the third-ranking cause of hospitalization among children under 15,[3] and the costs are enormous.[4] Each year, dozens of news stories claim or imply that air pollution plays a major role in whether a person develops asthma. However, data from a California study show that this relationship between air quality and asthma is not straightforward (Figure 4.2). Here, asthma prevalence increases

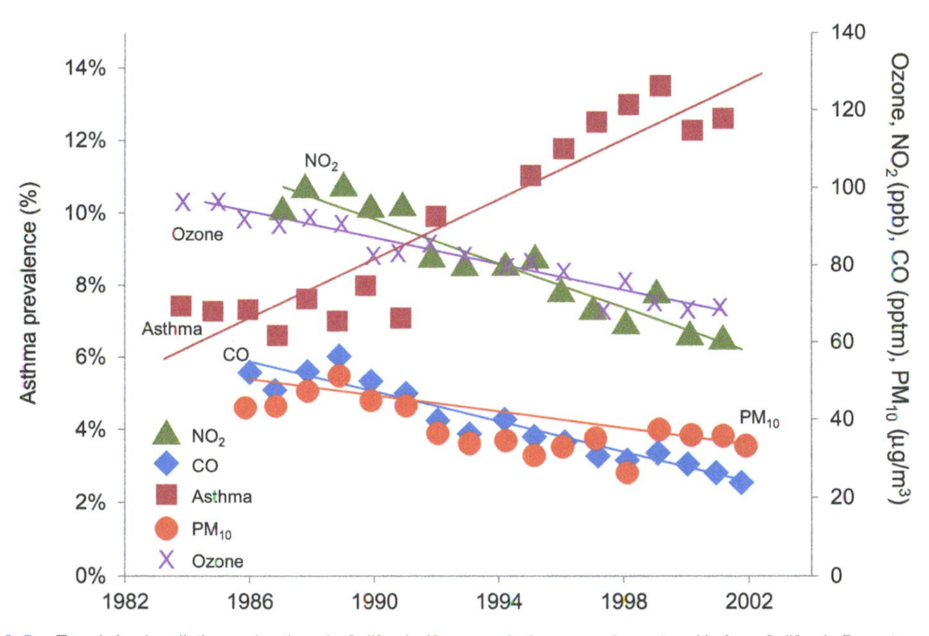

Figure 4.2 Trends in air pollution and asthma in California (Sources: Asthma prevalence trend is from California Department of Health Services; Air pollution trends are from California Air Resources Board).

[2] Centers for Disease Control and Prevention, Vital Signs, May 2011.

[3] DeFrances, C.J. et al. (2007), *Vital Health Statistics*, Vol. 12: p. 165.

[4] According to the CDC, asthma cost the U.S. about $3,300 per person each year from 2002 to 2007 in medical expenses, missed school and work days, and early deaths.

while the four monitored pollutants—ozone (O_3), fine suspended particulate matter (PM), nitrogen dioxide (NO_2), and carbon monoxide (CO)—all actually decrease over the two decade period. These four pollutants, along with sulfur dioxide (SO_2) and lead (Pb), are among the so-called **principal** or **criteria air pollutants** whose ambient levels are regulated by federal standards.

How many times have you heard the claim that you are breathing "some of the worstair pollution in the country"? Even my own local newspaper, the Fort Worth *Star-Telegram* (May 2, 2004), claimed that DFW has "some of the country's worst air," but is this really true? How can so many areas have some of the worst air pollution in the nation? And is the situation getting worse? In this chapter, we address these, and a number of other, questions relating to air quality. We also set U.S. air quality in a broader context globally.

DEFINING AIR QUALITY

There are two basic types of atmospheric pollutants: **primary pollutants** and **secondary pollutants** (Figure 4.3). Primary pollutants are compounds emitted directly into the air, such as SO_2, NO_2, suspended particulate matter (PM), and CO. They are emitted from **point sources**, such as

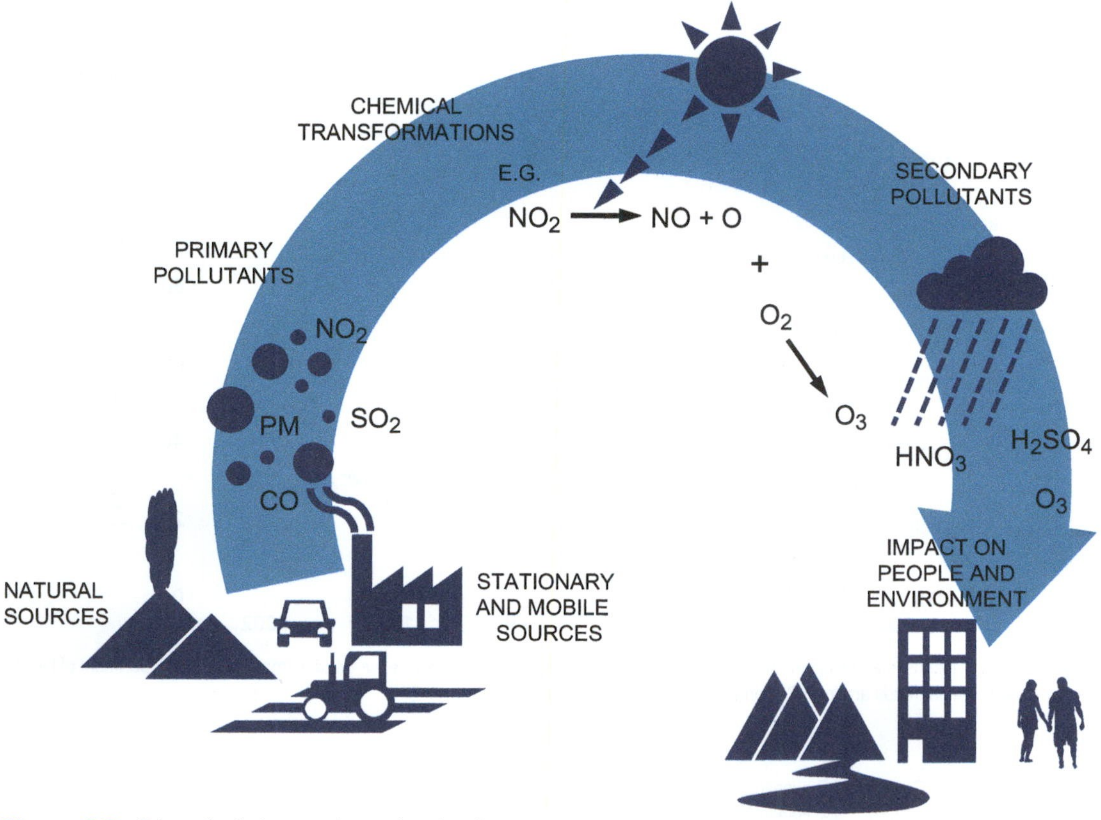

Figure 4.3 Schematic of primary and secondary air pollutants.

industrial stacks and coal-burning power plants (i.e., they are stationay), and **non-point sources**, such as motor vehicles (i.e., they are moving). Secondary pollutants are formed in the atmosphere through chemical reactions of the primary pollutants, usually involving sunlight. For example, ground-level ozone, or tropospheric ozone, which is often cited as being harmful to children and people with respiratory problems, is a secondary pollutant formed when oxides of nitrogen (NO_x) and volatile organic compounds (VOCs) react with sunlight to form O_3.

The **Clean Air Act**, established by Congress in 1970 and last amended in 1990, requires the EPA to set **National Ambient Air Quality Standards** (NAAQS) for pollutants considered harmful to the public's health and the environment. There are two types of NAAQS under the Clean Air Act. **Primary standards** set limits on pollutants in order to protect public health, especially the health of sensitive populations such as children, the elderly, and those that suffer from asthma. **Secondary standards** have limits that protect public welfare, things like protection against decreased visibility, damage to crops, and vegetation. The EPA assesses air quality based on standards for the six key air pollutants: CO, Pb, NO_2, O_3, SO_2, and PM.

The NAAQS are presented in Table 4.1. Setting these standards is a lengthy and often controversial process. For example, in 2005, the new eight-hour ozone standard replaced the one-hour standard but only after several years of litigation. The new standard is based on health studies that indicate long-term exposures to ozone are more harmful than shorter, one-hour exposure. The American Trucking Association, along with several other industries, sued the EPA over the new eight-hour rule, claiming that it unfairly targeted trucks and other diesel engines. They estimated that it would cost these businesses $46 billion a year to comply with the revised levels. In a landmark

Table 4.1 National Ambient Air Quality Standards for the United States.

Pollutant	Name	Primary Standards	Averaging Times
CO	Carbon monoxide	9 ppm (10 mg/m^3)	8-h (not to be exceeded more than once per year)
		35 ppm (40 mg/m^3)	1-h (not to be exceeded more than once per year)
Pb	Lead	1.5 μg/m^3	Quarterly average
NO_2	Nitrogen dioxide	0.053 ppm (100 μg/m^3)	Annual (arithmetic mean)
$PM_{2.5}$	Particulate matter	12.0 μg/m^3	Annual (arithmetic mean averaged over three years)
		35 μg/m^3	24-h
O_3	Ozone	0.075 ppm	8-h
		0.12 ppm	1-h (old standard—applies only in limited areas)
SO_2	Sulfur dioxide	0.03 ppm	Annual (arithmetic mean)
		0.14 ppm	24-h (not to be exceeded more than once per year)

(Source: www.epa.gov)

decision in 2001, the U.S. Supreme Court unanimously affirmed EPA's ability to set NAAQ standards that protect millions of people from the harmful effects of air pollution. On June 1, 2005, the new eight-hour standard was finally adopted.

On March 12, 2008, EPA further strengthened its NAAQS for ground-level ozone to a level of 0.075 parts per million (ppm), with the agency proposing to strengthen the standard further to 0.07 ppm, based on evidence about ozone's effects on public health and welfare. It does not sound like much, but a strengthened ozone standard of even 0.005 ppm will mean new requirements for emission controls on sources that emit nitrogen oxides and VOCs, such as vehicles, industrial facilities, and power plants. The revised standard will also likely cause significant portions of the country to fall into non-attainment, meaning they fail air quality based on the EPA limits. Compared to the previous 0.075 ppm standard, which had 28 areas classified as non-attainment, 241 counties would violate the new 0.07 ppm standard, according to the EPA. This is consequential, as counties and areas classified as non-attainment can suffer stringent penalties, including suspending federally supported highway and transportation projects. As expected, lawyers are challenging this new standard in a number of suits.

THE SIX CRITERIA POLLUTANTS

As noted above, the Clean Air Act of 1970 identified six common air pollutants of concern, called principal or criteria pollutants. Let us examine these in a bit more detail.

Nitrogen Oxides (NO_x)

NO_x is a general term for a group of highly reactive gases that contain nitrogen and oxygen in varying amounts. Most are colorless and odorless although NO_2, a more common pollutant, can often be seen as a reddish-brown layer over many urban areas. These oxides form when fuel is burned at high temperatures, for example, in motor vehicles and electric utilities. Although a primary pollutant, NO_x is also one of the main ingredients involved in the formation of ground-level ozone and acid rain, both of which are discussed in more detail in the following text.

Sulfur Dioxide (SO_2)

SO_2 belongs to the family of sulfur oxide gases (SO_x) that are easily dissolved in water. These gases are formed when fuel containing sulfur (e.g., coal) is burned. SO_2 dissolves in water vapor to form acid and other products that can be harmful to people and their environment. Power plants, especially those that burn coal, are by far the largest single contributor of SO_2 pollution in the United States, accounting for more than two-thirds of all SO_2 emissions nationwide. SO_2 reacts with other chemicals in the air to form tiny sulfate particles. Sulfates are major components of the fine particle pollution that plagues many parts of the country, especially communities nearby or directly downwind of coal-fired power plants. According to EPA studies, fine particle pollution from power plants causes more than 20,000 premature deaths a year.

Lead (Pb)

Pb is a metal found both naturally in the environment and in a number of manufactured products. The major sources of lead emissions have historically been vehicles, but industrial sources are also significant. Due to the phase out of leaded gasoline, metals manufacturing is now the major source of lead emissions, and the highest concentrations of lead are generally found near lead smelters.

Particulate Matter (PM)

Particle pollution is a mixture of very small particles and liquid droplets. Particle pollution is actually made up of a number of components, including ash, soil, and dust particles. Their impact, in terms of health, is directly linked to their size, with greatest concern focused on particles that are 10 μm in diameter or smaller. These particles have the ability to enter the lungs and blood stream and cause serious health effects.

Carbon Monoxide (CO)

CO is formed when carbon in fuel is incompletely burned. It is a primary component of vehicle emissions that, according to the EPA, contributes about 56% of all CO emissions nationwide. Thus, higher levels of CO generally occur in areas subject to heavy traffic congestion. In cities, 85–95% of all CO emissions may come from vehicle emissions. Other sources of CO include industrial processes and forest fires. At high levels, CO is poisonous and even lethal to healthy people.

Ozone (O_3)

O_3, consisting of three bound oxygen atoms, is the most problematic pollutant in terms of air quality in the United States. It occurs in both the **troposphere** (lower atmosphere) and the stratosphere (Figure 4.4). In the stratosphere, ozone is critical to life on Earth because it screens out harmful ultraviolet radiation from the sun, the focus of Chapter 5. However, in the troposphere, it is a human-made pollutant with potentially significant consequences for human health. Breathing is harmful at dosage levels of a few molecules per million air molecules. I have found the simple catch-phrase "good up high, bad nearby" is a useful way for students to keep the functioning of the two types of ozone clear in their minds.

Ozone formation is a complicated process. The majority of tropospheric ozone formation occurs when NO_x and VOCs react in the atmosphere in the presence of sunlight (refer back to Figure 4.3). NO_x and VOCs are therefore called **ozone precursors**, meaning the primary pollutants have to be emitted first before low-level ozone can form. Vehicle, industrial emissions, gasoline vapors, and chemical solvents are the major anthropogenic sources of these chemicals. Although these precursors often originate in urban areas, it is important to note that winds can carry those hundreds of miles, causing ozone formation to occur in less populated regions as well. Ground-level ozone is also the primary constituent of **smog**, which you may recognize as the reddish-brown haze that forms when air quality is particularly poor. The terms ozone and smog are often used

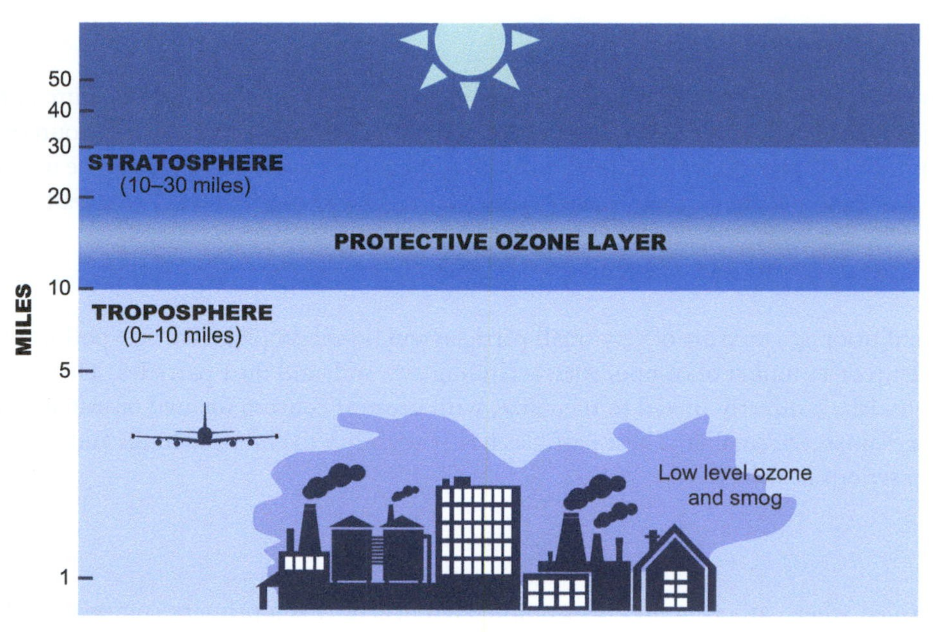

Figure 4.4 Distribution and characteristics of tropospheric and stratospheric ozone.

interchangeably for general use, but smog is more complex and comprises ground-level ozone, other gases, and particulate matter. Whereas ozone itself is colorless, the air can look clear even when high ozone concentrations are present.

The chemical reactions involved in tropospheric ozone formation are a series of complex cycles. In simple terms, NO_2 is first photolyzed (broken apart by sunlight) resulting in atomic oxygen (O). The atomic oxygen then combines with diatomic oxygen (O_2) resulting in a molecule of ozone (O_3):

$$NO_2 + UV \rightarrow NO + O \tag{4.1}$$

$$O + O_2 \rightarrow O_3 \tag{4.2}$$

The reactions involved in this process are illustrated with NO_2 in (4.1) and (4.2) and Figure 4.3, but similar reactions occur for VOCs as well. Because sunlight and hot weather are required to catalyze ozone to form in harmful concentrations in the air, it is often referred to as a summertime air pollutant. Figure 4.5, for example, shows the 8-hour average daily maximum ozone levels over Houston, Texas, on June 6, 2011. High ozone is common in this area of the United States (see also Figure 4.1) Many of the nation's largest refineries and petrochemical plants are located here and ozone routinely migrates from coastal areas into downtown Houston.

There is a great deal of evidence to show that high concentrations of ozone, created by high concentrations of pollution and daylight UV rays at the Earth's surface, can impair lung function and irritate the respiratory system. Breathing ozone can trigger a variety of health problems including chest pain, coughing, throat irritation, and congestion. It can worsen bronchitis, emphysema, and

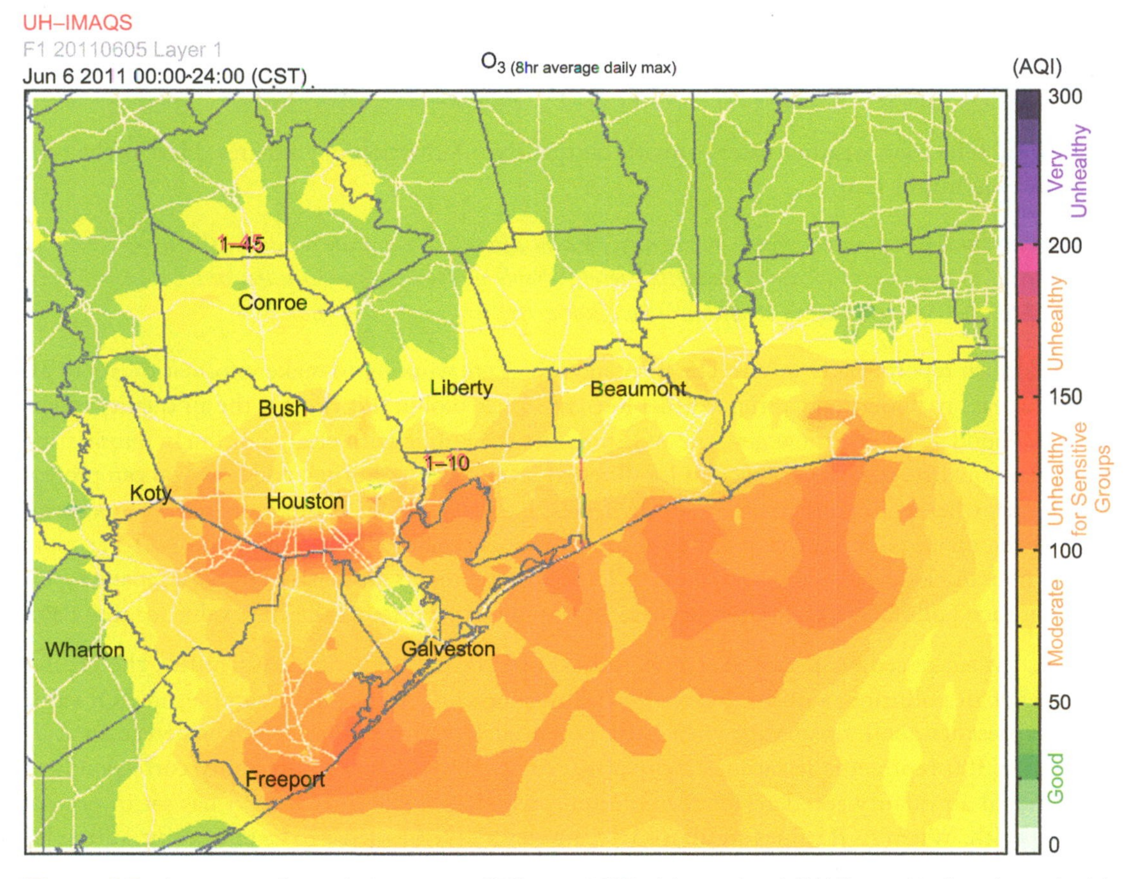

Figure 4.5 Contour map of tropospheric ozone over SE Texas and SW Louisiana on June 6 2011 (Source: http://www.imaqs.uh.edu/local_ozone.htm).

asthma. Repeated exposure can even permanently scar lung tissue. Ground-level ozone also damages vegetation and ecosystems. In the United States alone, ozone is responsible for an estimated $500 million in reduced crop production each year.[5]

Students often ask me why we cannot take all of this bad ozone and simply transport it up into the stratosphere where ozone protects us from harmful UV rays. The answer lies in the vast quantities needed. Because only 10% of the atmosphere's ozone occurs in the troposphere, the vehicle necessary to transport such enormous amounts of ozone vertically into the stratosphere does not exist. Even if it did, it would require so much fuel that the resulting pollution might undo any positive effect. Rather than seek such grandiose solutions, we need to focus on decreasing the emission of the pollutants that help create ozone in the troposphere.

[5] Source: AIRNow, a cross-agency U.S. Government web site using data from EPA, NOAA, NPS, and state and local agencies (http://www.airnow.gov) aimed at providing the public with easy access to national air quality information.

THE AIR QUALITY INDEX (AQI): A GUIDE TO AIR QUALITY AND YOUR HEALTH

Many local television weathercasters provide air quality information in your area. Here is the type of report you might hear:

> *Tomorrow will be a code orange ozone day. Warm afternoon temperatures and calm winds are expected to cause ozone to rise to unhealthy levels. People with respiratory disease, such as asthma, and children and the elderly should avoid outdoor activities.*

What do such reports really mean? The EPA and others are working to make information about air quality as easy to understand as the weather forecast. A key tool in this effort is the air quality index, or AQI. Found at www.airnow.gov, the AQI provides the public with clear and timely information on local air quality, the health concerns for different levels of air pollution, and how you can protect your health when pollutants reach unhealthy levels. Each day, the EPA monitors record concentrations of the major pollutants (ground-level ozone, particle pollution, CO, SO_2, and NO_2) at more than a thousand locations across the country. These measurements are converted into a separate AQI value for each pollutant. The highest of these AQI values is reported as the AQI value for that day.

The AQI is expressed on a scale that runs from 0 to 500; the higher the AQI value, the greater the level of air pollution, and the greater the health concern (Box 4.1). For example, an AQI value of 50 represents good air quality with little potential to affect public health, whereas an AQI value over 300 represents hazardous air quality. An AQI value of 100 generally corresponds to the national air quality standard for the pollutant, which is the level that EPA has set to protect public health. When AQI values are above 100, air quality is considered to be unhealthy, at first for

BOX 4.1

The Air Quality Index explained

To make it easier to understand, the AQI is divided into six categories:

Air Quality Index (AQI) Values	Levels of Health Concern	Colors
When the AQI is in this range:	*. . . air quality conditions are:*	*. . . as symbolized by this color:*
0–50	Good	Green
51–100	Moderate	Yellow
101–150	Unhealthy for sensitive groups	Orange
151–200	Unhealthy	Red
201–300	Very unhealthy	Purple
301–500	Hazardous	Maroon

certain sensitive groups of people (when a level orange alert is issued), then for everyone as AQI values get higher.[6] The color assigned to each AQI category makes it easier for people to understand quickly whether air pollution is reaching unhealthy levels in their communities. Air quality is also highly dependent on the state of the atmosphere and geographic location, as illustrated in Box 4.2.

National Trends in Criteria Levels

Let us now assess the temporal trends in air quality in the United States for the past two decades. Using a nationwide network of monitoring sites, the EPA has developed ambient air quality trends for the criteria pollutants discussed above. Figure 4.6 shows these national trends between 1990 and 2016, relative to their respective NAAQS. Two key points emerge from

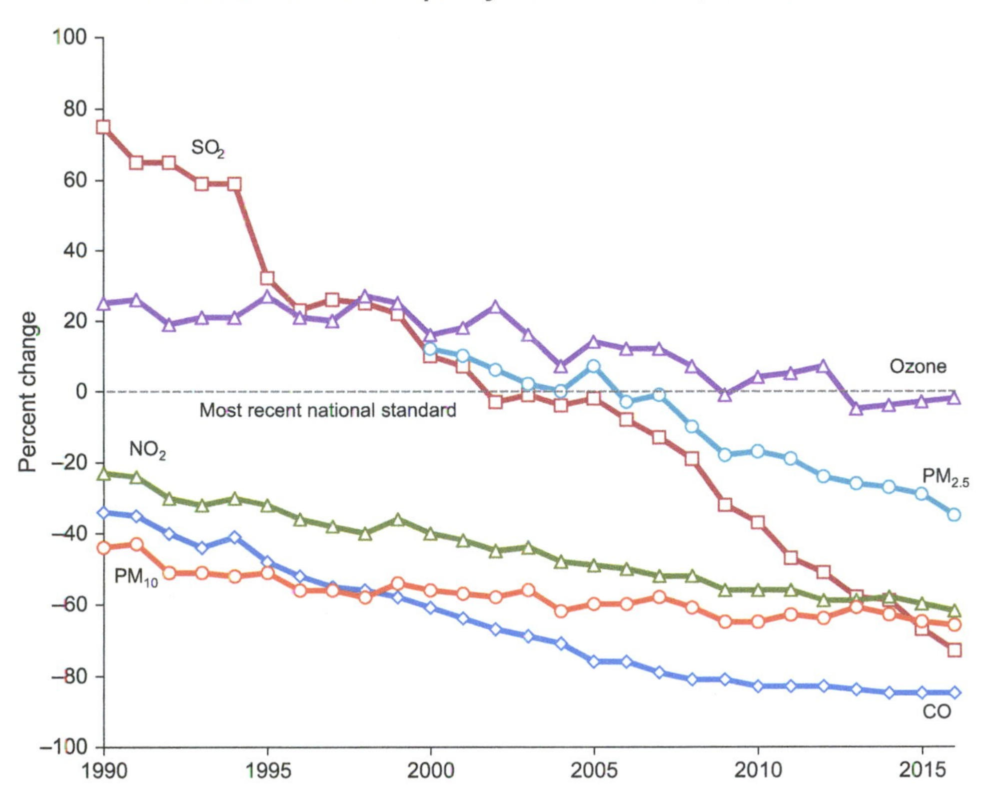

Figure 4.6 Comparison of national levels of the criteria pollutants to the most recent national ambient air quality standards, 1990–2017. National levels are averages across all monitors with complete data for the time. Note: Air quality data for $PM_{2.5}$ start in 1999. Lead is not plotted, as there are significant year-to-year changes in lead concentrations largely driven by changes in lead concentrations at monitoring sites near stationary sources. These year-to-year changes reflect changes in operating schedules and plant closings.

[6] At AQI values of 150 or above, everyone should limit prolonged outdoor exertion, especially children. When AQI levels reach 200 (that is, level purple or very unhealthy) people with respiratory disease, such as asthma, should avoid all outdoor exertion and limit exposure by staying inside (air conditioned spaces are best).

Air Pollution Meteorology

Pollution levels are highly dependent on the state of the atmosphere, specifically whether the atmosphere is **stable** or **unstable**. In a stable atmosphere, rising motion is suppressed whereas in an unstable atmosphere, vertical motion is enhanced. Stability and instability are determined by comparing the temperature of a rising parcel of air with the temperature of the surrounding atmosphere. The figure in this box shows a simple scenario where temperature decreases slowly with height. A parcel of air would rise and cool at a set rate along the solid line. Note that at any elevation, the temperature of the parcel is colder than the surrounding air. If released, the parcel will sink back to its original position—a stable atmosphere. In a **temperature inversion**, a condition in which the temperature of the atmosphere increases with altitude, cold air underlies warmer air at higher altitudes. Overnight radiative cooling of surface air often results in a nocturnal temperature inversion. This is often the case on clear, calm nights where the air in contact with the ground can cool quickly. In areas that lie in basins or valley bottoms, such as Los Angeles (see photograph) and Mexico City, the cold air drains down slope to collect in the low-lying areas leaving relatively warmer air aloft. During a temperature inversion, air pollution released into the atmosphere's lowest layer is trapped there and can be removed only by strong horizontal winds or rain.

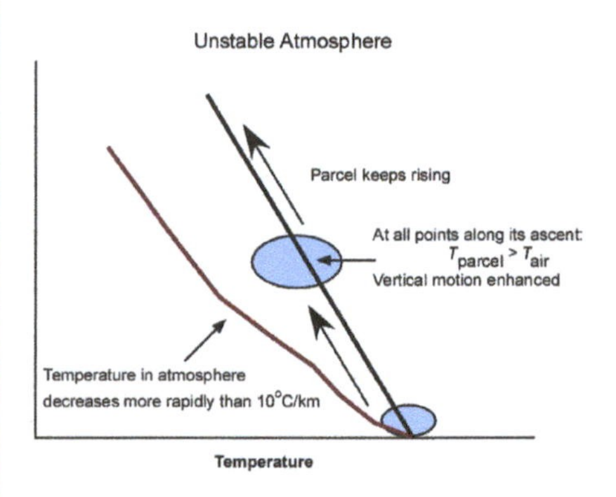

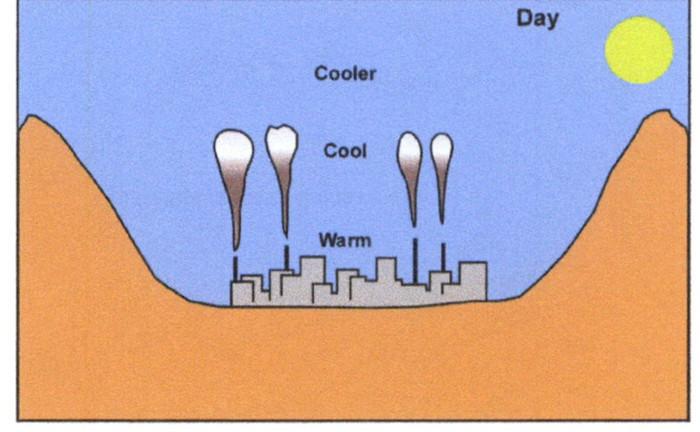

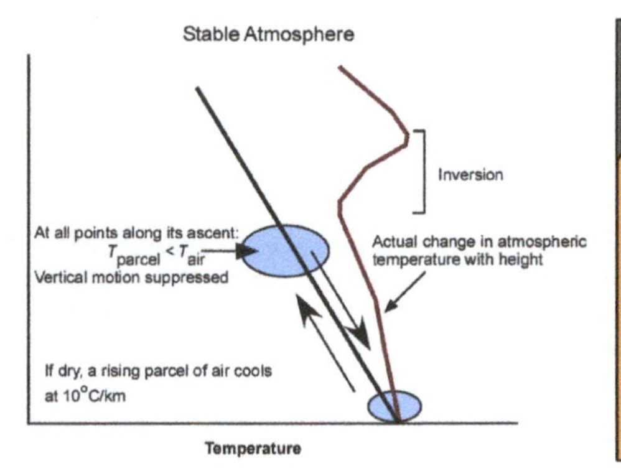

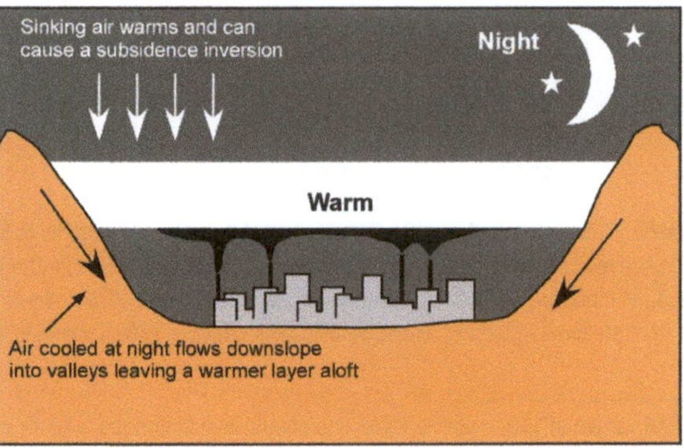

(Continued)

(Continued)

Strong, long-lived temperature inversions often accompany the dynamics of the large high-pressure systems depicted on weather maps. Descending currents of air near the center of the high-pressure system produce a warming (by compression), causing air at middle altitudes to become warmer than the surface air. Rising currents of cool air lose their buoyancy and are thereby inhibited from rising further when they reach the warmer, less dense air in the upper layers of a temperature inversion. Because high-pressure systems often combine temperature inversion conditions and low wind speeds, their long residency over an industrial area usually results in episodes of severe smog.

Los Angeles from the Griffith Observatory showing pollution trapped during a mid-morning inversion. Note the Hollywood sign at the extreme right.

this diagram. First, most pollutants show a steady decline throughout that time period, meaning that air quality has improved steadily across the United States since the Clean Air Act was amended more than two decades ago. The downward trend in air pollution has been especially evident over the past decade for SO_2 and the finest particulates ($PM_{2.5}$) and has had profound health benefits for the American people. Indeed, a recent peer-reviewed report by EPA suggests that by 2020, the Clean Air Act Amendments will prevent over 230,000 early deaths and 17 million lost work days annually.[7] The Amendments are also projected to result in a net improvement in U.S. economic growth and the economic welfare of American households, with benefits exceeding costs by a factor of more than 30 to 1. This net improvement in economic welfare occurs because cleaner air leads to better health and productivity for American workers and savings on medical expenses for air pollution-related health problems. The second trend evident in Figure 4.6 is that, despite improving air quality in the United States, ground-level ozone remains the most problematic in terms of the NAAQS in most areas. A look at Figure 4.7 shows that approximately 116 million people nationwide (or 36% of the population) live in counties with pollution levels above the NAAQS for ozone. Although control measures, such as catalytic converters, have reduced pollutant emissions per vehicle, the number of cars and trucks on the road and the miles they are driven has doubled in the past 20 years. Vehicles are now driven two trillion miles each year in the United States. With more and more cars traveling more and more miles, growth in vehicle travel may eventually offset progress in vehicle emission controls.

[7] The full report can be downloaded here: http://www.epa.gov/air/sect812/feb11/fullreport_rev_a.pdf

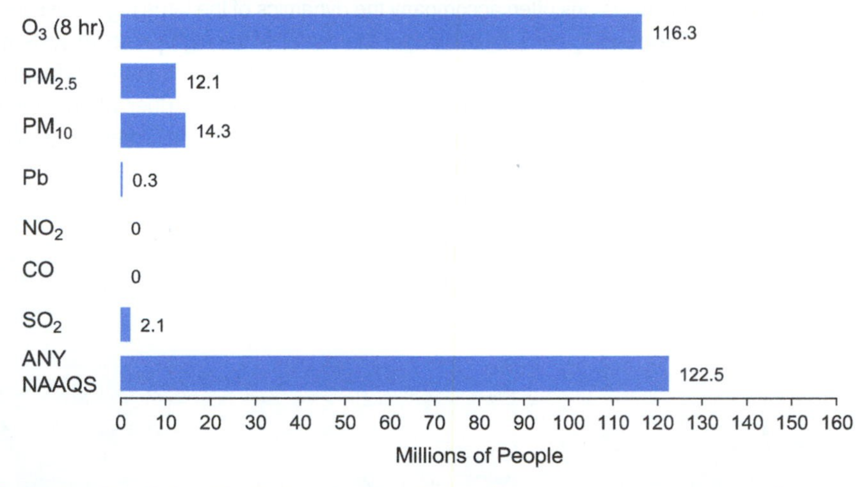

Figure 4.7 Number of people (in millions) living in counties with air quality concentrations above the level of the primary (health-based) NAAQS in 2017 (Source: http://www.epa/gov/ttn/naaqs/).

The decrease in average NO_2 and SO_2 concentrations over the 20 years can be attributed to a combination of factors including enforcement actions, tougher state laws, and reductions anticipated from EPA's Clean Air Interstate Rule (CAIR), a rule designed to cap SO_2 and NO_x emissions in states east of the Mississippi River. Satellite images unveiled by NASA in June 2014 show how air quality has improved as NO_2 declined in the United States between 2005 and 2011 (Figure 4.8). As you can clearly see, the most populated East Coast cities have seen the biggest drop. For example, New York saw a decrease of 32% over the monitoring period.

Sulfur dioxide is also on the decline as power companies begin to install scrubbers that will reduce SO_2 by as much as 90% at some of the dirtiest coal-burning facilities. Scrubbing is a loose term that describes an array of air pollution control devices that rely on a chemical reaction with a sorbent to remove pollutants. For SO_2 removal, these devices are usually called flue gas desulfurization (FGD) systems, or simply, scrubbers. "Wet" scrubbers, which use liquid to trap particles and gases in the exhaust stream, can reduce SO_2 by 90–95%. Large coal plants equipped with such scrubbers have shown that cleaner power is achievable and the investment in the cleanup of the oldest and dirtiest power plants should substantially reduce emissions that are a primary source of the fine particulate matter.

The overall momentum toward cleanup is clearly good news. As the graph in Figure 4.9 shows, between 1980 and 2017 in the United States, the gross domestic product increased 253%, vehicle miles traveled increased 190%, energy consumption increased 44%, and population increased by 58%. During the same time period, total emissions of the six principal air pollutants dropped by 73%. It is a remarkable accomplishment, achieved through regulations, voluntary measures taken by industry, partnerships among federal, state, local, and tribal governments, and environmental organizations.

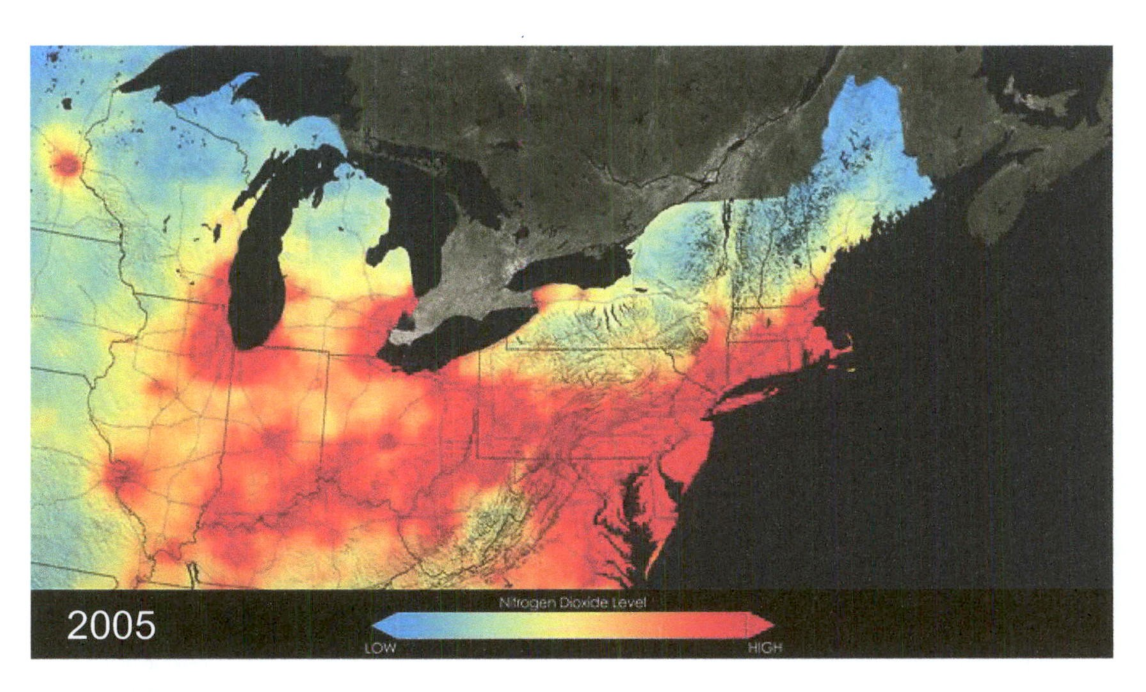

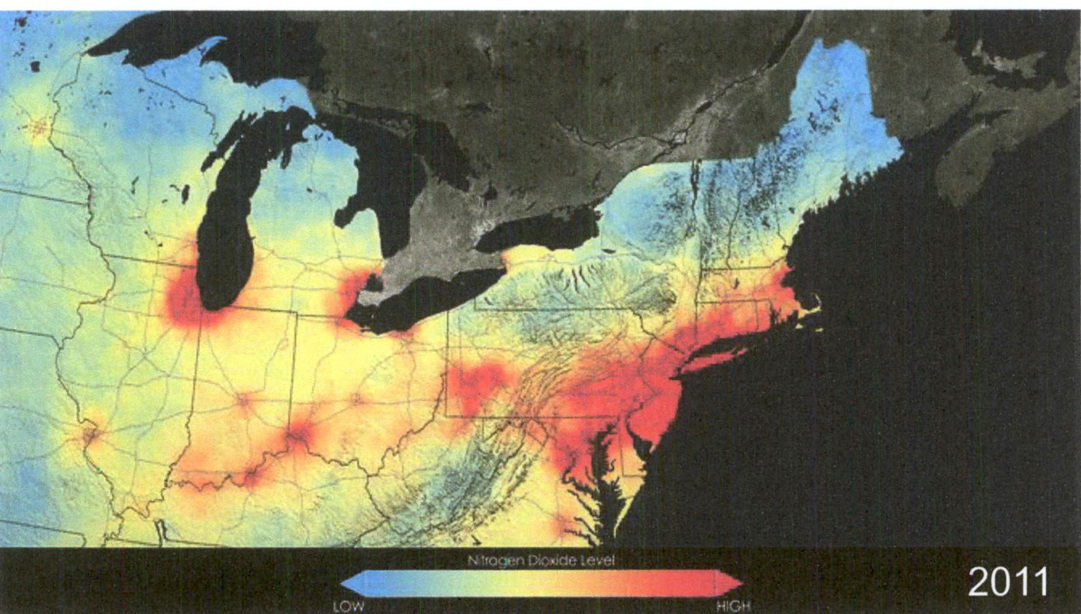

Figure 4.8 Maps showing the decrease in the criteria pollutant NO_2 between 2005 and 2011. The areas of red and orange on the map signify higher concentrations of NO_2 (Source: NASA Goddard Scientific Visualization Studio).

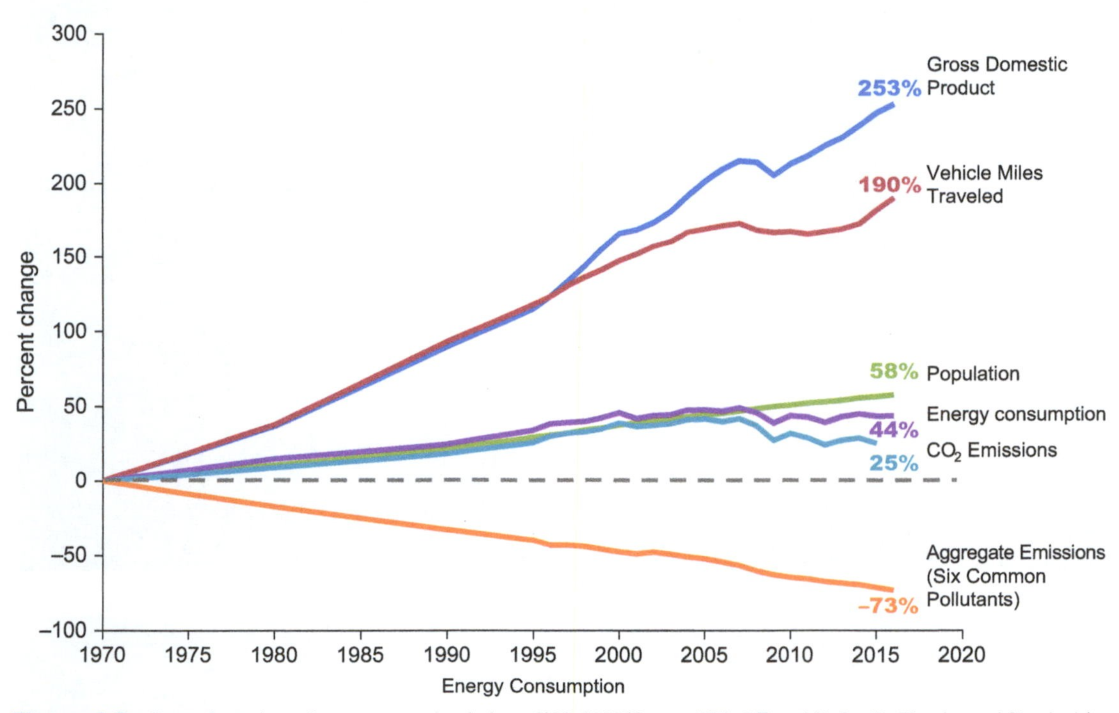

Figure 4.9 Comparison of growth measures and emissions, 1980–2017 (Source: EPA, Office of Air Quality Planning and Standards).

Yet in other parts of the world, the air quality situation is not nearly quite as rosy. The World Health Organization (WHO) estimates that 91% of the world's population lives in places where air quality exceeds healthy limits.[8] Chinese cities often receive the most media attention for atrocious air pollution, and rightly so, but cities in India are actually far more polluted. The WHO's database of more than 4,300 cities showed the ten most polluted cities in the world are all in India (Figure 4.10). New Delhi's air was among the worst in the world with an annual average of 153 $PM_{2.5}$ micrograms per cubic meter (see Box 4.3 that covers my own experience with New Delhi's air). If you compare that to Table 4.1 and the U.S. air quality standard of 12 $\mu g/m^3$, it means that, put in context, New Delhi's air pollution is about 13 times higher than the U.S. limit and almost 4 times higher than our most polluted air which is found in Fresno, CA. Only three of China's cities came in the top 20, with Beijing now coming in at number 57 (thanks to a string of effective pollution control measures) but still nearly five times above the estimated WHO safe limit.

This latest WHO report has garnered much attention because it shows that air quality in most cities worldwide fails to meet guidelines for safe levels, putting people at additional risk of serious respiratory disease and other long-term health problems. In many parts of the world, air pollution is simply getting worse. Reliance on fossil fuels, especially coal-fired power plants,

[8] Source: http://www.who.int/airpollution/en/

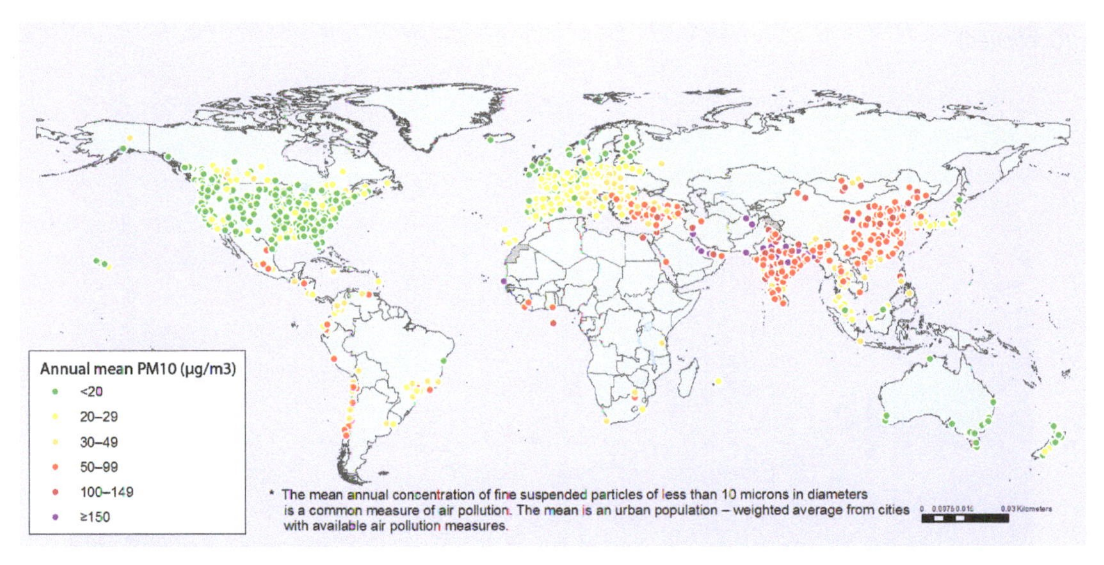

Figure 4.10 Exposure to particulate matter with an aerodynamic diameter of 10 micrometers or less in 1,600 urban areas, 2008–2015 (Source: World Health Organization—www.who.int).

BOX 4.3

Air Quality in New Delhi, India

In November 2017, I visited New Delhi for a geography conference. I had always wanted to go, and the fifth most populous city in the world (~22 million people) certainly did not disappoint! The sights and sounds, the hustle and bustle, the smell of spices—all quite extraordinary, as was the air quality!

India has seen rapid development in the past two decades, but it has come at the cost of increasing pollution. In addition, weak industrial regulation means that factories generally do not follow pollution-control measures.

When I landed in the early hours of November 6, air quality in Delhi had dropped precipitously, so much so that United Airlines, among several, cancelled flights into the capital for the week. By Wednesday, November 8, Delhi had become what one government minister called " a gas chamber." Daily PM_{10} values for 15 official recording stations across the city *averaged* 879 micrograms per cubic meter ($\mu g/m^3$), 15 times the level deemed safe by the World Health Organization. For $PM_{2.5}$, November 8 averaged 616 $\mu g/m^3$, almost 25 times the safety limit! At one station fairly close to my hotel, the AQI exceeded 999, the equivalent of smoking 45 cigarettes a day. Visibility was less than five feet. It was like walking through a soupy mixture of polluted air. I could feel my eyes and throat burning. The government declared a state of emergency and closed all schools for the week. Stores quickly ran out of filtration masks; I ended up using the eye mask from my airline amenity kit to cover my mouth to aid my breathing!

(Continued)

(Continued)

Photo from my hotel room on the morning of November 8, 2017. © Mike Slattery.

The dense smog caused a number of car crashes, including a 24-car motorway pileup that we drove past on our way to Agra to visit the Taj Mahal. The source of the tiny particulates in the air was combustion, released in this region by burning coal, running diesel engines, open fires, and most importantly at this time of year, burning crop stubble in neighboring states as farmers clear their fields. The state of the atmosphere exacerbated the problem, with north-westerly winds bringing in pollutants from the stubble-burning regions and cool temperatures and high humidity causing the air to stagnate. North India's topography also acts as a basin that traps pollution, making it impossible for the millions of people in the region to escape the toxic air.

The smog festered in the city for another two days and then began to clear just as I left the capital and headed south. However, the vast majority of New Delhi's residents, many of whom are poor and homeless, do not have the luxury of escaping the city. As if to add insult to injury, a new study on the impact of air pollution on life expectancy suggests that people in Delhi could live six years longer if India just met its national $PM_{2.5}$ standard (Source: https://epic.uchicago.in/impact/new-tool-shows-pollutions-impact-india/).

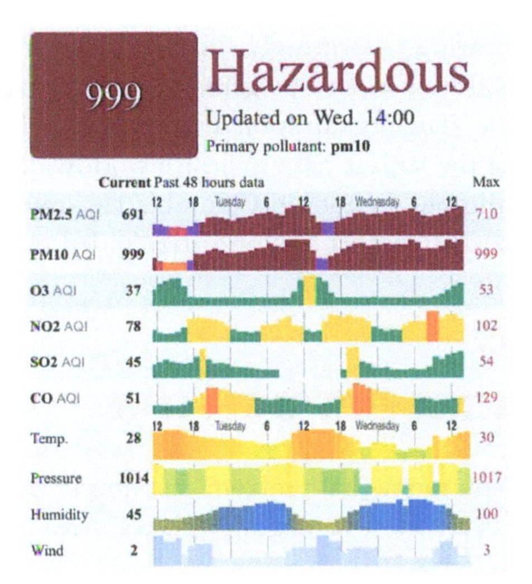

Air quality readings from a monitor nearby my hotel on November 8.

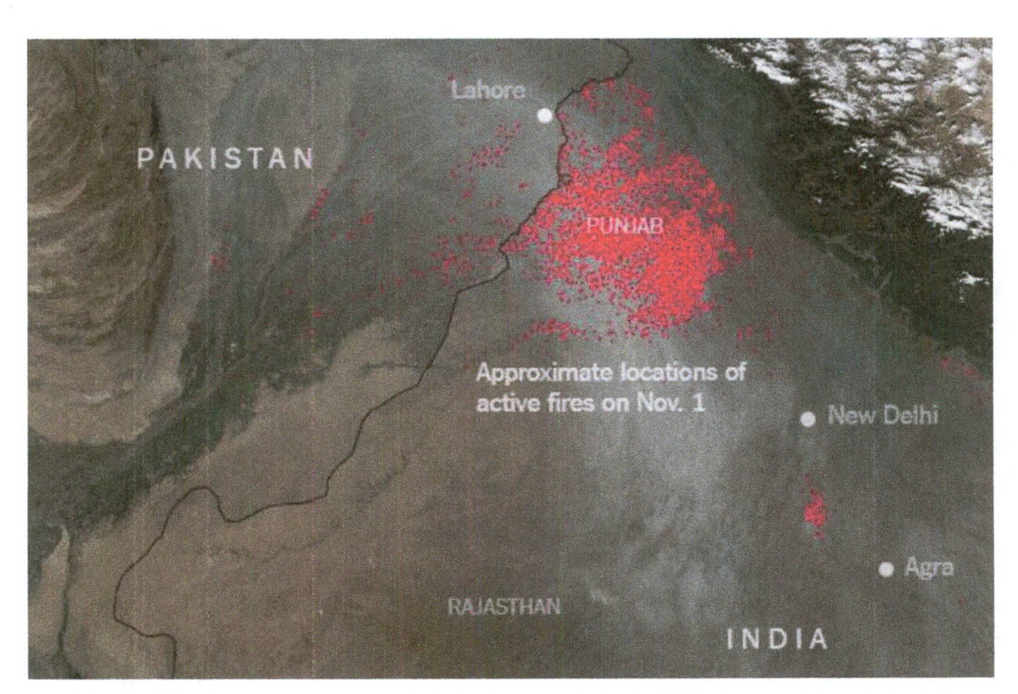

Farmers in neighboring Punjab and Haryana setting fire to paddy stubble in their fields after cultivating the crop as part of the slash and burn. (Source: NASA Rapid Response Team, GSFC)

and dependence on private transport motor vehicles are two major factors, along with the use of biomass for cooking and heating. Alarmingly, in 2018, the WHO issued new information estimating that outdoor air pollution was responsible for the deaths of some 4.2 million people under the age of 60 worldwide. The organization also emphasized that indoor and outdoor air pollution combined are among the largest risks to health worldwide. But some cities are making notable improvements. For example, Bogotà in Columbia has improved air quality by promoting "active" (i.e., non-motorized) transport and prioritizing dedicated networks of urban public transport, walking, and cycling.

NO_X, SO_X, AND ACID DEPOSITION

Acid deposition is a serious environmental problem that affects large parts of the United States and much of Europe. Its effects are most damaging to lakes, streams, forests, and the plants and animals that live in these ecosystems.

Acid deposition results primarily from the transformation of SO_2 and NO_x into secondary pollutants such as sulfuric acid (H_2SO_4) and nitric acid (HNO_3). The reaction for SO_2 can be written as follows:

$$SO_2 + OH \rightarrow HOSO_2 \quad (4.3)$$

which is then followed by:

$$HOSO + O_2 \rightarrow HO_2 + SO_3 \quad (4.4)$$

The sulfur trioxide molecule (SO_3) is then converted rapidly into sulfuric acid (H_2SO_4) in the presence of water:

$$SO_3 + H_2O \rightarrow H_2SO_4 \quad (4.5)$$

Nitric acid (HNO_3) is formed by the reaction of OH with nitrogen dioxide, as follows:

$$NO_2 + OH \rightarrow HNO_3 \quad (4.6)$$

The result of these reactions is a solution of **sulfuric acid** and **nitric acid**. This transformation of SO_2 and NO_x to acidic particles occurs as these pollutants are transported in the atmosphere over distances of hundreds to thousands of miles.

Acidic particles are deposited via two processes: wet deposition and dry deposition. Wet deposition refers to **acid rain**, the process by which acids are removed from the atmosphere in rain, fog, and snow. Dry deposition takes place when particles such as sulfates and gases (such as SO_2 and NO_x) are deposited on or absorbed onto surfaces. Dry deposited gases and particles can then be washed from these surfaces by rainstorms. About half of the acidity in the atmosphere falls back to Earth through dry deposition.

Acid rain is measured using a scale called pH. Because acids release hydrogen ions (H^+), the acid content of a solution is based on the concentration of hydrogen ions. The lower a substance's pH (i.e., the smaller the number on the pH scale), the more acidic it is. The pH scale ranges from 0 to 14 (Figure 4.11). A pH of 7 is neutral, a pH less than 7 is acidic, and a pH greater than 7 is basic. It is important to appreciate that each whole pH value below 7 is 10 times more acidic than the next higher value. For example, a pH of 4 is 10 times more acidic than a pH of 5 and 100 times (10 times 10, or 100 times) more acidic than a pH of 6. The same holds true for pH values above 7, each of which is 10 times more alkaline—another way to say basic—than the next lower whole value. For example, a pH of 10 is 10 times more alkaline than a pH of 9.

In the 1950s, it was discovered that fish were disappearing from lakes and waterways in southern Scandinavia. Today, some 14,000 Swedish lakes are affected by **acidification**, with widespread damage to plant and animal life as a consequence. The damage is not only extensive in large parts of Scandinavia but also occurs in parts of the United Kingdom and the Alps. Several regions in the United States have also been identified as containing surface waters sensitive to acidification. They include the Adirondacks and Catskill Mountains in New York State, the mid-Appalachian Highlands along the East coast, the upper Midwest, and mountainous areas of the Western United States (Figure 4.12). In areas like the Northeastern United States, some lakes now have a pH value of less than 5. Rainfall is most acidic in the Northeast, a pattern caused by the large number of cities, the dense population, and the concentration of power and industrial plants in the region (Figure 4.13). Most lakes and streams naturally have a pH between 6 and 8, although some lakes are naturally acidic even without the effects of acid rain.

Lakes that have been acidified cannot support the same variety of life as healthy lakes. As a lake becomes more acidic, crayfish and clam populations are the first to disappear, then various types of fish. At pH 5, most fish eggs cannot hatch. Some species of fish, such as smallmouth bass, walleye,

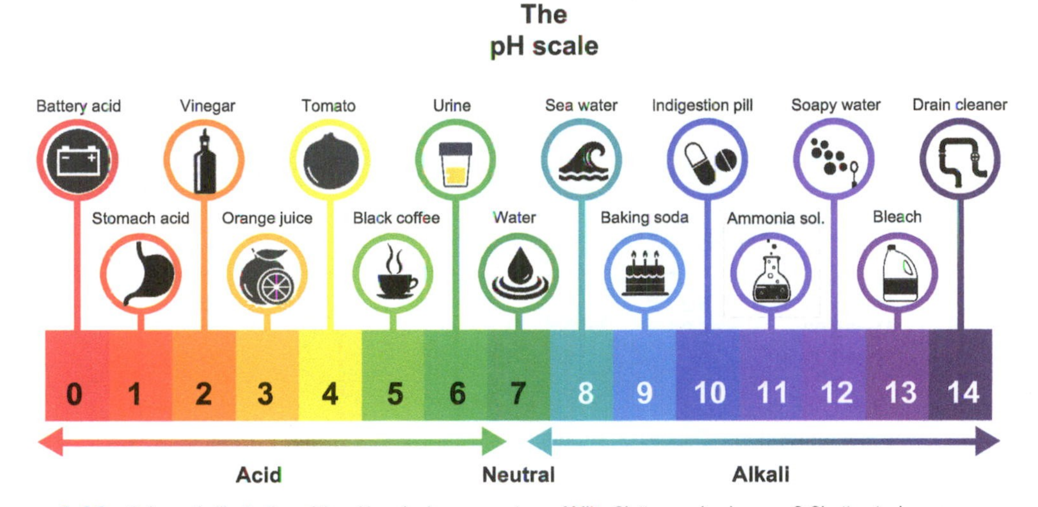

Figure 4.11 Schematic illustration of the pH scale. Image courtesy of Mike Slattery, using images © Shutterstock.com

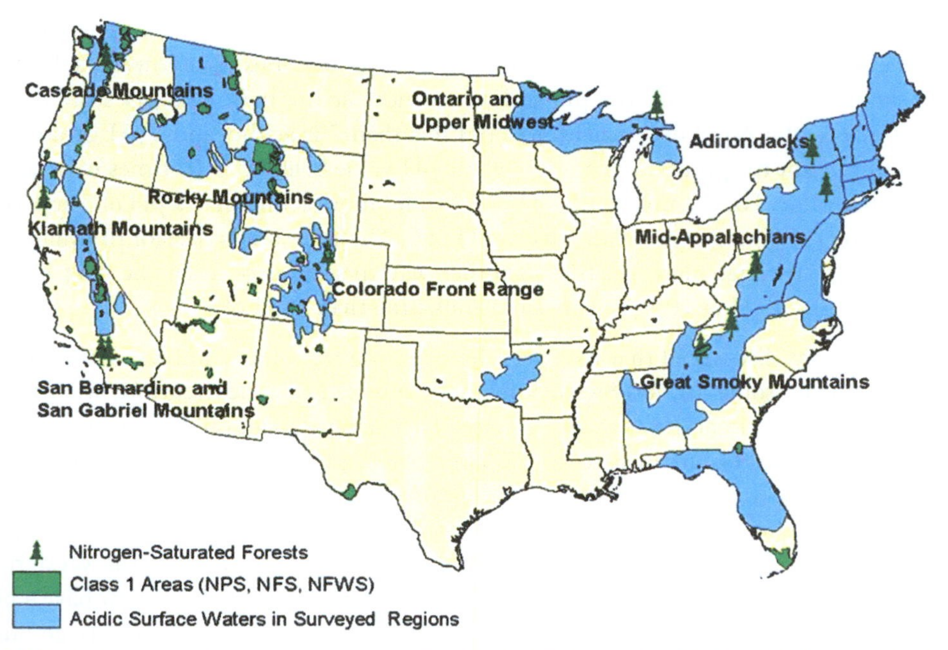

Figure 4.12 Acidic surface waters and nitrogen-saturated forests in the U.S. (Source: U.S. EPA Acid Rain Program). Note that Class I Areas are those areas protected under the Clean Air Act from air pollution damage (NPS = National Park Service; NFS = National Forest Service; NFWS = National Fish and Wildlife Service).

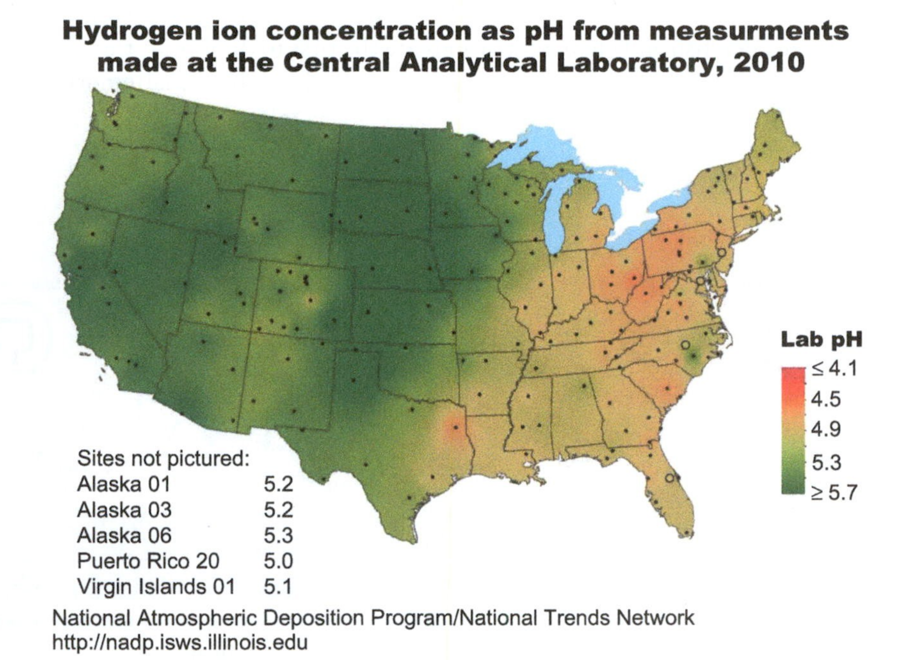

Figure 4.13 Hydrogen ion concentration (pH) in rainfall across the U.S., 2010 (Source: National Atmospheric Deposition Program—wwwnadp.sws.uiuc.edu/ntn/annualmapsbyyear.aspx).

brook trout, and salmon, are more sensitive to acidity than others and tend to disappear first. One of the first signs of acid stress is the failure of females to spawn. Sometimes, even if the female is successful in spawning, the hatchlings or fry are unable to survive in the highly acidic waters. Many types of plankton, minute organisms that form the basis of the lake's food chain, are also affected. However, the lakes do not become totally dead. Some life forms actually benefit from the increased acidity. Lake-bottom plants and mosses, for instance, thrive in acid lakes.

Acid rain also affects forest ecosystems. The term **waldsterben**, or dying of the forest, was first coined in Germany in the early 1980s to describe forest decline in the Black Forest (Figure 4.14). Acid rain, acid fog, and acid vapor damage the surfaces of leaves and needles, reduce a tree's ability to withstand cold, and inhibit plant germination and reproduction. Consequently, tree vitality and regenerative capability are reduced. More importantly, prolonged exposure to acid rain causes forest soils to lose valuable nutrients. It also increases the concentration of aluminum in the soil which interferes with the uptake of nutrients by the trees. Lack of nutrients causes trees to grow more slowly or stop growing altogether. The more visible damage, such as defoliation, may show up later. Trees exposed to acid rain may also have more difficulty withstanding other stresses, such as drought, disease, insect pests, and cold weather. Acid rain has been implicated in forest and soil degradation in many areas of the eastern United States, particularly high elevation forests of the Appalachian Mountains from Maine to Georgia that include areas such as the Shenandoah and Great Smoky Mountain National Parks.

Figure 4.14 Tree death in the Black Forest, Germany because of acid rain in 1996. Note the new forest growing around the acidified trees (Source: iStockphoto.com with copyright Michael Fernahl).

Not all lakes and forests that are exposed to acid rain become acidified. The ability of any ecosystem to withstand acidification depends on its neutralizing capability—or **buffering capacity**. This is largely determined by the region's geological conditions and soils. For example, in areas where there is abundant limestone rock (e.g., Kentucky and Texas), lakes are better able to neutralize acid. In areas where rock is mostly granite (for example, New England, western Montana, and Idaho), the lakes have a more difficult time neutralizing acid. If finely ground limestone calcium carbonate (or $CaCO_3$) is added to water, it raises the pH and increases resistance to acidification. The liming of lakes and waterways is carried out on a large scale in Sweden and Norway. In Sweden, around 7,500 lakes and 7,000 miles of waterways are now limed each year, a very expensive undertaking. Unfortunately, much of the eastern United States—where most of the acid rain falls—has a lot of granite rock and therefore a very low capacity for neutralizing acids.

The ability of soils to buffer acidity depends on the nature of the soil itself and type of bedrock beneath the forest floor. Midwestern states like Nebraska and Indiana have soils that are well buffered. Places in the mountainous Northeast, like New York's Adirondack and Catskill Mountains, have thin soils with low buffering capacity. As with lakes, the acidification process in soils can be countered by liming. Lime acts like a filter in the upper layer of the forest soil, where it can capture and neutralize future acid deposition. The effect of the added lime penetrates slowly into the soil, at roughly 0.5 inches per year, but it persists for a long time in the future. Therefore, the liming of soil can help counter the acidification of surface water in the long term. A dosage of 3–5 tons of lime per hectare is estimated to protect soil from acidification for 20–30 years with current levels of acid deposition in southern Sweden.

Mercury in the Environment

Mercury contamination of the environment is one of the most important ecological problems humans face in the 21st century, yet it is under-reported in the media. Mercury is a teratogen[9] that interferes with neurological development. The risk to humans and wildlife occurs as mercury is transported to watersheds and accumulates in the aquatic food chain (Figure 4.15). Fortunately, the concentrations of all forms of mercury in most natural waters are very low. However, certain aquatic bacteria methylate, or transform, inorganic mercury into **methylmercury** that greatly increases the bioavailability and toxicity of mercury.

Methylmercury biomagnifies to high concentrations in aquatic food webs. **Biomagnification** is defined as the increasing concentration of a contaminant with increasing **trophic level** in a food web. Organisms at the base of the food web such as phytoplankton absorb methylmercury directly from the water while consumers, including fish, are primarily exposed to methylmercury through their diet. Because mercury bioaccumulates from trophic level to trophic level, concentrations of methylmercury in fish can exceed those in ambient surface water by a factor of 10^6–10^7.

Bioaccumulation of mercury has been more intensively studied in fish than in other aquatic organisms because gamefish often contain very high concentrations mercury, and fish are the

[9] A teratogen is defined as a drug or other substance capable of interfering with the development of a fetus, causing birth defects.

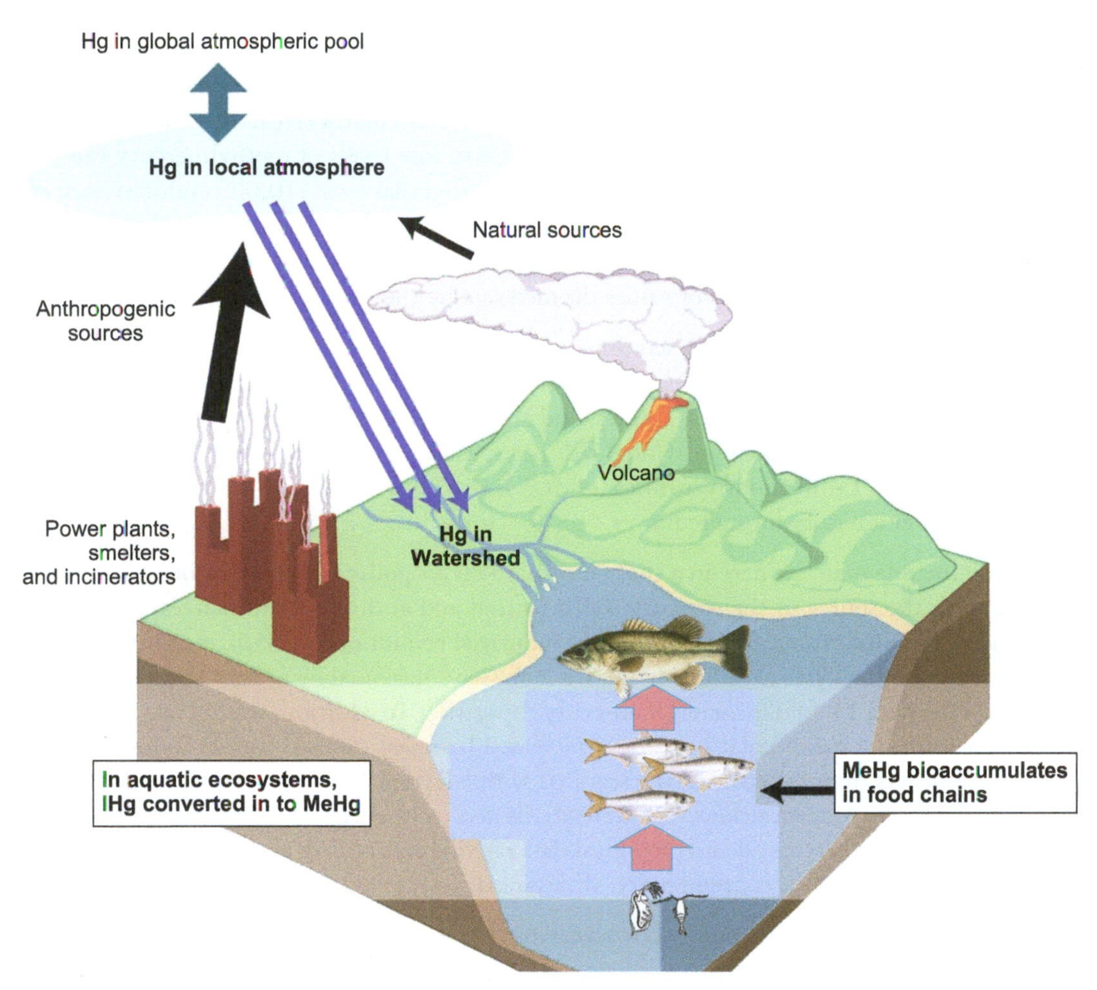

Figure 4.15 The mercury cycle. Most of the Hg contamination observed today originates from atmospheric deposition. Although there are natural sources of Hg emissions to the atmosphere, like volcanoes, two-thirds of Hg emissions are human-caused with the majority of these originating from coal-burning power plants. Power plants release inorganic and elemental forms of Hg to the atmosphere that is relatively non-toxic. This Hg can be deposited near power plants or remain suspended in the atmosphere circulating around the globe in a "global pool of Hg" that is capable of contaminating even the most remote regions. Once this inorganic form of Hg is deposited on the Earth and is washed into lakes and rivers, it is chemically transformed by naturally occurring, aquatic microbes to methyl Hg (MeHg)—a toxic form that bioaccumulates in the food chain and contaminates fish, putting humans and wildlife at risk around the globe.

primary pathway of mercury to humans. Concentrations of methylmercury in muscle tissue or whole fish typically increase with increasing age, body size, and trophic position. Consumption of fish is then the main pathway of methylmercury exposure for birds. The biomagnification of mercury in aquatic food webs leads to high concentrations in fish-eating birds, and methylmercury can adversely affect adult bird survival, reproductive success, and behavior. Methylmercury in bird diets can cause teratogenic effects on birds and is passed from mother to the eggs. According to several scientific reviews, the embryos of birds are much more sensitive than the adult to methylmercury exposure.

The human health risks of dietary exposure to methylmercury underlie the concerns about mercury contamination of aquatic food webs. Human health risks to even low doses of methylmercury can include damage to the nervous systems. Fetuses are particularly sensitive to methylmercury consumed by pregnant women, and prenatal exposure to low levels of methylmercury can cause developmental and cognitive problems. It has been reported that over 410,000 children born each year in the United States have been exposed in the womb to methylmercury levels that are associated with impaired neurological development. Eight percentage of U.S. women of childbearing age have blood Hg levels in excess of values deemed safe by the EPA.

To help reduce the risk of mercury exposure, fish consumption advisories regarding mercury contamination are issued by the EPA. A consumption advisory is a recommendation to limit consumption to specified quantities, species, and sizes of fish and is issued based on the analysis of muscle tissue of at least three individual fish per species. As of December 2011, 17.7 million lake acres (or 42%), 1.4 millionriver miles (or 36%), and 36% of the nation's contiguous coastal waters were under advisory for mercury.

Why do we include mercury here in a chapter focused on air pollution? The reason, quite simply, has to do with the source of the mercury. Both natural and anthropogenic sources contribute to mercury in the environment. However, since the industrial revolution (c. 1850), mercury deposition rates (as revealed by analyses of sediment and ice cores) have increased by a factor of 3–4, with some regions experiencing 11-fold increases in mercury deposition. In addition, studies have shown that mercury levels in the fur of polar bears have increased by seven times since pre-industrial times. Analyses of feathers from two fish-eating seabirds sampled from 1885 to 1994 showed long-term increases in mercury concentrations that were attributed to increases in global trends in mercury contamination. These geological and biological data provide compelling evidence that the problem of mercury contamination is contemporary, widespread, and linked to anthropogenic sources.

Mercury is released into the environment by several human activities including gold production, nonferrous metal production, incinerators, cement production, improper disposal of consumer products, and coal-burning power plants. Mercury emissions in the United States have declined over the past decade due to federally mandated reductions in Hg emissions in medical waste incinerators and municipal incinerators.[10] However, unlike incinerator emissions, emissions from coal-burning power plants have remained largely unchanged, and their relative contribution to total U.S. emissions has increased over the past decade. Currently, the largest single anthropogenic source of environmental mercury is emissions from coal-burning power plants. However, the coal industry continues to deny any apparent causal relationship between mercury contamination of the environment and emissions from the power plants. For example, the American Coalition for Clean Coal Electricity, a non-profit group that represents the interests of the coal industry, claims in several company reports that power plants are not the major source of mercury emissions in the United States, that local deposition of mercury from power plantsis not prevalent, and that there are currently no mercury advisories on Texas' power plant lakes.[11] This view is clearly at odds with the general consensus among the scientific community.

[10] http://www.epa.gov/mercury/control_emissions/emissions.htm

[11] Source: www.cleancoalusa.org/

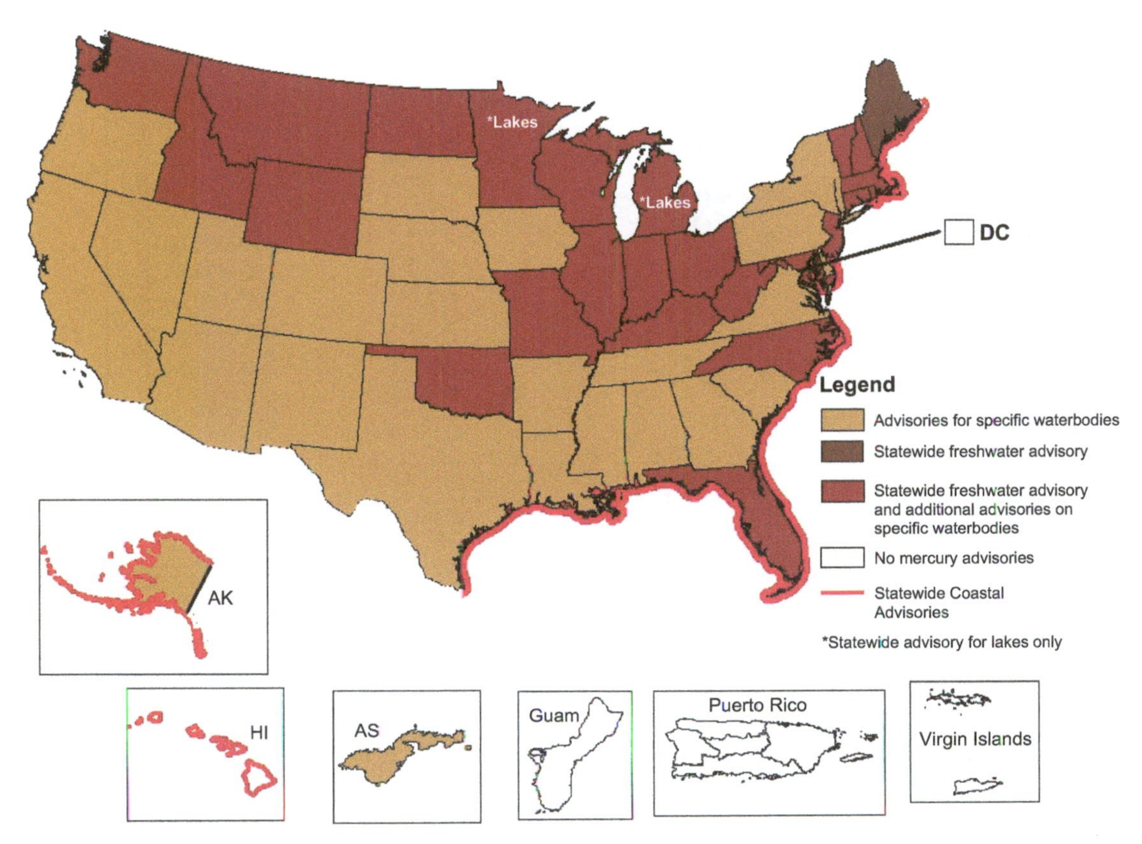

Figure 4.16 Fish consumption advisories for mercury, 2010 (Source: http://water.epa.gov/scitech/swguidance/fishshellfish/fishadvisories/upload/nlfa_slides_2011.pdf). Note: This map depicts the presence and type of fish advisories issued by the states for mercury as of December 2010. Because only selected waterbodies are monitored, this map does not reflect the full extent of chemical contamination of fish tissues.

Figure 4.16 shows the extent of mercury advisories across the United States. Many states have advisories for the entire state (as shown in red), while others have advisories only for specific water bodies. However, maps like this can be misleading because they only really give a partial view of the extent of the problem. For example, in my home state of Texas, there is widespread contamination of fish in the eastern third of the state (Figure 4.17), but this picture has only emerged after extensive sampling over many years. Atmospheric modeling from coal-fired power plants and other major mercury sources in the region (Figure 4.18) clearly shows that, under the predominant transport winds, there is intense mercury deposition in the both immediate vicinity of the sources (as shown by the red, orange, and yellow colors on the map) and regionally. Here, deposition rates are in the order of 1.5–3 times higher than mercury deposition in the western United States where, according to the EPA, mercury transported in from Asia (specifically, China) is the predominant source. Simply put, there is a very strong and compelling correlation between regional mercury levels in fish and the emissions from coal-fired power plants and other point sources in the region.

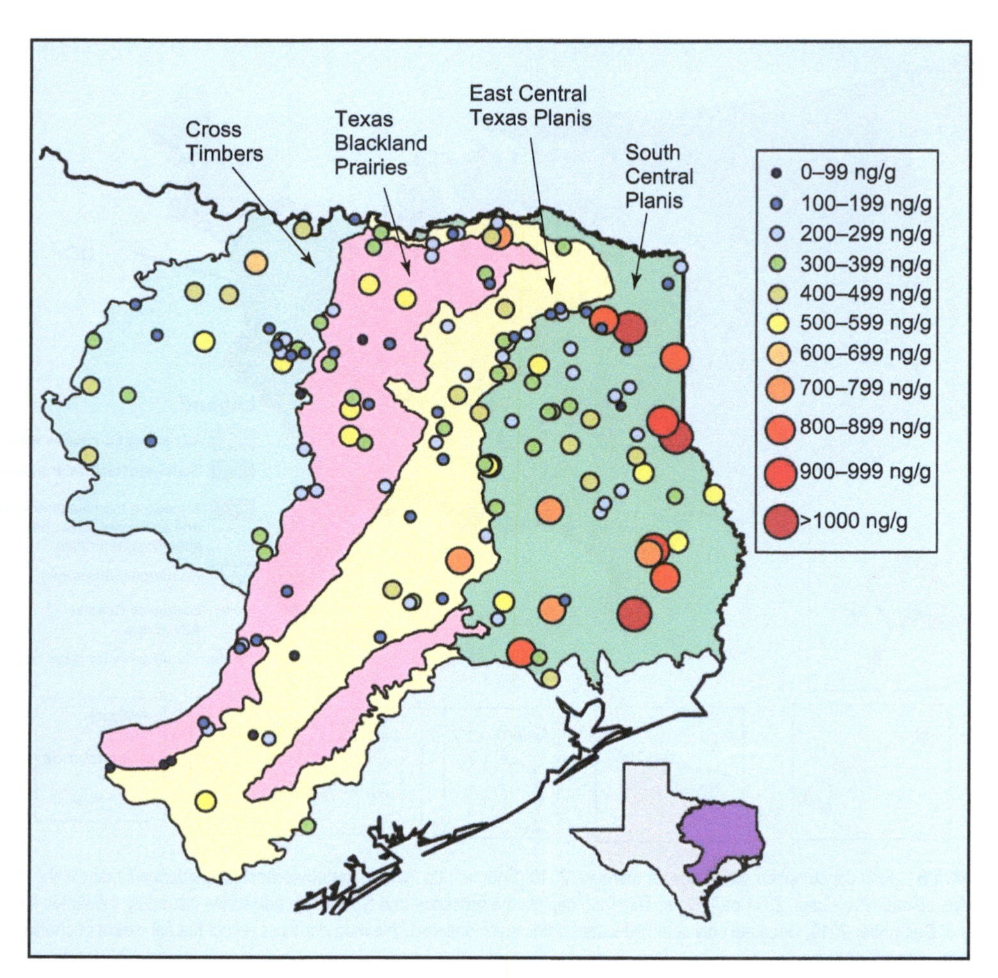

Figure 4.17 Estimated total mercury concentrations in largemouth bass from reservoirs in four ecoregions of North Texas. Each point represents the average mercury concentration in a water body. The redder the color and the larger the point, the higher the concentration of mercury in the fish. Shades of blue are considered safe for unlimited human consumption by the U.S. EPA (< 300 nanograms/gram of fish tissue). Every other color point represents fish with Hg concentrations high enough that U.S. EPA would recommend issuing a fish consumption advisory to warn people to limit their consumption of fish (Source: Drenner, R.W. et al. (2011), *Environmental Toxicology*. 30, 1–5).

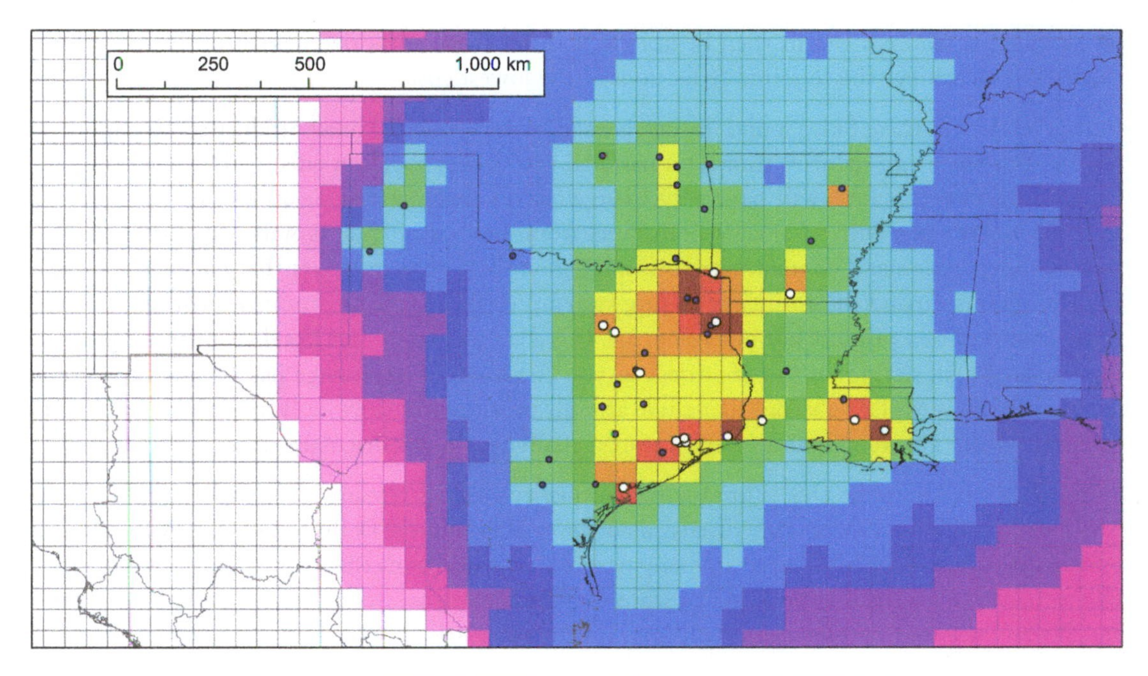

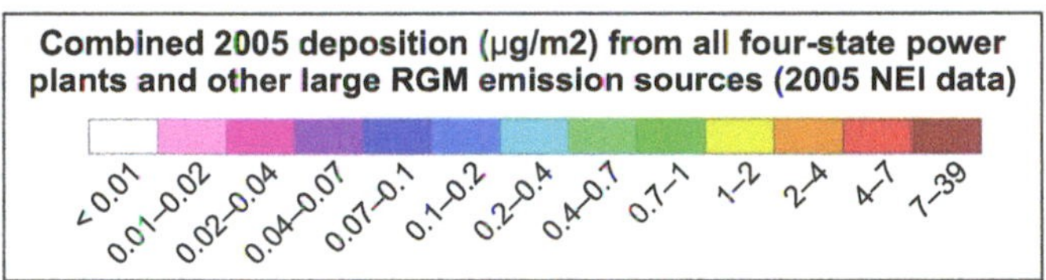

Figure 4.18 Annual deposition of mercury from 51 major emitters in TX, OK, AR and LA. Each square represents an area of about 50 × 50 km (Source: Slattery, unpublished data).

CONCLUDING THOUGHTS

We have made great strides in cleaning up our air in the United States. The national trends showing that declining levels in the criteria pollutants are evidence that we are moving in the right direction. The problem is that, as a society, we continue to grow, consume, and develop. These actions will continue to put pressure on our environment, particularly our air. Ozone remains a significant challenge, and under-regulated toxins, such as mercury, are rapidly becoming a concern in certain sensitive areas.

The current emphasis of clean air legislation is on national **cap-and-trade** programs like the Clear Skies initiative. According to the EPA, this legislation would create a mandatory program that would dramatically reduce power plant emissions of SO_2, NO_x, and mercury by setting a national cap on each pollutant. By 2018, SO_2 and NO_x emissions would be cut by approximately 70% from year 2000 emissions. A federally enforceable emission limit (or cap) for each pollutant would be established. Sources would then be able to transfer these authorized emission limits among themselves to achieve the required reductions at the lowest cost.

Initiatives such as Clear Skies sound impressive and, if implemented, may produce some important results. But requiring utilities to meet a national cap, where they can effectively trade pollutant emissions between plants, will have very little effect in areas such as Texas and its surrounding regions, where the addition of new pollutants from new power plants will very likely lead to increased deposition and contamination in the region. The bottom line is that national cap-and-trade programs expand the pollution trading system, so while some communities will get cleaner, many communities will lose out on cleaner air. It is a bit like trashing a hotel room and then leaving a wad of cash sitting on the front desk as you check out. The damage has already been done. Such cap-and-trade plans also fail to include a single measure to reduce or even limit the growth of carbon dioxide. This remains one of our greatest challenges and is discussed in detail in Chapter 6.

While the situation in the U.S. is improving, air quality globally remains an enormous challenge, as illustrated by Figure 4.10 and my discussion of air quality in New Delhi in Box 4.3. Air pollution remains at dangerously high levels in many parts of the world with the poor especially vulnerable. Exposure to fine particles in polluted air leads to diseases such as stroke, heart disease, lung cancer, and chronic respiratory infections, including pneumonia. It is simply shocking that, in 2018, air pollution causes 1 in 9 deaths worldwide.[12]

[12] Source: http://www.who.int/airpollution/en/

5

Stratospheric Ozone Depletion

INTRODUCTION

In the preceding chapter, we covered tropospheric ozone—a low-level, human-formed pollutant that degrades air quality and has toxic effects on humans and vegetation. We now turn our attention to **stratospheric ozone**, which is widely referred to as the "ozone layer" and contains about 90% of all atmospheric ozone. Many people picture this layer as a surface or shield that protects us from harmful ultraviolet (or UV) **radiation.**[1] Ozone in the **stratosphere** does not exist like some sort of blanket; rather, it is dispersed throughout the stratosphere (Figure 5.1), reaching its highest concentrations at an altitude of about 14 miles (22 km) above Earth's surface (refer back again to Figure 4.4). At these altitudes, the actual amount of ozone in the atmosphere is minute. The pressure exerted by ozone at peak concentrations is about 30 millipascals, whereas average sea level pressure (i.e., the pressure exerted by all gases in the atmosphere) is over 101 million millipascals. This means that only 10 or less of every million molecules of air are ozone molecules.

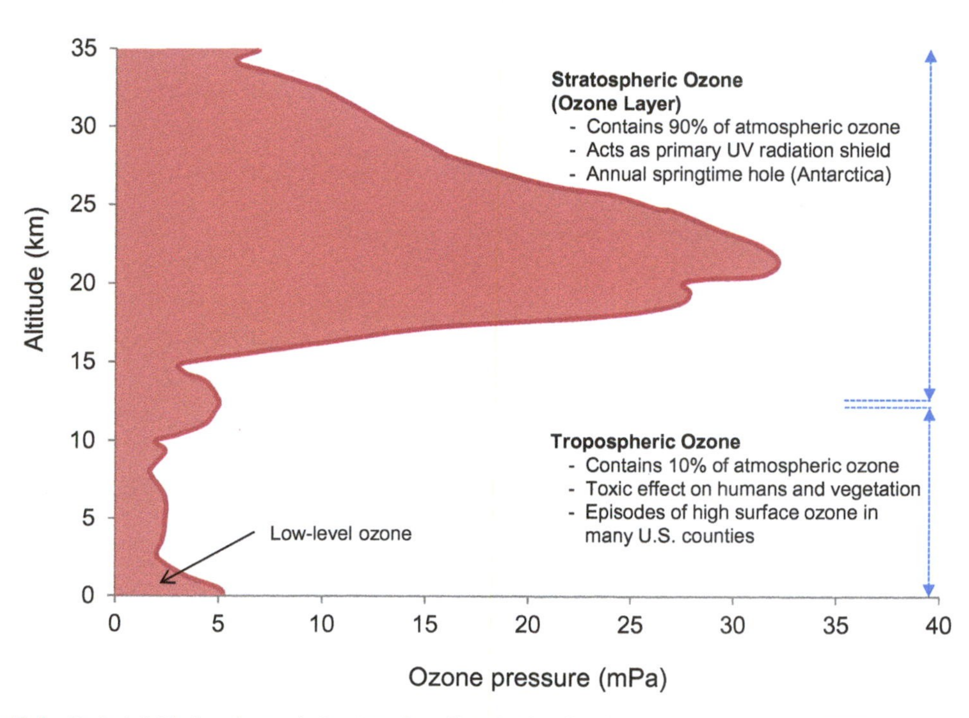

Figure 5.1 Vertical distribution of ozone in the atmosphere. Note that 25 mPa of atmospheric pressure = 0.00025 millibar; mean sea level pressure exerted by all gases in the atmosphere = 1013 millibar.

[1] Turn to Figure 6.5 (Chapter 6) for a first look at the electromagnetic spectrum and where UV radiation fits in relative to visible light energy emitted from the sun.

Ozone[2] is a relatively simple molecule, consisting of three bound oxygen atoms. Its chemical notation is O_3, and it is chemically identical to the O_3 found in the troposphere. However, ozone in the stratosphere has very different environmental consequences for humans and other life forms than ozone in the troposphere. At Earth's surface, breathing it is lethal at dosage levels of a few molecules per million air molecules. This is why ozone at the surface is a pollutant. Yet, ozone high in the atmosphere (the stratosphere) screens out biologically harmful solar ultraviolet radiation, keeping it from reaching the surface. Such ultraviolet radiation is destructive to genetic cellular material in all living organisms (plants, animals, and humans). Without the ozone layer high up in the atmosphere, life on the surface of Earth simply would not exist.

The key issue with stratospheric ozone is that scientists have observed a long-term downward trend in the stratospheric ozone on a global scale over the past several decades. The thinning of the ozone layer means that more harmful UV rays enter our atmospheric system. Of greater concern is the substantial thinning in ozone that occurs over Antarctica each spring, an occurrence that is commonly referred to as the Antarctic ozone hole. Springtime Arctic ozone losses have also been observed in recent years, which all beg the question: Is the thinning of stratospheric ozone a natural phenomenon, or are humans somehow involved? And if we are to blame, to what extent can we reverse this trend of decreasing ozone in the stratosphere?

FORMATION, MEASUREMENT, AND FUNCTION OF OZONE

The first point to appreciate with stratospheric ozone is that it is formed naturally in the middle to upper stratosphere of the tropics. It is here that the sun provides enough of the very energetic extreme ultraviolet (EUV) radiation. Ozone is generated in a two-step splitting-plus-combination process. In the first step, ultraviolet radiation from the sun breaks apart a diatomic oxygen molecule (O_2) to form two separate oxygen atoms (2O), as shown in reaction (5.1). In the second step, the O atoms undergo a binding collision with other O_2 molecules to form ozone molecules. In the overall process, three oxygen molecules react to form two ozone molecules. As a chemical equation, this process is represented by:

$$O_2 + UV \rightarrow O + O \tag{5.1}$$

$$O + O_2 \rightarrow O_3 \tag{5.2}$$

These reactions occur continually wherever ultraviolet sunlight is present in the stratosphere (Figure 5.2).

The next step in the lifecycle of an ozone molecule (Figure 5.2) shows that the ozone spends most of its life absorbing UV radiation. This absorption process occurs when the UV ray breaks the ozone (O_3) molecule into a diatomic oxygen molecule (O_2) and a single oxygen atom (O), followed by the

[2] For the rest of this chapter, you may assume that "ozone" refers to stratospheric ozone unless otherwise specified.

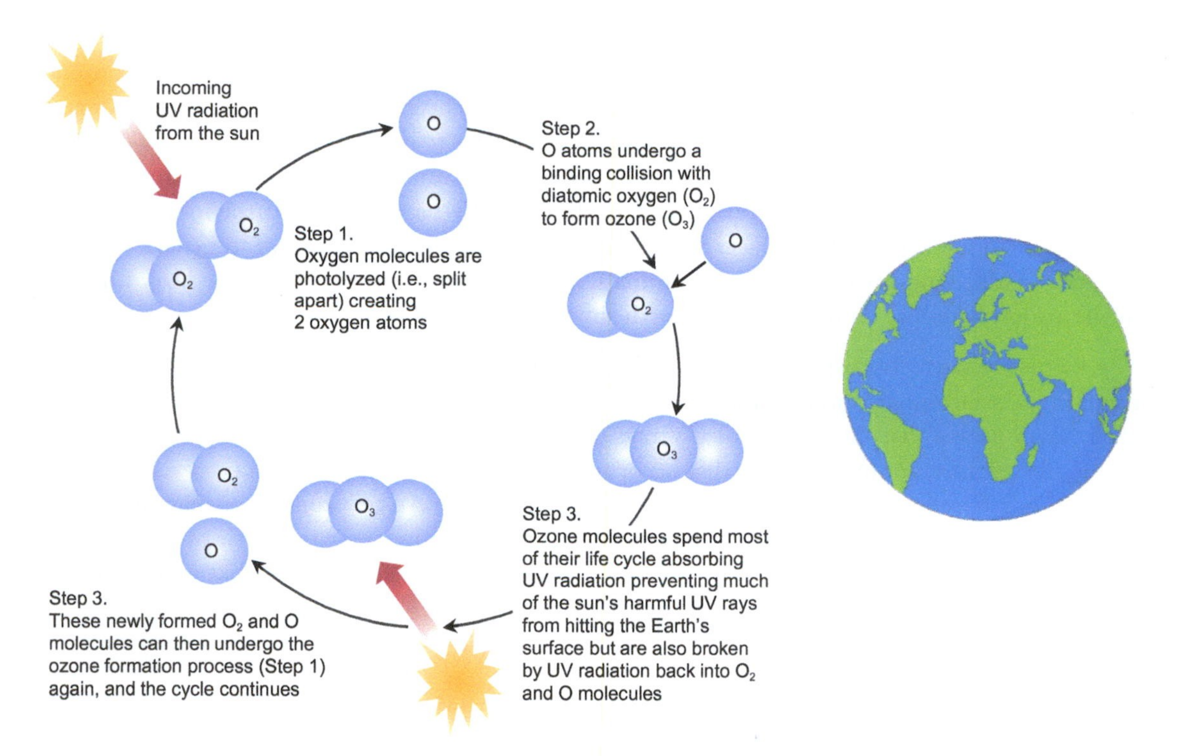

Figure 5.2 Schematic of ozone creation and destruction in the atmosphere. Source: Mike Slattery; image of Earth © Shutterstock.com.

recombination of the single oxygen atom with another diatomic oxygen molecule to reform ozone (Figure 5.2). In this process, UV radiation is converted into heat energy. Again, we can represent this chemically as

$$O_3 + UV \rightarrow O_2 + O \tag{5.3}$$

This process of absorption is an extremely efficient process since ultraviolet radiation is effectively screened out before it reaches Earth's surface. Ozone then reforms through reaction (5.2) above resulting in no net loss of ozone. Finally, it is worth noting that the absorption of solar UV energy via reaction (5.3) above actually leads to a warming of the stratosphere where temperatures begin to increase with increasing altitude. Nevertheless, overall temperatures in the stratosphere are still significantly colder than at Earth's surface.

Now, imagine a column of air extending from Earth's surface into outer space. If all of the O_3 in this column were to be compressed and spread out evenly over the area, it would form a slab approximately 3 mm thick (0.12 inches, about the thickness of two stacked pennies). In those terms, O_3 is very thin indeed (Figure 5.3), but in space, it is best not to envision the ozone layer as a distinct, measurable band, or layer. Instead, think of it in terms of parts per million (ppm) concentrations in the stratosphere (the layer 6–30 miles above Earth's surface). Ozone is measured in Dobson Units (DU), named after G.M.B. Dobson, one of the first scientists to investigate atmospheric ozone.

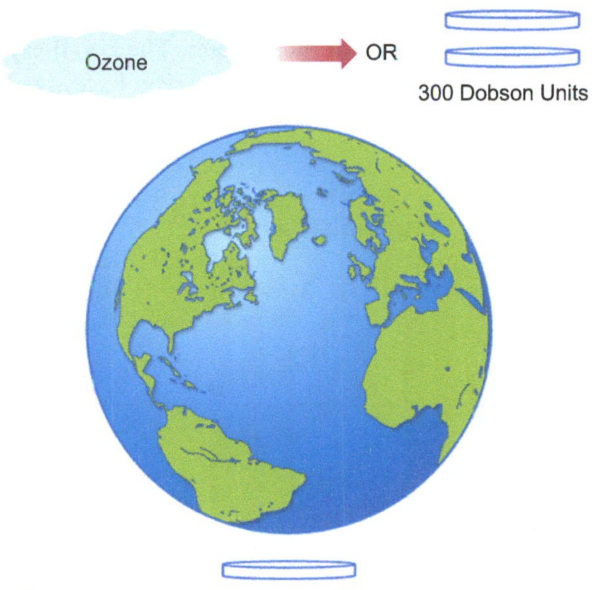

Figure 5.3 Ozone thickness and the Dobson unit. Source: Mike Slattery; image of Earth © Shutterstock.com.

He designed the Dobson Spectrometer—the standard instrument used to measure the total amount of ozone in a column extending vertically from Earth's surface to the top of the atmosphere. One Dobson Unit is defined to be 0.01 mm thickness; thus, the compressed ozone slab would be 300 DU, approximately the global average of total ozone. This measurement of ozone is directly related to the amount of UV light reaching the surface. It is therefore a measure of UV exposure received at the surface. The less total ozone in the column means more UV light penetrates, hence the faster you get sunburned.

Ozone concentrations are typically about 260 DU near the tropics and higher elsewhere. At first, that may seem somewhat counter-intuitive. Surely, one would expect that total ozone levels would be highest over the tropics and correspondingly lower in the polar region given the greater intensity of solar ultraviolet radiation in equatorial regions (which would essentially speed up the reaction in equation 1). The fact is, the natural processes of ozone formation and destruction are located predominantly in the upper stratosphere the tropics. However, the puzzle of overall lower ozone in the tropics is explained by the atmospheric circulation, which transports high ozone from the tropics poleward and downward to the lower stratosphere of the high latitudes, thereby altering the distribution of ozone from what we would expect (i.e., highest over the tropics) to what is actually observed (i.e., highest in colder air masses over higher latitudes).

Ozone, like carbon dioxide, is a **selective gas**, which means that the gas absorbs and emits radiation only at certain wavelengths. Stratospheric ozone is considered good for humans and other life forms because it absorbs ultraviolet radiation from the sun. If not absorbed, UV-A and UV-B, the highest intensity portion of the UV spectrum, would reach Earth's surface in amounts that are harmful to a variety of life forms (Figure 5.4). In humans, as exposure to UV-B increases, so does the risk of skin cancer, cataracts, and a suppressed immune system. The UV-B exposure before adulthood and cumulative exposure are both important factors in the risk. Excessive UV-B exposure also can damage terrestrial plant life, single-cell organisms, and aquatic ecosystems. Other UV radiation,

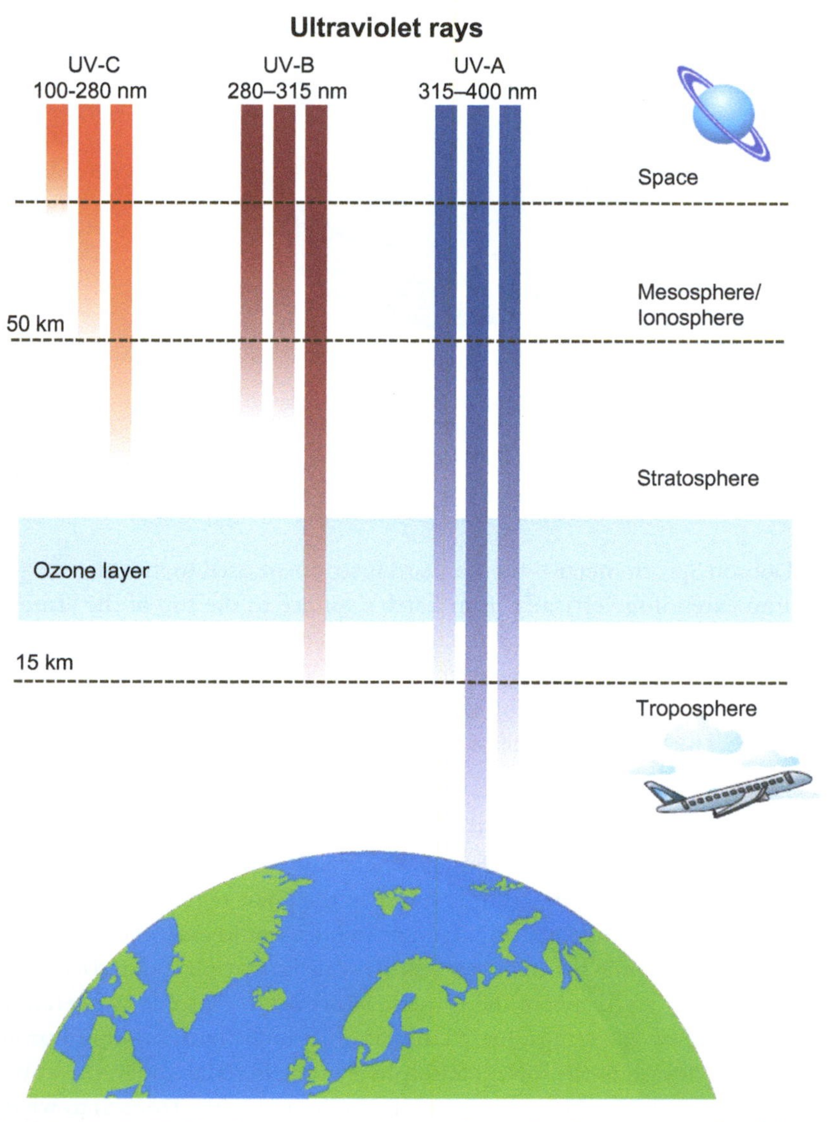

Figure 5.4 Absorption of UV radiation in the atmosphere by ozone. Source: Mike Slattery; image of Earth © Shutterstock.com.

UV-A, which is not absorbed significantly by ozone, causes premature aging of the skin. Sun screens have been developed by commercial manufacturers to protect human skin from UV radiation. The labels of these sun screens usually note that they screen both UV-A and UV-B. Why not also screen for UV-C radiation? From Figure 5.4, we can see that when UV-C encounters ozone in the mid-stratosphere, it is quickly absorbed so that none reaches Earth's surface.

OZONE DEPLETION OVER THE POLES

The first systematic measurements of stratospheric ozone were conducted by a research group from the British Antarctic Survey (BAS) at the South Pole in the late 1950s. Dramatic loss of ozone was first noticed in the early 1970s. In 1974, chemists, F. Sherwood Rowland and Mario Molina of the University of California at Irvine, theorized that **chloroflurocarbons** (CFCs), a family of chemical compounds developed back in the 1920s as a safe, non-toxic, non-flammable alternative to dangerous substances like ammonia for purposes of refrigeration and spray can propellants, might attack the ozone layer. They suspected that CFCs' non-reactivity, the very quality that made them so useful, would allow them to drift, intact, 10–15 miles up above Earth, into the stratosphere. Here, the chemists predicted, short-wavelength ultraviolet radiation could break off a chlorine atom (Cl) from a CFC molecule (Figure 5.5). This highly reactive freed chlorine atom would grab onto an ozone molecule and split it, forming a chlorine monoxide (ClO) radical, which would, in turn, combine with an O atom to form O_2. This would free up the Cl and set off a chain reaction that would destroy ozone molecules in the stratosphere, steadily weakening the ozone layer. The reactions shown in Figure 5.5 are written as

$$Cl + O_3 \rightarrow ClO + O_2 \tag{5.4}$$

$$ClO + O \rightarrow Cl + O_2 \tag{5.5}$$

Rowland and Molina's prediction that CFCs would significantly deplete the ozone layer in the coming decades garnered much negative attention at the time from both the scientific community and industry. CFC producers and users continued to expand the chemical's use, and industry fought hard against any limitations on its production. But in the early- to mid-1980s, scientists had found overwhelming proof of Rowland and Molina's theory: the Antarctic ozone hole[3], a severe annual depletion of the ozone above the South Pole (Figures 5.6 and 5.7). By 1984, ozone had dropped to below 200 DU compared to the 300–320 DU values of the 1960s. The loss of ozone appeared to begin soon after the end of the polar winter darkness and proceeded very rapidly for several weeks into mid-October. The speed and scale of depletion caught scientists by surprise, prompting suggestions that the instruments measuring stratospheric ozone were faulty. However, once the depletion was verified through a series of NASA satellite images, the Antarctic ozone hole garnered worldwide attention as a symbol of humankind's potential to

[3] The ozone hole is not technically a hole where no ozone is present but is actually a region of exceptionally depleted ozone. Any area where the concentration drops below 220 Dobson Units is considered part of the ozone hole.

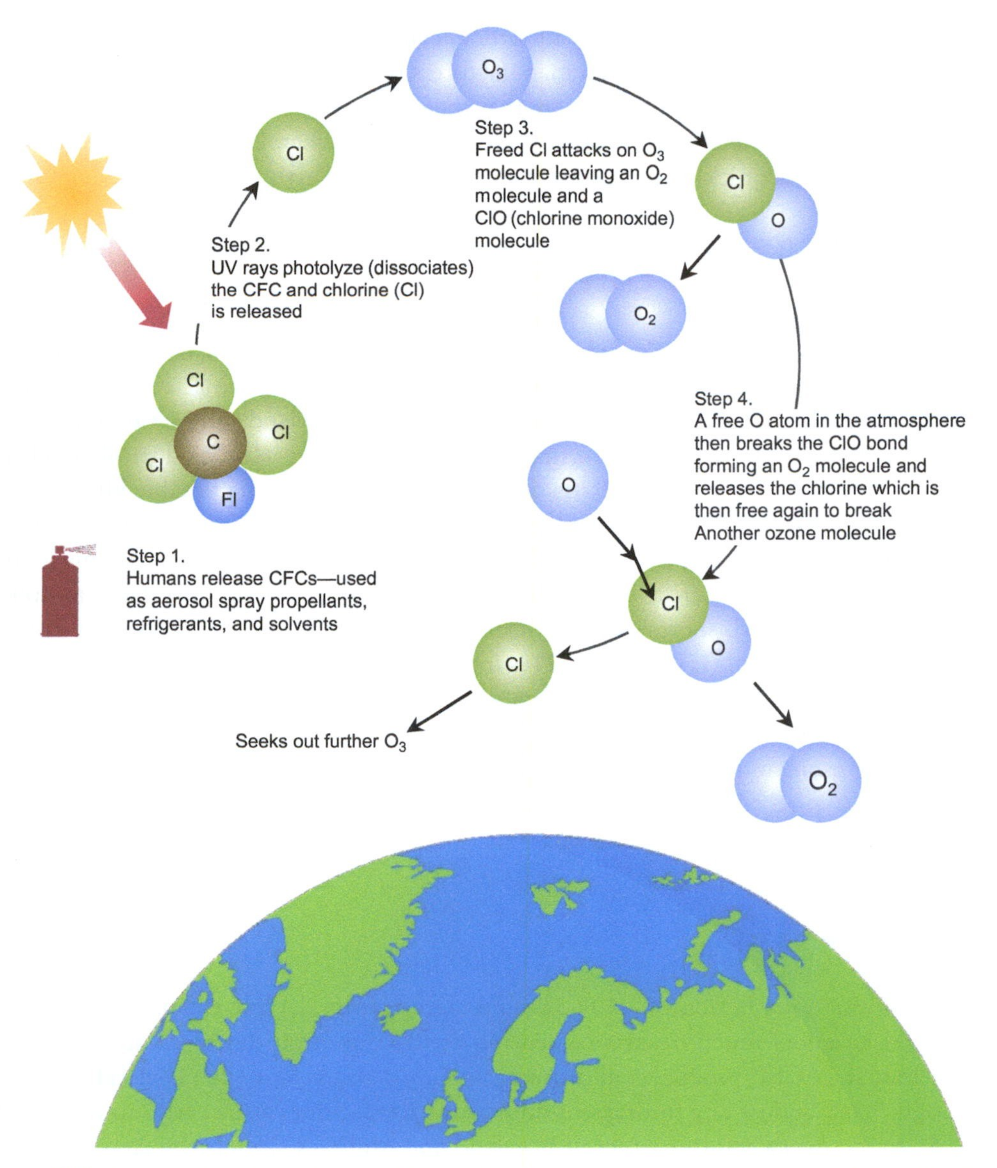

Figure 5.5 Schematic illustrating the life cycle of CFCs and how their breakdown products destroy ozone. Source: Mike Slattery; image of Earth © Shutterstock.com.

cause unanticipated damage to Earth's fragile atmosphere. Rowland and Molina, whose alarming theory was attacked mercilessly by the chemical industry, were vindicated in 1995 when they won the Nobel Prize for their discovery. This marked the first Nobel Prize ever presented in the environmental sciences.

Ozone in the atmosphere is now mapped using a range of satellite-borne instruments used to gain a global perspective of ozone levels.[4] The total ozone mapping spectrometer (TOMS) instrument

[4] Source: www.ozonewatch.gsfc.nasa.gov/

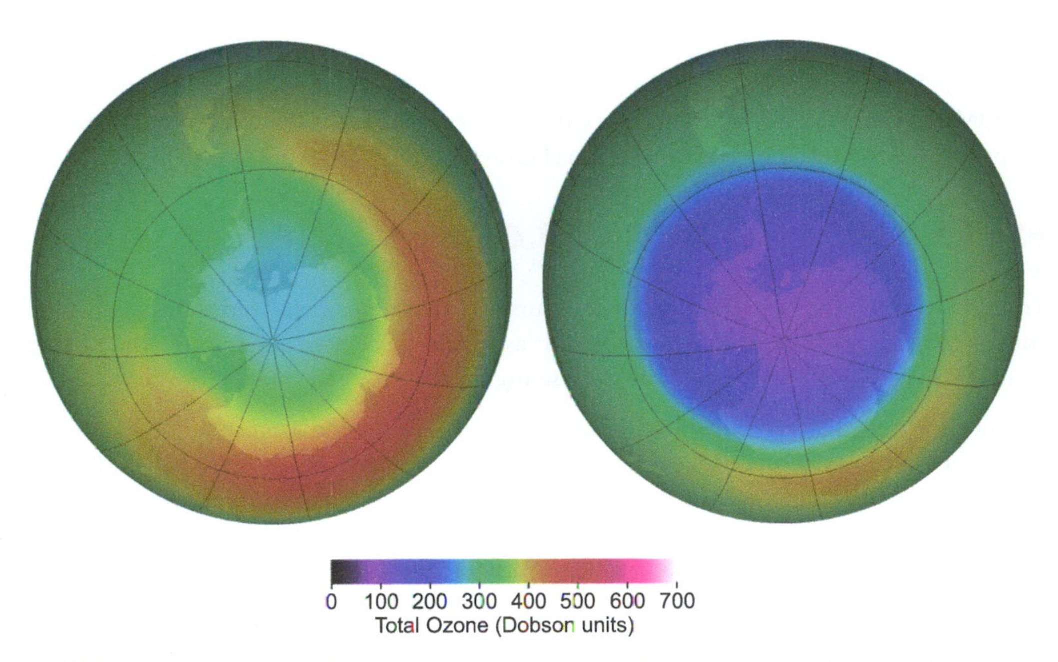

Figure 5.6 Average size of the ozone hole (for October) over Antarctica in 1979 (left) and 2011 (right) (Source: NASA Total Ozone Mapping Spectrometer—http://jwocky.gsfc.nasa.gov/).

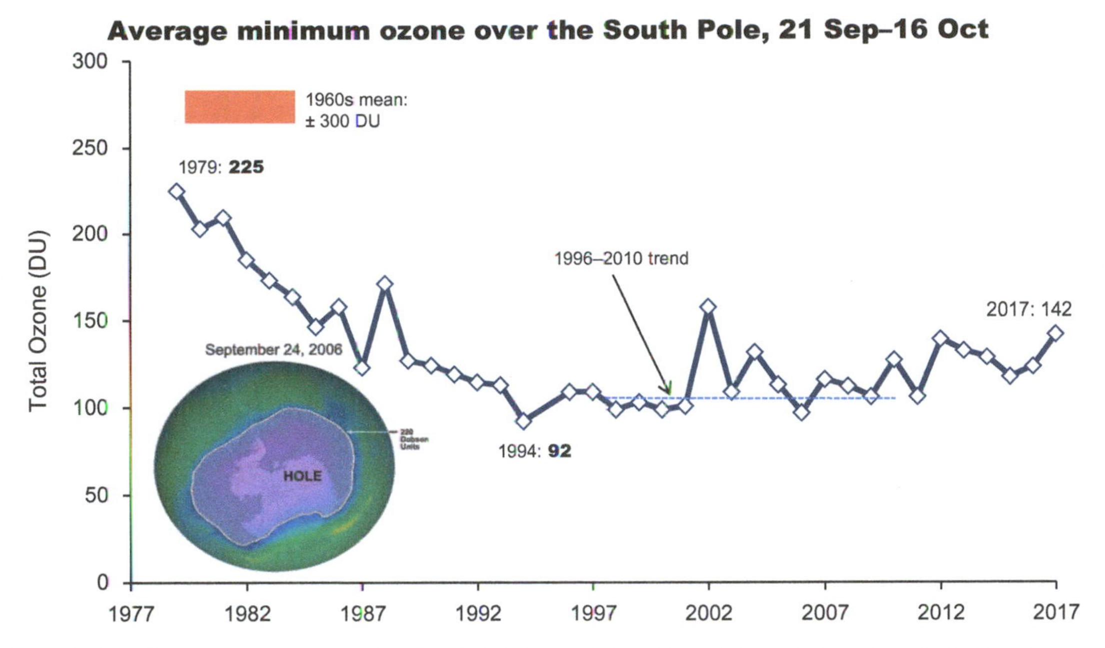

Figure 5.7 Trend in Antarctic ozone levels, 1979–2017 (Source: NASA Total Ozone Mapping Spectrometer—http://jwocky.gsfc.nasa.gov/). The ozone hole area is determined from total ozone satellite measurements. It is defined to be that region of ozone values below 220 Dobson Units (DU) located south of 40°S. Values below 220 DU represent anthropogenic ozone losses over Antarctica.

measured ozone levels from the back-scattered sunlight, specifically in the ultra-violet range, up until 2005.[5] Data from 2005 onward are provided by the aptly named ozone monitoring instrument (OMI) aboard a NASA satellite.[6] Recent measurements from OMI show that the ozone hole now approximates the size of North America (Figure 5.8).

THE RECIPE FOR OZONE LOSS

In trying to understand how the ozone loss occurs and the things that need to happen to destroy so much ozone so quickly, it helps to think of it as a recipe. We need several ingredients to make the ozone loss occur. We will now look at these ingredients one at a time.

It is now accepted that chlorine (and bromine) compounds in the atmosphere cause the ozone depletion observed in the ozone hole over Antarctica and the North Pole. Nearly, all of the chlorine in the stratosphere, where most of the depletion has been observed, comes from human sources. As shown in Figure 5.5, chlorine (Cl) is part of the gaseous CFC compound that, once released, is transported up into the stratosphere and subsequently broken down by sunlight. This chlorine then destroys ozone that allows more UV radiation to pass through the atmosphere to Earth's surface. However, as with many environmental processes, the reactions described here are not quite that simple. When CFCs break down in the atmosphere, active (or free) chlorine (Cl), the form of chlorine required for ozone destruction, is not immediately released (i.e., Step 2 in Figure 5.5). What

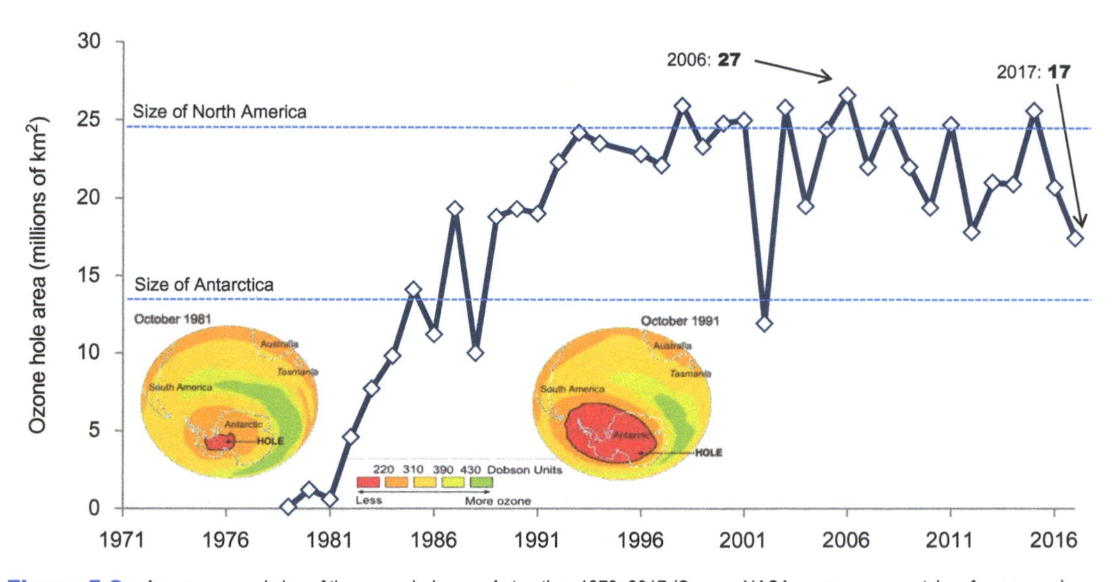

Figure 5.8 Average annual size of the ozone hole over Antarctica, 1979–2017 (Source: NASA—www.ozonewatch.gsfs.nasa.gov).

[5] Source: www.ozoneaq.gsfc.nasa.gov/earthProbeOzone.md

[6] Source: www.aura.gsfc.nasa.gov/instruments/omi.html

happens in reality is that the chlorine breaks down into several inorganic carriers (or reservoirs) of chlorine, the two most important of which are hydrogen chloride (HCl) and chlorine nitrate ($ClONO_2$). These chlorine reservoir species HCl and $ClONO_2$, and their bromine counterparts, are then converted into more active forms of chlorine, but these reactions are unusual. They cannot take place in the atmosphere unless certain conditions are present, most importantly, the presence of a very unique type of cloud, called a **polar stratospheric cloud** (or PSC, as shown in Figure 5.9). So what exactly are these PSCs, and how do they form?

Figure 5.10 shows schematically what happens over Antarctica during winter.[7] As winter arrives, a vortex of winds develops around the pole and isolates the polar stratosphere. During the winter polar night, sunlight does not reach the South Pole, and the air within the **polar vortex** gets extremely cold. When temperatures drop below −78 °C (−109 °F), nitric acid (HNO_3) and sulfur-containing gases condense with water vapor to form solid and liquid PSC particles (Figure 5.11). At even lower temperatures, ice particles also form. Although these thin clouds are not the same as clouds that you are used to seeing in the sky, which are composed of water droplets, they are crucial for ozone loss to occur.

Figure 5.9 As colorful as any aurora, iridescent polar stratospheric clouds (PSCs) glitter in the low light of the spring sunrise at McMurdo Station in Antarctica in late August or early September 2003. Photo courtesy Seth White, UNAVCO.

[7] Remember, Antarctica has six months of daylight darkness. During the winter (March through August), Antarctica is tilted away from the sun, causing it to be dark. The lowest temperature ever recorded in Antarctica was −129°F. The warmest temperature ever recorded in Antarctica was 59°F. The average summer temperature is 20°F. The average winter temperature is −30°F.

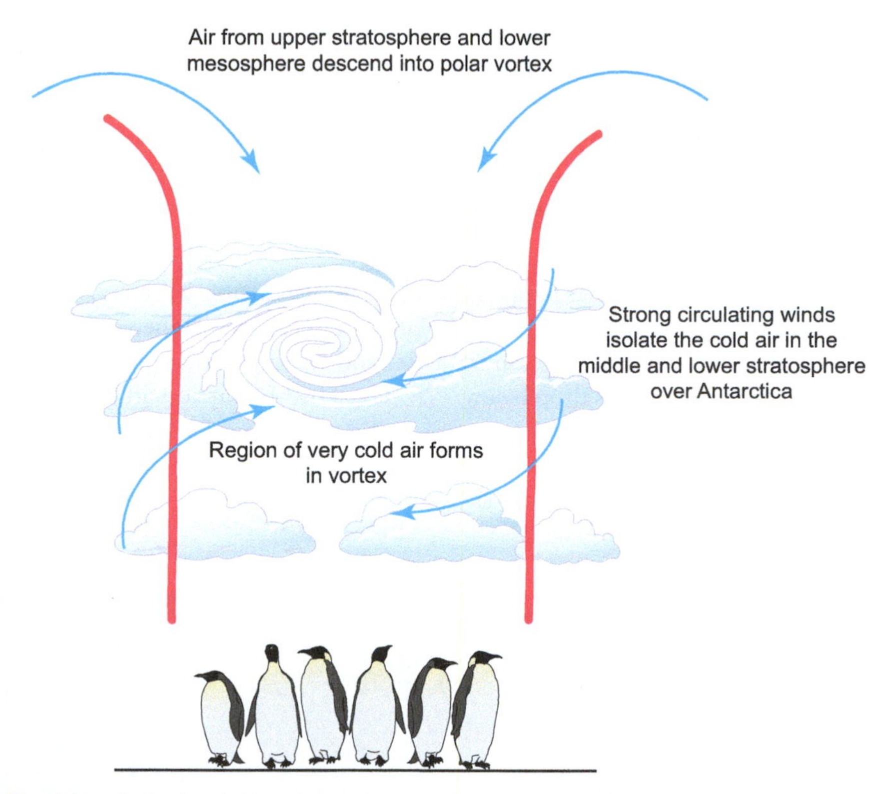

Figure 5.10 Schematic showing what happens over Antarctica during winter. During the winter polar night, sunlight does not reach the South Pole resulting in the development of a strong circumpolar wind in the middle to lower stratosphere. These strong winds are known as the "polar vortex," which has the effect of isolating the air over the polar region.

The chlorine reservoir species HCl and $ClONO_2$, which have been transported poleward by the global winds, now collect on the PSCs, as shown in Figure 5.12, and react as follows:

$$HCl + ClONO_2 \rightarrow Cl_2 + HNO_3 \quad (5.6)$$

Molecular chlorine (Cl_2) then comes off the PSC while HNO_3 remains on the PSC to settle out of the stratosphere later on. These reactions in Equation 6 are referred to as "dark reactions" because they occur during the polar winter (i.e., 24 hours of darkness). Note, however, that we still only have formed Cl_2 from the reactions. To destroy ozone requires atomic or **free chlorine** (Cl), as shown in Figure 5.5. Unfortunately for the ozone layer, molecular chlorine is easily photodissociated (split by sunlight) as follows:

$$Cl_2 + UV \rightarrow Cl + Cl \quad (5.7)$$

This reaction is the key to the timing of the onset of the "opening" of the ozone hole and the speed at which it depletes. During the polar winter, molecular chlorine builds up in the stratosphere (Equation 6). Then, when the sunlight returns to the polar region during the southern hemisphere

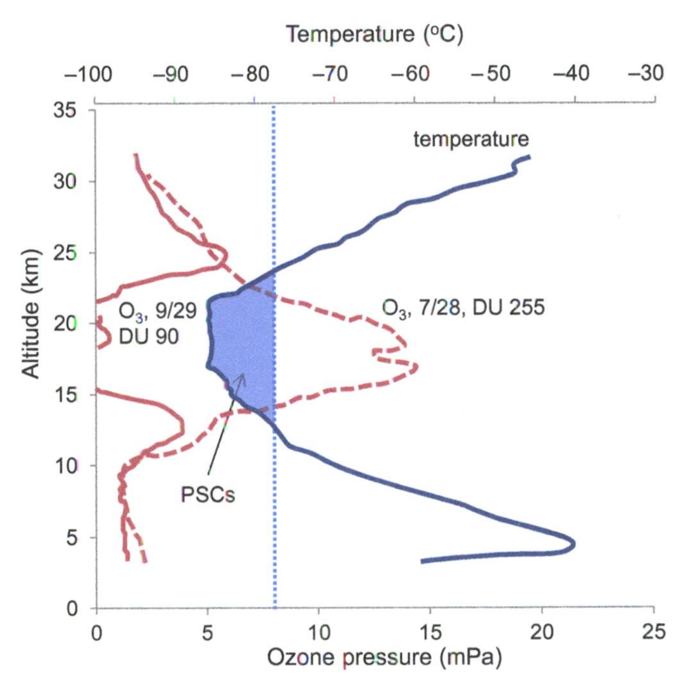

Figure 5.11 Vertical profiles of O_3 and temperature, South Pole, (Source: Rowland, F.S. (2006), *Philosophical Transactions of the Royal Society B*, Vol. 361: p 769–790).

spring (during August), the Cl_2 is rapidly split into free chlorine atoms that lead to the sudden loss of ozone (Figure 5.13). Notice that the ozone minimum occur toward the end of September/early October, during which time the sun is fully above the horizon even at the geographic South Pole. This rapid depletion of ozone occurs via **catalytic cycles**, as shown in Figure 5.5.

Over its lifetime in the stratosphere, an individual chlorine atom can destroy about 100,000 ozone molecules. The dramatic fall in ozone is thus the result of the speed at which the catalytic reaction takes place once the sun rises above the horizon. The ozone hole grows throughout the early spring (September) until temperatures warm and the polar vortex weakens, ending the isolation of the air in the polar vortex. As air from the surrounding latitudes mixes into the polar region, the ozone-destroying forms of chlorine disperse. The ozone layer then stabilizes until the following spring.

We should note that the same ingredients or conditions necessary for the destruction of ozone that we see in Antarctica apply similarly to the loss of ozone in the Arctic stratosphere during winter. While there have been significant—even severe—losses of ozone recorded in the past several years over the Arctic, there is not a symmetrical "hole" of similar magnitude, extent, and duration centered over the North Pole. One of the reasons for this is that the range of minimum temperatures found in the Arctic is much greater than in the Antarctic. In some years, PSC formation temperatures are not reached in the Arctic, and significant ozone depletion does not occur.

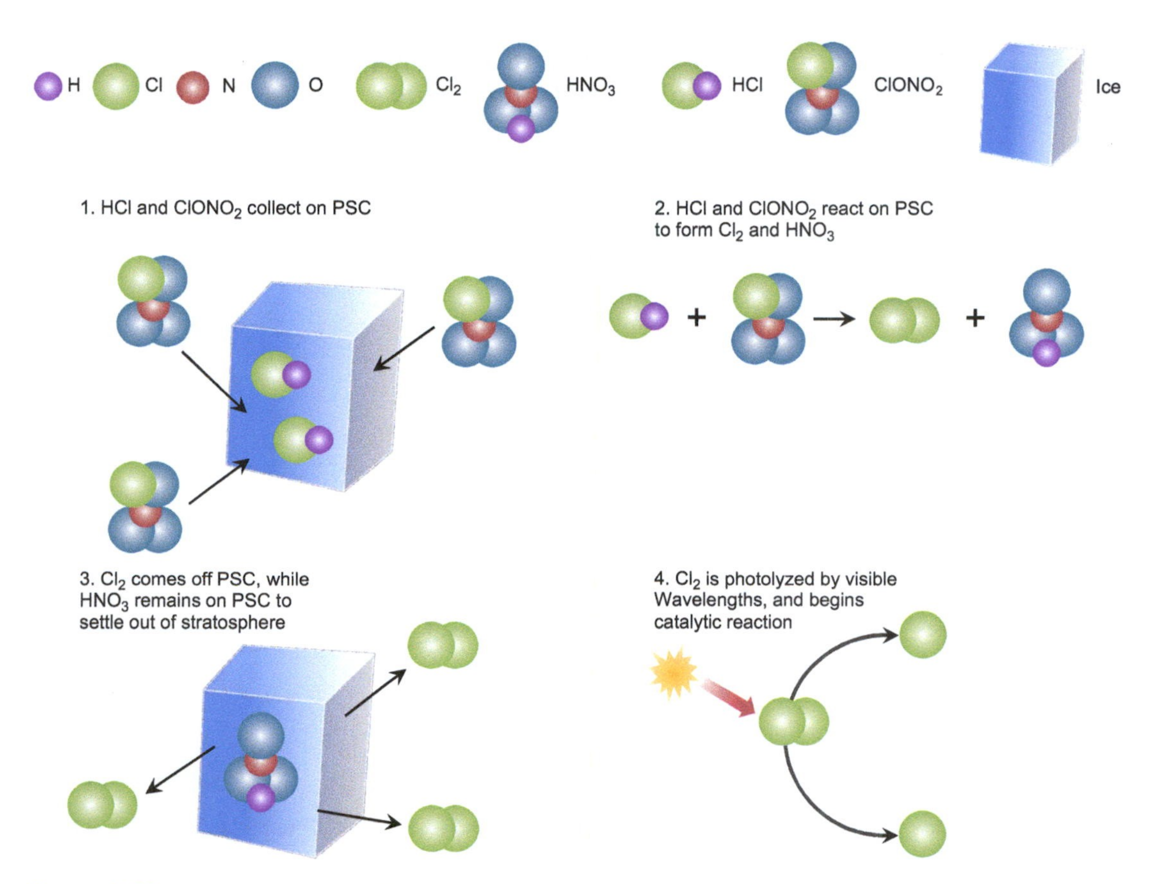

Figure 5.12 Schematic showing how CFC breakdown products collect on polar stratospheric clouds, ultimately freeing up chlorine to destroy ozone in a catalytic reaction. Source: Mike Slattery.

WHAT IS BEING DONE?

Concern for the health of the stratospheric ozone layer led to an international agreement in 1987, the landmark **Montreal Protocol**, which restricted CFC production. Twenty-seven nations signed the Protocol, which required them to agree to a 50% reduction of CFC production by 1999. As the evidence of the damage to the ozone layer accumulated, nations realized that this agreement would be insufficient and the Montreal Protocol was strengthened with several so-called amendments and adjustments. These revisions added new controlled substances, accelerated existing control measures, and scheduled phase outs of the production of certain gases.

The 1990 London Amendments to the Protocol, signed by over 80 nations, called for a phase-out of the production of the most damaging ozone-depleting substances in developed nations by 2000 and in developing nations by 2010. This was a radical advancement over the Montreal Protocol. New scientific evidence indicated that even this action would not be soon enough to stop some destruction of the ozone layer. So once again, the international community met to revise CFC

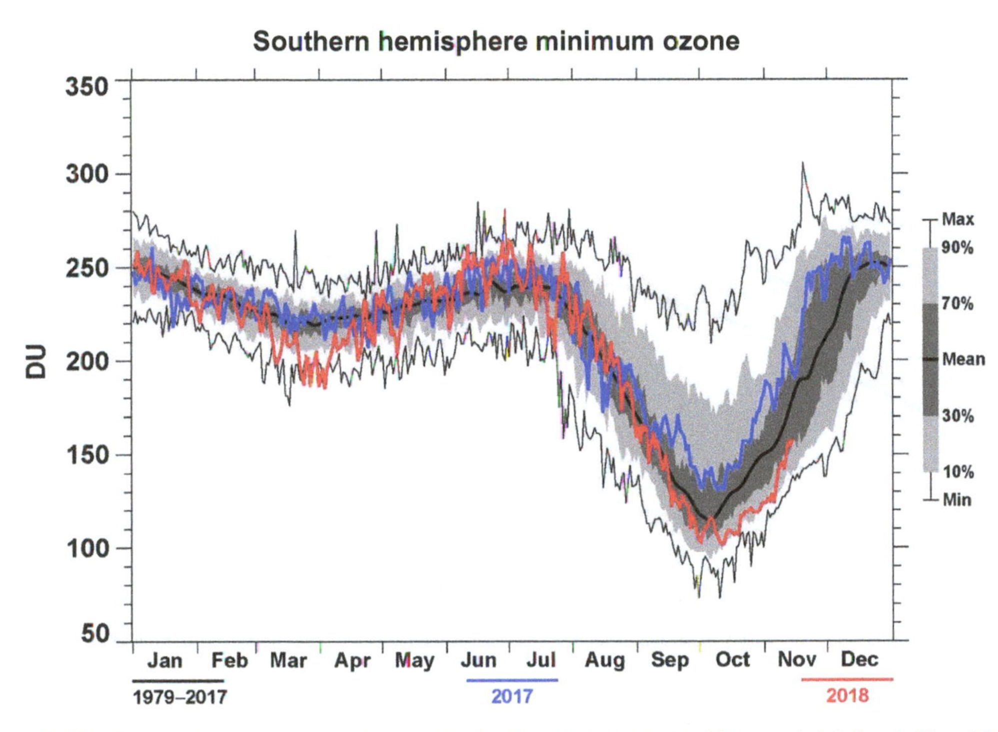

Figure 5.13 Thinning of the ozone layer as determined from total ozone measurements. The ozone hole beings to thin rapidly in August and reaches its minima (and largest geographic area) in the middle of September to early October period. The blue line shows data plotted for 2017; the red line includes data up to November 2018. The other lines and shadings show data with respect to the climatological period indicated. The solid black line shows the mean over the period of record, 1979–2013. The light gray shading indicates the 10th and 90th percentile values, while the dark shading shows the 30th and 70th percentiles for the climatology (Source: http://ozonewatch.gsfc.nasa.gov/meteorology/SH.html).

policy. The result was the Copenhagen Amendment of 1992, in which the international community agreed to the complete phase out of CFCs in developed nations by 1996 and a reduction or phase-out of HCFCs (a less destructive replacement chemical) by the year 2030. The protocol has undergone a further six revisions, culminating in the 2016 Kigali revisions which have been agreed upon but not adopted as of writing.

Thus far, the Montreal Protocol appears to have been successful in slowing and now reversing the increase of ozone-depleting gases in the atmosphere. In the latter half of the 20th century up until the mid-1990s, the effective chlorine content in the stratosphere steadily increased.[8] This long-term increase in effective chlorine then slowed, reached a peak in late 1996 and has now begun to decrease (Figure 5.14). This small and continuing decrease means that the potential for stratospheric ozone depletion has begun to lessen as a result of the Montreal Protocol. The decrease in effective chlorine is projected to continue throughout the 21st century if all nations continue to comply with the provisions of the Protocol.

[8] Effective chlorine is defined here as the sum of hydrochloric acid (HCl) and chlorine nitrate ($ClONO_2$).

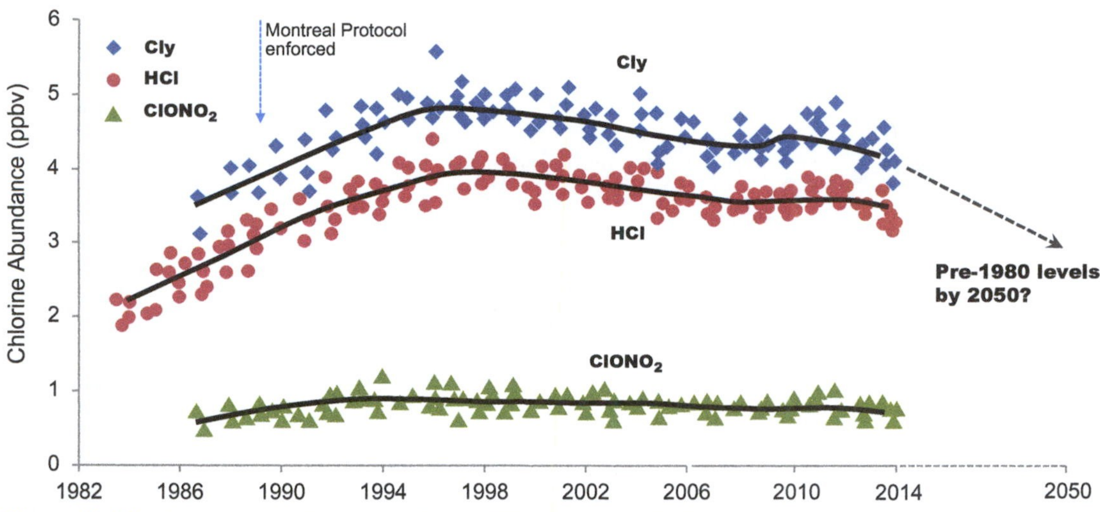

Figure 5.14 Time series of monthly mean total column HCl (red circles) and $ClONO_2$ (green triangles) for the Northern Hemisphere. The blue diamonds represent Cly which is the summation of the corresponding HCl and $ClONO_2$ data points. The tick black lines show the best fit trend lines (Source: Zerefos, C. et al. (2009), *Proceedings for the Symposium of the 20th Anniversary of the Montreal Protocol*, Springer (with data to 2014 updated from author's online database).

Today, CFCs and many other ozone-depleting chemicals have been largely eliminated, and the ozone layer has begun to show signs of recovery to the point that it may return to 1980 levels between 2050 and 2070.[9] Satellite-mounted instruments show that the amount of chlorine in the stratosphere is on the decline as indicated in Figure 5.14. The current consensus among the scientific community is that as chlorine concentrations continue to decrease, Antarctic and Arctic ozone amounts should begin to recover. While experts estimate that it will be the middle of the 21st century before the effects completely disappear, the ozone hole over Antarctica, which reached record size in late 2006, will remain with us for decades. Part of the reason for this is that CFC molecules have atmospheric lifetimes measured from many decades to a century. Thus, there will be significant quantities of these compounds present in the atmosphere throughout the 21st century. However, what is encouraging is that the ozone loss appears to be approximately at a balance point (1996–2010 trend line in Figure 5.7), although with normal year-to-year fluctuations, it will probably be another decade or more before we can truly say that ozone recovery is fully underway.

CONCLUDING THOUGHTS

The story of ozone and CFCs has evolved greatly over the last 30 years as our understanding of the stratosphere has expanded. Our ability to monitor the stratosphere, investigate its phenomena, and assess its future has dramatically improved because of the investments by government, industry,

[9] Source: Douglass et al. (2014), *Physics today*, Vol. 67, p. 42.

and the academic community. Forecasting the future is always a tricky process (ask any weather forecaster), but we are now able to largely determine the stratospheric effects of new chemicals and technologies, and thereby heal and preserve the ozone layer for future generations and for the ecological health of the planet.

The disturbing discovery of the hole in the ozone layer set the stage for what has been called an environmental triumph: the Montreal Protocol of 1987. Despite entrenched opposition from the chemical industry, a great deal of squabbling in the scientific community, and the challenge of devising a global agreement that would satisfy the requirements of both developed and developing nations, the world ultimately came together to address a clear environmental danger. The political and scientific processes seem to have worked together in helping to resolve the ozone/CFC problem. Indeed, this pact to phase out the use of CFCs and restore the ozone layer was eventually signed by every country in the United Nations—the first UN treaty to achieve universal ratification. Such unparalleled cooperation had a major impact. The late Secretary-General of the United Nations, Kofi Annan, was quoted as saying that "perhaps the single most successful international agreement to date has been the Montreal Protocol."[10] It is interesting to note how particularly useful the set of model-based predictions from scientists of the consequences of a particular set of actions were in this process. Many people are hopeful that the success of the Montreal Protocol can be replicated to attain international action on global warming. As we shall see in Chapter 6, optimism is, at best, guarded, because banning CFCs involved only one specific group of industrial chemicals. The causes of global warming and abrupt climate change are more numerous and complex, and are presenting much more forbidding scientific, economic, and political challenges.

[10] Source: https://www.un.org/en/events/ozoneday/background.shtml

6

Global Climate Change

"I'm starting to get concerned about global warming."

INTRODUCTION

If you do a Google™ search on the phrase **Climate Change**,[1] these are two of the first images you find: pollution pouring into the atmosphere from smokestacks and the ever-present polar bear, precariously making her way across a sea of melting ice (Figure 6.1). The latter, in particular, has become the global symbol of the movement to curb greenhouse gas (GHG) emissions. This is perhaps not surprising since polar bears are well known to the public, and they make a big impression (which is aided by how ridiculously adorable they are as cubs). The potential impacts of climate change are also easy to visualize in connection with polar bears: their habitat is literally melting away.

Figure 6.1 Smokestacks at a coal-burning power plant (top); a polar bear roaming the icy waters of Svalbard, an archipelago in the Arctic (bottom) (Source: www.istock.com).

[1] In this chapter, I use the terms **climate change** and **global warming** with very specific meaning. Global warming describes the average global surface temperature increase largely from human emissions of greenhouse gases. Climate change includes global warming and everything else that increasing greenhouse gas amounts will affect, such as changes to precipitation patterns and sea level. Within scientific journals, this is still how the two terms are used.

Through the language of catastrophe and imminent doom, climate change is now widely reported in the media as one of the greatest problems facing humanity, and we (i.e., humanity) are to blame. Unfortunately, the science underpinning climate change has become obscured and appropriated by many different special interest groups in an attempt to promote their own causes. Climate change, for all intents and purposes, has been transformed from a physical phenomenon, measurable, and observable by scientists, into a social, cultural, and political phenomenon. This has led to what some call "climate porn" or the tendency of some sections of the scientific community and the media to sensationalize climate change data in evermore apocalyptic terms.[2] Here is an example. In May, 2014, the White House released a report on climate change in the United States.[3] The study's verdict? The impacts of a changing climate were already impacting our infrastructure and economy, in effect, threatening our future. The media covered the report extensively, but a quick glance at two headlines reveals disbelief and fear and are typical of the alarmist language now used to frame the debate:

Landmark Report Warns Time Is Running Out To Save United States From Climate Catastrophe

Thinkprogress.org

Brace Yourself: Top Ten Terrifying Impacts of Climate Change

ABCNews.com

I find headlines like these particularly unhelpful when trying to engage in a civil discourse about climate change.

One of the challenges in trying to engage the public in the topic of climate change, especially here in the U.S., as that we invariably end up in one of two boxes: either you are a believer in climate change, or you are a denier. This happens to me all the time. Once people find out what I do and what subject I teach, I get the inevitable question: "So, do you believe in this climate change thing?" The so-called believers ascribe to the notion that Earth's climate is rapidly warming, that the cause is a thickening of CO_2 enhanced by human activities, and that a rise in global temperatures will have devastating consequences. The deniers or skeptics do not believe that there is any credible evidence that humankind's activities are the cause of climate change, if indeed that is even happening at all. In the United States, this dichotomy between believers and deniers/skeptics is most frequently aligned along party lines. The believers (generally on the political left) are perceived as pro-environment and favoring immediate action, while the deniers and skeptics (generally on the political right) are perceived as wanting to drag their heels for fear of hurting the economy. A survey conducted by the well-respected Pew Research Center in October, 2013 found that while 67% of Americans believed that there is solid evidence that the Earth has been getting warmer over the past few decades, there were sharp partisan divides: 88% of Democrats said that there was solid evidence of warming versus

[2] See two thoughtful articles at the Institute for Public Policy Research (http://www.ippr.org/press-releases/111/2500/climate-porn-turning-off-public-from-action) and the Tyndall Center for Climate Change Research (http://www.tyndall.ac.uk/sites/default/files/wp98.pdf) on this topic.
[3] http://nca2014.globalchange.gov/

only 50% of Republicans.[4] Just 25% of Tea Party Republicans said that there is solid evidence of global warming.

Progress in taking action to curb GHG emissions has been additionally hindered by the fossil fuel industry, with large multinational corporations (MNCs) like ExxonMobil regularly publishing papers that minimize the impacts of climate change.[5] Both MNCs and domestic corporations in the oil and gas sector are throwing millions of dollars at (primarily Republican) lobbyists and politicians who represent the interests of the fossil fuel industry.[6] Further exemplifying the political polarization of climate change, the former vice president Al Gore lay blame squarely on President George W. Bush and his administration for not signing up to the Kyoto protocol. In *An Inconvenient Truth*, the third highest grossing documentary film of all time, he asks, "Are we going to be left behind as the rest of the world moves forward? There are only two advanced nations in the world that have not ratified Kyoto and we are one of them."[7] President Bush and fellow Republicans were (and arguably still are) portrayed in the media as climate skeptics, even though in 1997 the U.S. Senate voted 95-0 during a Democratic administration against ratification of the **Kyoto Protocol**. President Bush never supported ratification primarily because of the strain he believed that the treaty would put on the economy, not helped by the fact that China and India were exempt from Kyoto's carbon reduction targets.

President Trump has upped the ante even further, vowing to roll back President Obama's efforts to fight climate change. He dropped climate change from the administration's list of global threats and, to global condemnation, withdrew the U.S. from the **Paris Climate Agreement**. His Tweets on climate change have proven to be divisive. As a candidate, he called **global warming** "an expensive hoax" (29 January 2014), one "created by and for the Chinese in order to make U.S. manufacturing non-competitive" (6 November 2012). During a particularly cold spell in 2017, he Tweeted "In the East, it could be the COLDEST New Year's Eve on record. Perhaps we could use a little bit of that good old **Global Warming** that our Country, but not other countries, was going to pay TRILLIONS OF DOLLARS to protect against. Bundle up!" (28 December 2017).

Of all the topics covered in this book, climate change is undoubtedly the most controversial, complex, and politically divisive. How, then, are we supposed to respond to an issue such as climate change with any sense of objectivity? How do we make an informed decision about what our course of action should be without the confusion and noise of political and media bias? In this chapter, I present current scientific understanding agreed on by the majority of climate scientists on what climate change is, how it could potentially affect our world, and what is currently being done politically and socially to prepare for its effects. First, as always, we turn first to the scientific data.

---

[4] http://www.people-press.org/2013/11/01/gop-deeply-divided-over-climate-change/

[5] According to a study by the Union of Concerned Scientists, between 1998 and 2005, ExxonMobil dispersed roughly $16 million to organizations that were challenging the scientific consensus view. After heavy criticism from the press and environmental groups in late 2006 and early 2007, ExxonMobil began distancing itself from these organizations.

[6] Source: http://www.opensecrets.org/lobby/indusclient.php?id=E01 and http://www.opensecrets.org/industries/indus.php?Ind=Eource

[7] The Kyoto Protocol is the international treaty on climate change designed to get signatory nations to commit to reduce their emissions of greenhouse gases.

CO_2 AND THE GREENHOUSE EFFECT

We begin our discussion of climate change at an elevation of 11,000 ft on the northern slopes of Mauna Loa, the spectacular volcano on the Big Island of Hawaii. Here you will find the Mauna Loa Observatory (MLO), a research station where scientists have been monitoring our atmosphere since the 1950s. They record changing levels of atmospheric gases, including CO_2.

Charles David Keeling, a scientist who developed the first instrument capable of measuring CO_2 in air samples, started the MLO measurements in the late 1950s. The data show that CO_2 levels have been rising steadily throughout the record (Figure 6.2), a graph that has become known as the **Keeling Curve**. This data is the longest instrumental record of atmospheric CO_2 in the world and is considered an extremely reliable indicator of current trends in CO_2 levels. The data show that the atmospheric concentration of carbon dioxide has increased from approximately 315 ppm in 1958 to 407 ppm in 2017.

The CO_2 record shown in Figure 6.2 raises several important questions. First, is this trend in increasing CO_2 levels significant? That is, is there some driving force (or cause) behind this increase, or is it just part of a longer-term (say several hundred or thousand years) natural variation in CO_2 levels? Second, are the levels themselves noteworthy (i.e., what does a concentration of 407 ppm mean)? Third, why should we focus on CO_2 as opposed to other atmospheric gases that are potentially more harmful? Finally, what does this all have to do with a warming global climate?

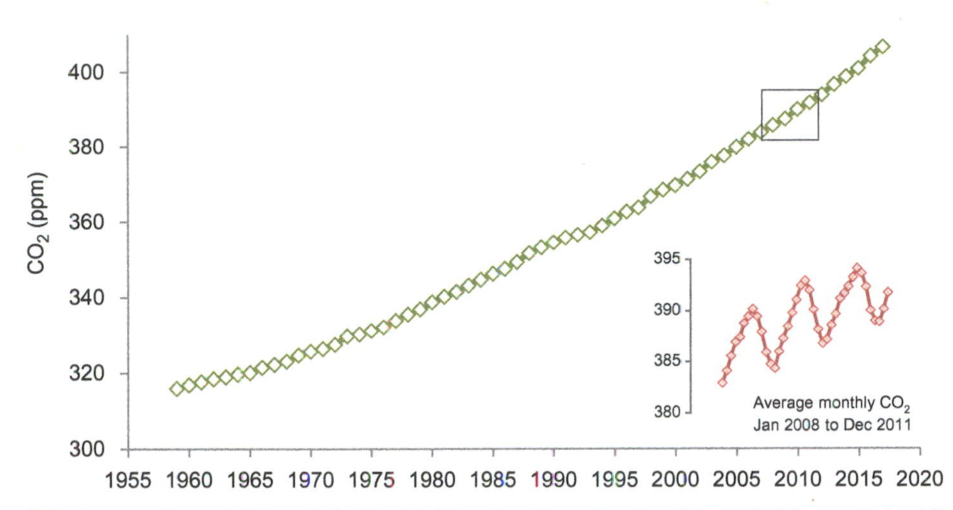

Figure 6.2 Instrumental record of atmospheric CO_2 at the Mauna Loa Observatory, Hawaii, 1956–2017 (Source: National Oceanic and Atmospheric Administration, http://www.esrl.noaa.gov/gmd/ccgg/trends/). The inset shows the monthly mean CO_2 record for 2008–2011. The seasonal cycle is due to the vast land mass of the Northern Hemisphere, which contains the majority of land-based vegetation. The result is a decrease in atmospheric carbon dioxide during northern spring and summer, when plants are absorbing CO_2 as part of photosynthesis. The pattern reverses, with an increase in atmospheric carbon dioxide during northern fall and winter. The yearly spikes during the cold months occur as annual vegetation dies and leaves fall and decompose, which releases their carbon back into the air.

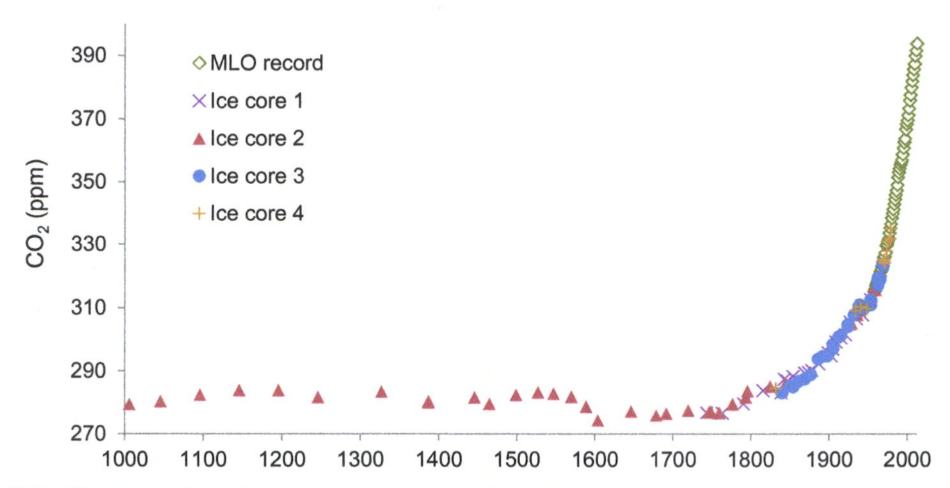

Figure 6.3 CO_2 concentration in the atmosphere over the last 1,000 years based on both direct measurements on Mauna Loa, Hawaii (green diamonds, as shown in Figure 6.2) and sampling of gases trapped in ice cores in Antarctica. Core 1 is from the Siple Station, West Antarctica and cores 2–4 are from Law Dome, East Antarctica (Source: Carbon Dioxide Information Analysis Center—http://cdiac.ornl.gov/).

In order to address the first question regarding the significance of increasing CO_2 levels, we have to put the past 60-plus years of CO_2 data in a much wider context. In Figure 6.3, the MLO record is shown alongside CO_2 concentrations measured from relatively shallow **ice cores** obtained from drilling expeditions in Antarctica.[8] These cores provide an important means for determining atmospheric gas concentrations thousands of years ago. The principle is very simple: as annual snowfall settles and is compacted under subsequent snow layers, tiny bubbles of air become trapped within the ice. These bubbles actually contain samples of what the atmosphere was like at different times in Earth's history. By extracting the air trapped inside these bubbles, we can measure what CO_2 concentrations used to be in our atmosphere. How neat!

As you can see in Figure 6.3, CO_2 concentrations are approximately 270–275 ppm for the 750 years or so preceding the Industrial Revolution. Levels then begin to rise throughout the 19th and 20th centuries and then accelerate dramatically during the past 60 years, the period of the MLO record. Overall, CO_2 in the atmosphere has risen by almost 40% since the Industrial Revolution. Visually and, more importantly, statistically, the increase in CO_2 shown in Figure 6.3 is significant and makes for a compelling argument that we are now in unchartered territory in terms of atmospheric CO_2. Note that the CO_2 measurement for 1978 from the ice cores (333 ppm) lines up very well with the 1978 instrumental measurement from the MLO (335 ppm), indicating that the ice core data corroborates the historical record.

However, the secrets from the ice cores do not stop there. Scientists have now sampled ice from the Antarctic ice cap to a depth of over 3,000 m and have been able to extend the record back several

[8] The deepest ice cores retrieved from Antarctica that provide the most comprehensive record of CO_2 are those at the Russian station at Vostok and the European Project for Ice Coring in Antarctica (EPICA), both of which are >3,000 m deep. These data are shown in Figure 6.4.

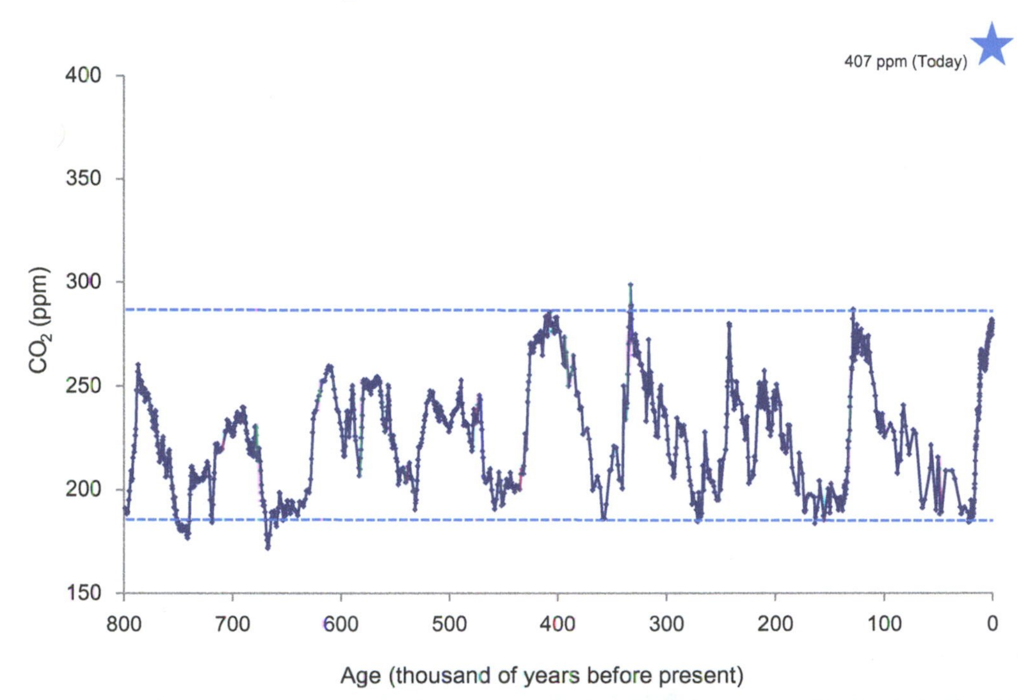

Figure 6.4 CO_2 concentration in the atmosphere over the last 800,000 years based on ice core data. The dashed lines show the long-term upper and lower bounds of the CO_2 record indicating how CO_2 is "phase locked" over the last 800,000 years relative to the last 60 years (Source: Luthi, D. et al. 2012. *Nature*, 453: 379–382).

hundreds of thousands of years (Figure 6.4). The results show atmospheric CO_2 concentrations of approximately 180–200 ppm during ice ages, increasing to approximately 280 ppm during warmer periods, known as the **interglacials** (those periods in between glaciations). During the **Holocene**, a geologic period stretching back 10,000 years, CO_2 levels fluctuated between about 260 and 280 ppm, a narrow range. The dramatic rise during the latter half of the 20th century, shown in Figure 6.2, clearly lies outside anything previously recorded. Therefore, it seems highly improbable that the recent increase in CO_2 from 280 to 407 ppm is part of some natural background variability. The overwhelming consensus among the scientific community is that it is driven largely by the release of CO_2 during the combustion of fossil fuels. Emissions from such sources are constantly adding CO_2 to the atmosphere at rates that significantly exceed those supplied by natural sources, such as volcanic activity. Most of this CO_2 has been released since 1945 (see Box 6.1).

CO_2 levels of 407 ppm look and sound impressive, but at this concentration, it only represents 0.04% of the atmosphere (Table 6.1). Nitrogen and oxygen outweigh all other gases in the atmosphere; they are literally the heavyweights of the atmosphere. Why then is CO_2 the focus of such attention? Surely at such low concentrations, adding just a few more parts per million of a particular gas, even during a relatively short time period of 60 years, will not have much effect, right? The truth is that CO_2 plays a disproportionate role relative to its concentration in affecting climate. It is a **GHG (Greenhouse Gas)**, a term I am sure many of you have encountered, but

BOX 6.1

The carbon cycle

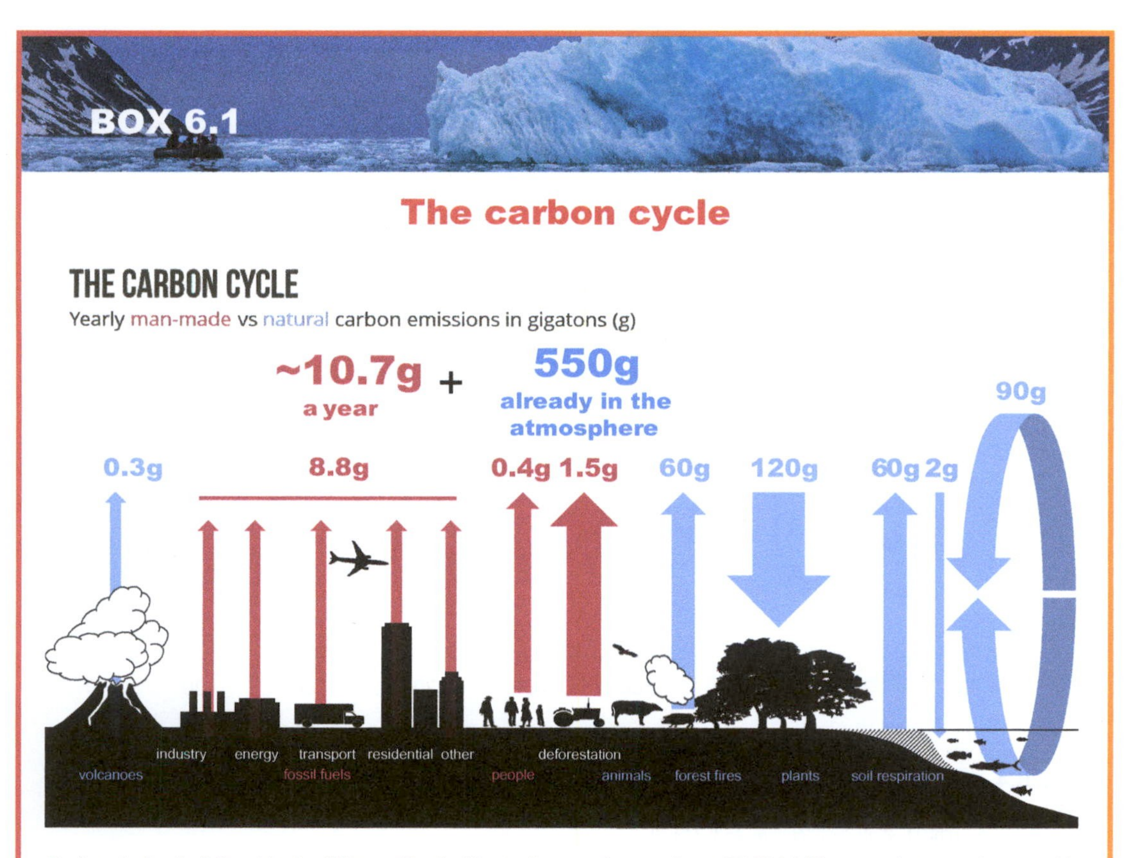

Carbon is the building block of life on Earth. Most of our carbon—about 65,500 billion metric tons—is stored in rocks with the rest stored in the ocean, atmosphere, soil, plants, and fossil fuels. Carbon flows between each of these so-called "reservoirs" in an exchange called the carbon cycle, which has both slow and fast components.

Over millennia, the carbon cycle maintains a balance that prevents all of Earth's carbon from entering the atmosphere (as is the case on Venus) or from being entirely stored entirely in rocks. This balance helps keep Earth's temperature relatively stable, like a thermostat. Scientists have discovered that this thermostat works over the timescale of hundreds of thousands of years; we refer to this as the slow carbon cycle. From an environmental perspective, we are more concerned with the fast carbon cycle that operates over shorter time periods—tens to a hundred thousand years. Over these timescales, Earth's temperature can vary and, in fact, our planet swings between ice ages and warmer interglacial periods. The diagram above is a schematic of the fast carbon cycle and shows the movement of carbon between land, atmosphere, and oceans. Blue numbers represent natural fluxes while red are anthropogenic contributions in gigatons of carbon per year. Note that ocean absorbs about two gigatons of carbon more from the atmosphere than it gives off to the atmosphere. Marine organisms use that extra amount of carbon which is eventually incorporated into deep-sea deposits and sediments. So the net level of carbon in the ocean remains roughly the same every year. The bottom line is that humans are adding about 10.7 gigatons of carbon into the atmosphere each year mostly due to fossil fuel burning and land use changes such as deforestation. (Diagram adapted and redrawn from U.S. DOE, Biological and Environmental Research Information System, and McCandless, D. 2009. *Information is Beautiful*, Collins.)

Table 6.1 Composition of the atmosphere.

Gas	ppm	(%)
Nitrogen (N_2)	780,840	78.1
Oxygen (O_2)	209,460	20.9
Argon (Ar)	9,340	0.9
Carbon dioxide (CO_2)	407	<0.04

Not included in above-mentioned dry atmosphere: water vapor (~0.25%) over full atmosphere; typically 1–4% near surface.

what does that really mean? Well, GHGs such as CO_2, methane (CH_4), and water vapor (H_2O) occur naturally in our atmosphere and regulate the atmospheric thermostat—i.e., they keep our planetary temperatures livable. To understand how these gases operate, and how temperature may respond to increased GHG concentrations in the 21st century, it is important to understand first how energy enters, is processed by, and eventually exits our atmosphere.

The Electromagnetic Spectrum

In Chapter 4, we talked about UV light, which describes a type of energy, or **radiation** (i.e., emitted by the sun). UV, like all radiation from the sun, comes to Earth in the form of waves, just like energy is moved through the ocean via swell and waves. UV light is an example of **shortwave radiation**, and it can be graphed on a spectrum with other different types of radiation, known as the **electromagnetic (EM) spectrum** (Figure 6.5). This is a very important diagram, and it illustrates several key concepts about radiation.

Most of the waves that arrive from the sun have very short **wavelengths**, defined as the distance between the crest of two waves. Gamma rays, ultraviolet rays, and visible light are all classified as **shortwave radiation** and constitute about 90% of the radiation coming from the sun. It is also important to understand that the shorter the wavelength, the higher the intensity of the radiation.

Now the sun is not the only body that emits radiation. Everything on Earth emits radiation. As you read this book, everything around you is constantly emitting radiation. This is **long wave radiation** or **infrared radiation** (the same radiation emitted by heat lamps) and is much less intense than the incoming solar shortwave radiation. On the electromagnetic spectrum, these waves all lie to the right of visible light (Figure 6.5). The radiation emitted by Earth lies wholly within the middle infrared bands.

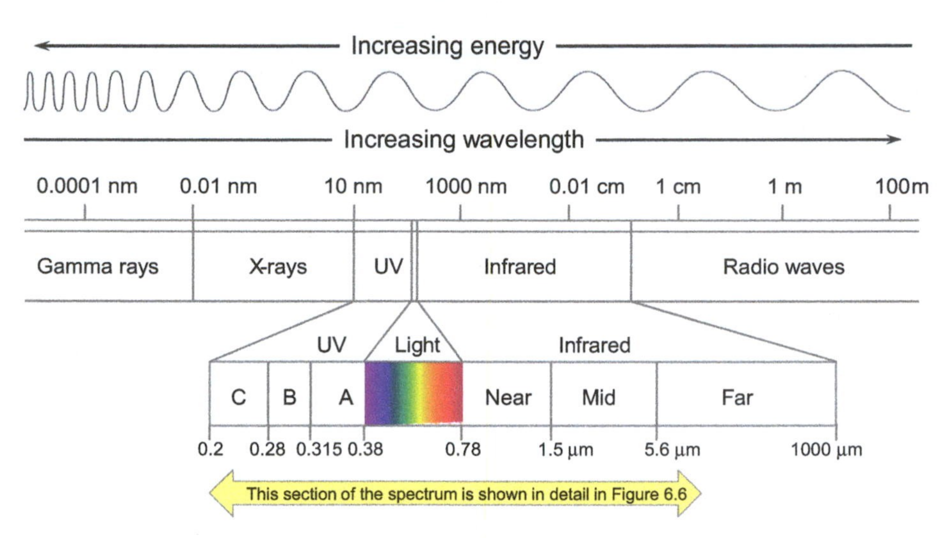

Figure 6.5 The electromagnetic spectrum. Note that radiation emitted by the sun includes all wavelengths to the left of the mid infrared band.

The type of radiation emitted from an object is also based on its temperature. As you increase the temperature of an object, the wavelength decreases. For example, Earth as an object has an average temperature of about 59 °F, whereas the sun's average temperature is almost 10,000 °F! This enormous thermal difference accounts for the difference in wavelengths emitted by each object: the sun is dominated by short, intense UV, and visible wavelengths, while the Earth emits longer, meandering infrared waves.

The Earth–Energy Balance and the Greenhouse Effect

Now, Earth is located approximately 93 million miles from the sun, although this distance varies slightly because Earth's orbit around the Sun is elliptical (i.e., oval in shape). Still, the amount of solar radiation that reaches Earth annually remains relatively constant. This consistent amount of radiation being supplied to Earth's atmosphere in any given year is balanced by an amount of radiation given off by our Earth-atmosphere system into outer space. This is called the **energy balance** and is written simply as

$$I – O = \Delta S \tag{6.1}$$

where I is the energy input, O the energy output, Δ = the Greek letter meaning change, and S, the storage of total energy within Earth's atmosphere.

We can quantify the energy balance a little more by studying what happens to radiation when it hits Earth's atmosphere (Figure 6.6). As solar radiation passes through the atmosphere toward

Earth's surface, some is reflected off clouds and Earth's surface itself (about 30%), which is known as the **albedo**, or reflectivity, and some is absorbed by the atmosphere (about19%), while the rest (about 51%) strikes Earth's surface and warms it. This is probably a bit surprising since it is hard to believe that only half of the sun's radiation actually hits the ground. However, once it does, Earth's surface heats up. Now what happens? Well, some of the heat from the Earth's surface is transported into the atmosphere by convection currents (think rising air), some is transported in water vapor as latent or hidden heat that gets released in the atmosphere once the water vapor condenses into clouds, and some is simply radiated directly from the Earth's surface as long wave radiation.

Now, the atmosphere, specifically its composition, plays a critical part in the energy balance story. The atmosphere contains different so-called **GHGs**, such as CO_2 and water vapor. These gases are **selective absorbers** of radiation, and how they work is illustrated in Figure 6.7. Although this is a complex diagram, it is absolutely critical to understanding how the atmosphere works relative to climate change. Notice in Figure 6.7 that CO_2 is transparent to the incoming shortwave radiation from the sun. It simply does not "see" it. Water vapor (i.e., H_2O) behaves much the same way, although it does start absorbing some of the sun's energy in the near infrared (i.e., to the right of

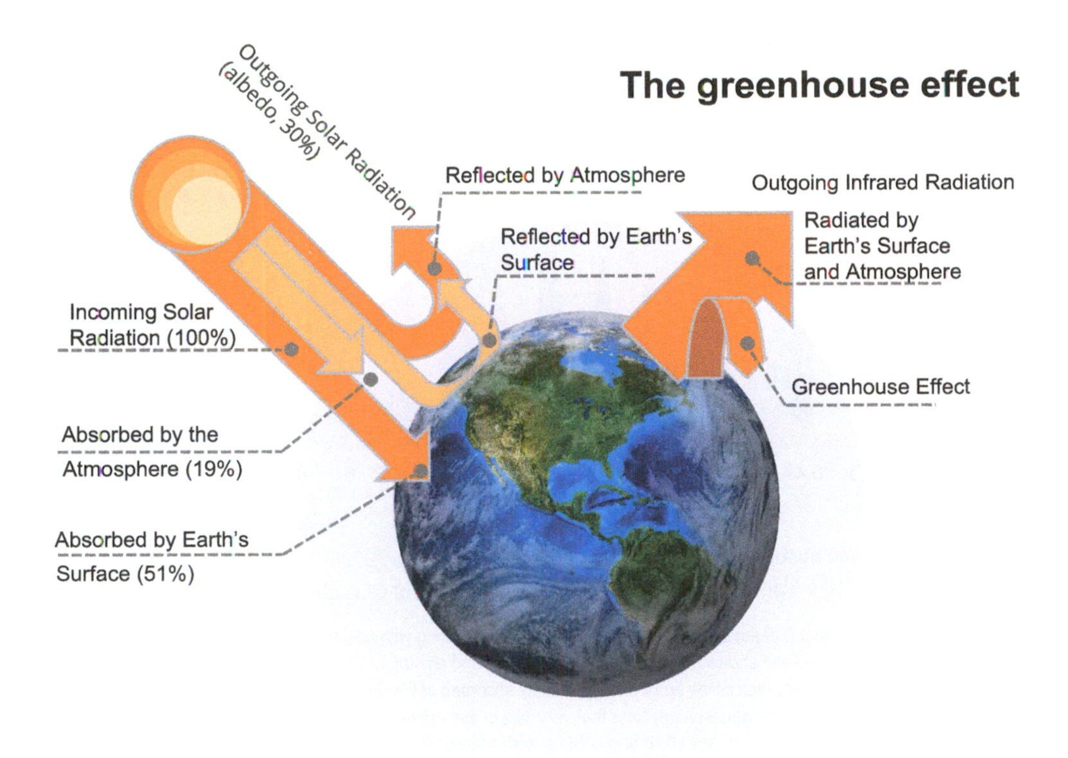

Figure 6.6 Earth's greenhouse effect which keeps our planetary temperature 33 degrees C warmer than it otherwise should be. Source: Mike Slattery; image of Earth © Shutterstock.com.

the visible spectrum). However, CO_2 (and the other GHG molecules) are really able to "see" long wave radiation given off by the Earth and they easily absorb it (notice the 100% absorption peaks at 4.5 μm, 5–7 μm, and again beyond 15 μm). This atmospheric absorption of longwave radiation given off by the Earth by the GHGs and the re-emission of that radiation back toward the Earth is called the **greenhouse effect.** It reduces the amount of longwave radiation emitted directly back into space and warms the Earth's surface and lower atmosphere. Put another way, we get our heat

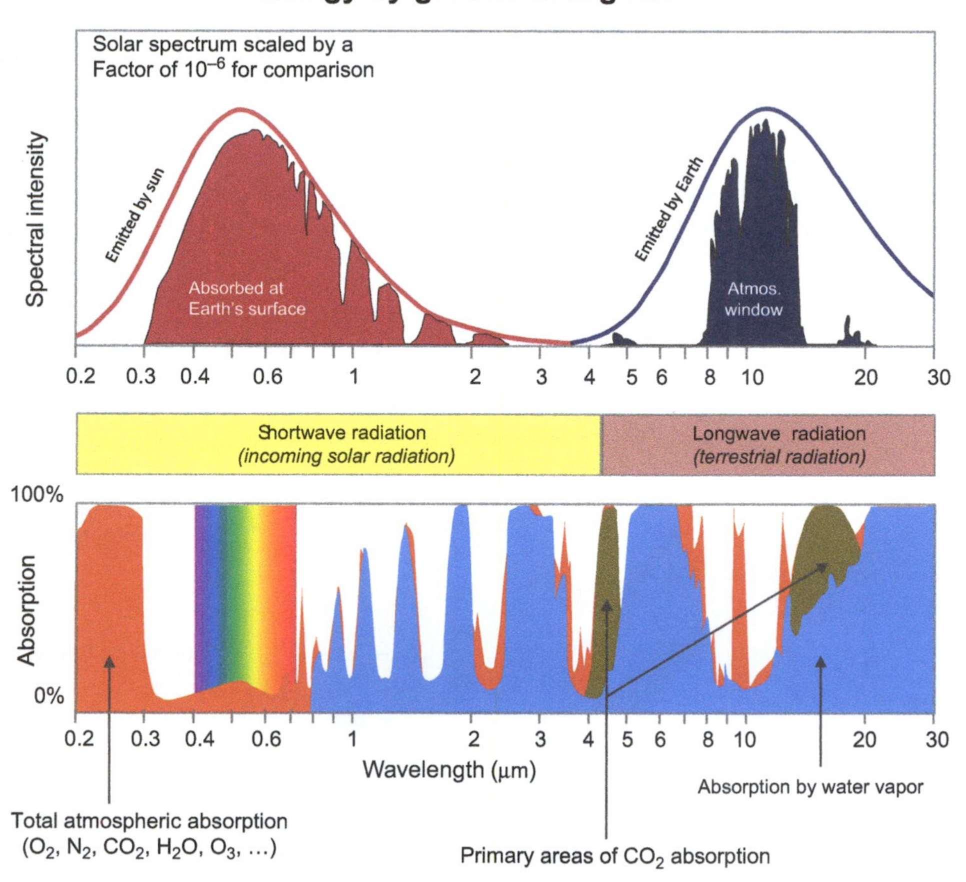

Figure 6.7 The energy spectra of the sun (red line) and Earth (blue line) along with absorption of radiation in the atmosphere by greenhouse gases. The solid red line shows radiation emitted by the sun and arriving at the top of Earth's atmosphere while the red shading under the curve shows that portion of the sun's energy actually absorbed at the Earth's surface. The unshaded area under the sun's curve is thus the energy absorbed by the atmosphere. Note that very little of the visible spectrum is absorbed, while the absorption to the left of 0.3 microns (solid orange) is predominantly UV-B and UV-C absorbed by ozone. The blue curve shows the energy emitted by Earth (longwave radiation) while the blue shading under the curve is the part of Earth's radiation that escapes directly back to space, called the atmospheric window. As with the sun's energy curve, the unshaded area is the energy emitted by the Earth that is absorbed by the atmosphere. As shown below the emission curves, some of that absorption is accomplished by water vapor (light blue) which is particularly effective across wavelengths. CO_2 accounts for two distinct absorption peaks, one at about 4.5 microns and a second, larger peak at approximately 15 microns. All of this atmospheric absorption causes the greenhouse effect.

from two sources: the sun, which is obviously pretty important in driving the entire system, but also the atmosphere.

Just to illustrate how important this effect is in regulating Earth's temperatures, let us imagine for a moment that we did not have an atmosphere containing CO_2, water vapor, and other selective absorbers surrounding our planet. What would the average temperature of Earth be? The answer will probably surprise you: −0.7 °F, or about −18 °C. In other words, Earth would be frozen and life would not exist. Because we have an atmosphere and GHGs within it, our Earth is a much more comfortable 59 °F, or 15 °C. That is a difference of 33 °C!

If this still does not seem convincing, do a little research on Venus which sits about 67 million miles from Earth. Venus is similar to Earth in terms of size and mass, but its surface temperature is about 460 degrees Celsius, hot enough to melt lead! The Venusian atmosphere, which is 96.5% CO_2 (a stark contrast from Earth's 0.04%) has a strong greenhouse effect. The physics behind this are not up for debate.

In summary, scientists have known for *centuries* that certain gases in the atmosphere, like CO_2, *prevent* Earth's heat from escaping directly into space. We also know that emissions from human activities, like cutting down forests and burning fossil fuels, which we can measure directly, are *thickening* the layer of these gases in the atmosphere. We can also measure this directly. Bringing this back to the energy balance in Equation 1, we are reducing the output (O) on an annual time frame which, with a constant input of solar radiation (I), means that the change in storage (ΔS) must be positive. This means that *more* heat remains within our Earth-atmosphere system thereby warming up our planet. This is the very definition of **anthropogenic climate change** or global warming: a human-caused acceleration of the greenhouse effect. Thousands of scientists worldwide agree that this is happening.

CHANGING GLOBAL TEMPERATURE

Figure 6.8 shows the record of global average temperatures over the past 130 years. Figure 6.9 shows surface temperature anomalies for every country in 1900 and again 2016. Together, these two diagrams are central to the climate change debate and show that the planet's average near-surface atmospheric temperature has risen by ±1.4 °C (±2.5 °F) during the period of record. It is important to stress here that this is the **instrumental temperature record**, where temperature is measured by ground-based thermometers. The five warmest years in the global record have all come in the 2010s whilst the 20 warmest years on record have all come since 1995.[9] Note that, in Figure 6.8, the warming actually occurs during two periods: 1910–1945 and 1976 to the present day. Scientists explain the early 20th century warming by a combination of factors, including GHGs and natural forcing, such as decreased volcanic activity, which allows increased radiation to reach the ground, as well as greater **solar irradiance**.[10] Scientists also agree that in the second half of the century, increased concentrations of GHGs, specifically CO_2, are largely the cause of this warming.

[9] Source: www.climatecentral.org

[10] Sunlight, including light, infrared, ultraviolet, and any other wavelength of electromagnetic radiation the sun gives off.

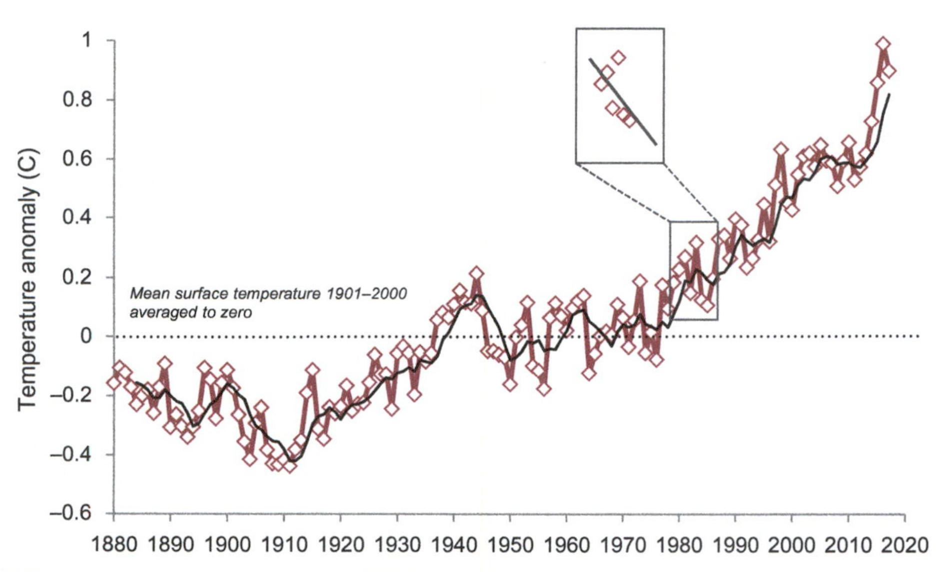

Figure 6.8 Surface temperature anomalies since 1880. The term temperature anomaly means a departure from a reference value or long-term average. A positive anomaly indicates that the observed temperature was warmer than the reference value, while a negative anomaly indicates that the observed temperature was cooler than the reference value. The black trend line is the 5-year running mean (Source: www.ncdc.noaa.gov/cmb-faq/anomalies.php).

I have highlighted six data points in Figure 6.8 (1980–1985) to illustrate an important point with regard to analyzing and interpreting the data in the instrumental temperature record. During these six highlighted years temperatures actually fall (albeit with quite some variability) and could lead one to conclude the onset of a cooling trend. Several other groupings of data points in Figure 6.8 could also be "cherry picked" in the same way to support a different (and perhaps preconceived) conclusion. This certainly happens and is grossly dishonest. What climate scientists look for are statistical trends over long periods of time. With regard to global temperatures, the trend in Figure 6.8, along with the 116-year anomaly chart shown in Figure 6.9, is self-evident and statistically significant.

If we accept that Earth's atmosphere has warmed over the past century, as Figure 6.8 shows, the next question relates to the significance of the warming relative to the long-term climatic record. Like the 60-year MLO CO_2 record shown in Figure 6.2, the 20th century temperature record must be set within a broader context. The problem, of course, is that there are no instrumental records going back hundreds (let alone thousands) of years. Our only course of action is to turn to so-called **proxies**—variables that independently may not be of any enormous interest but from which a variable of interest, in this case temperature, can be obtained. Tree-ring widths are a well-documented example of such a temperature proxy. Dendrochronologists (i.e., tree-ring scientists) use the width and other characteristics of tree rings to infer temperature. Generally, the ring pattern reflects the climatic conditions in which a tree grew, with wide rings reflective

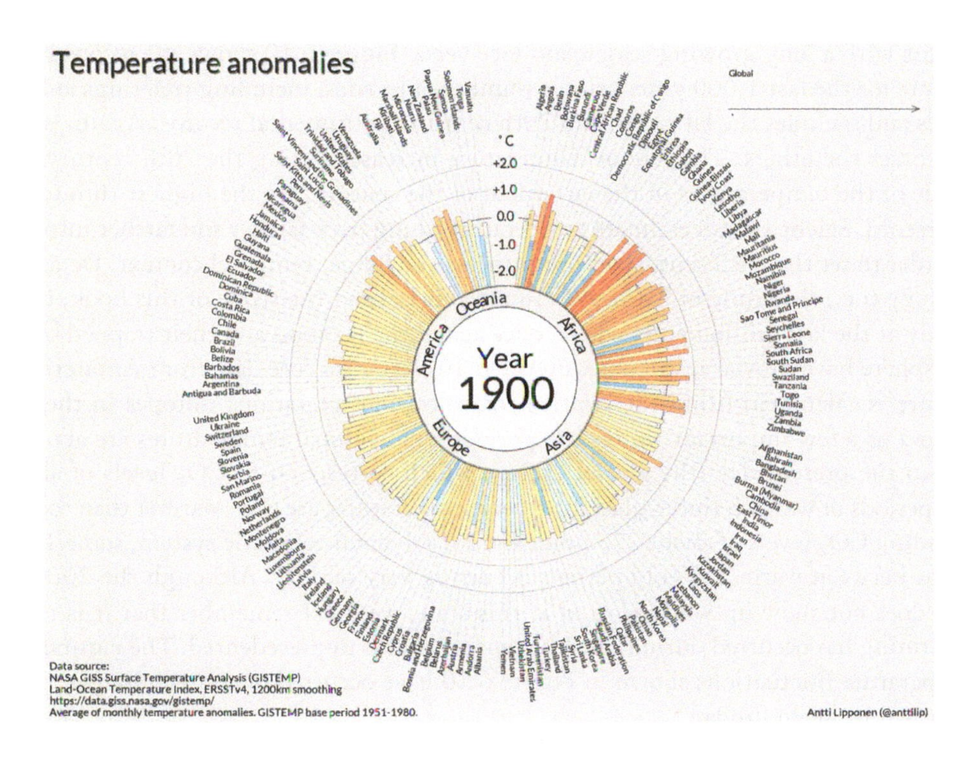

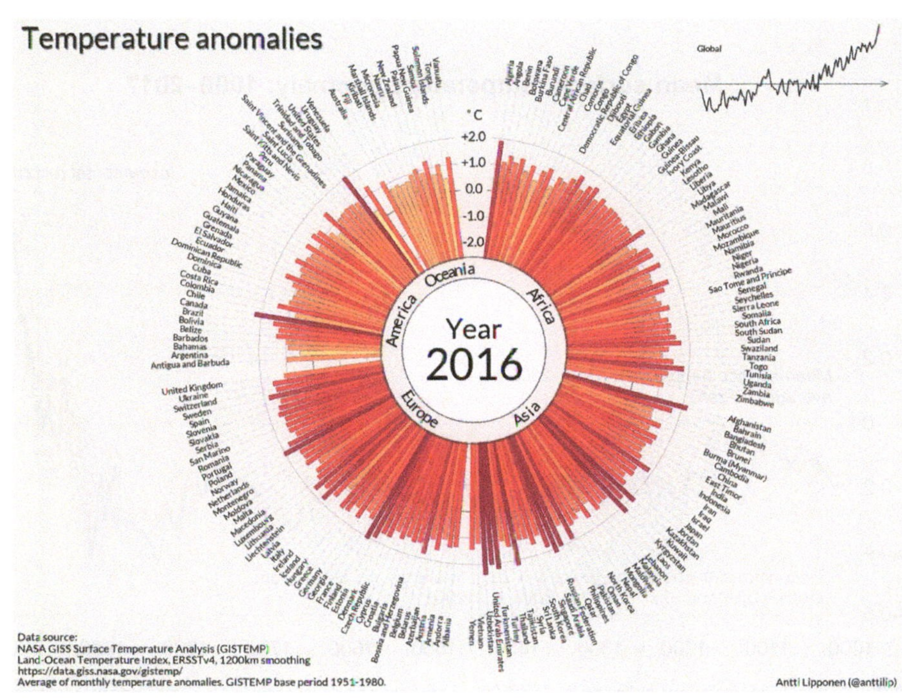

Figure 6.9 Surface temperature anomalies for each country in 1900 and 2016, showing the extent of global warming across the globe.

of wet years with a long growing season and vice versa. Figure 6.10 shows the reconstruction of temperature for the last 1,000 years using a number of proxies, including tree rings and historical records and includes the late 19th and 20th century instrumental records. Again, each set of data reinforces the others. The rate of temperature increases during the 20th century, and the magnitude of the temperatures in the latter half of the century, are the highest throughout the climatic record. Scientists can estimate temperature going back farther and farther into geologic time in order to set the 20th century record into an even longer temporal context. Detailed paleoclimatology (i.e., the study of historical climates) is beyond the scope of this book, but let us look briefly at the last million years where, once again, the ice cores and their trapped bubbles of the atmosphere have proven invaluable. Figure 6.10 shows ice core data from Antarctica. Here, temperature is calculated using the relative concentrations of various isotopes in the ice. The curve shows us a few important things: (1) periods where global temperatures are about −6 °C colder than the present day (the glacial cycles), which correspond to CO_2 levels of about 200 ppm; (2) periods of warmth (inter-glacials) where temperatures are even warmer than today, with corresponding CO_2 levels of about 275 ppm; and (3) a dynamic climatic system, sometimes with transitions between warm and cold periods occurring very rapidly. Although the 20th century warming does not show up when plotted at this scale, we must remember that it is the *rate* at which warming has occurred during the past century that is unprecedented. The naturally occurring temperature fluctuations shown in Figure 6.10 have occurred over much longer timescales than what we are seeing today.

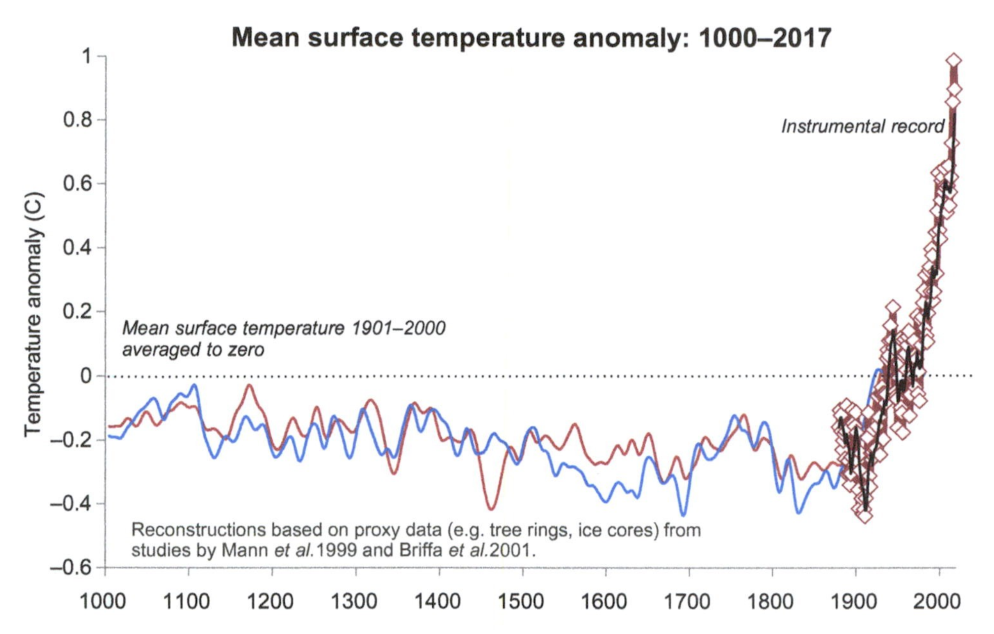

Figure 6.10 Surface temperature anomalies for the last 1,000 years based on direct sampling (i.e., the surface temperature record from 1880 shown by the pink diamonds) and proxy evidence (red and blue lines from two different studies) (Source: National Oceanic and Atmospheric Administration and National Climatic Data Center—www.ncdc.noaa.gov).

Given the preceding discussion, I think we can state with considerable certainty that:

1. Atmospheric CO_2 has increased by almost 40% since the Industrial Revolution.
2. Current levels of CO_2 are unprecedented over both human history and over a longer-term **geologic timescale** (at least the last 800,000 years).
3. The increase in CO_2 is *most likely* due to human activity, primarily through the combustion of fossil fuels.
4. CO_2 is one of several GHGs that keep our planetary temperatures much warmer than they otherwise would be (i.e., the greenhouse effect).
5. Average global temperatures have risen by about 1.4 °C (2.5 °F) during the 20th century, an unprecedented rate of change over both human history and longer-term geologic timescales.

Not even the most committed global warming skeptic disputes the physics behind the greenhouse effect and that increased amounts of CO_2 in the atmosphere will magnify this effect. However, it is here that the more difficult and controversial questions begin: Are the increased amounts of CO_2 the *primary* cause of the observed atmospheric warming? Will CO_2 continue to rise during the 21st century, and if so, at what rate? How will the atmosphere respond? Will global temperatures begin to level-off, find some equilibrium with the changes in atmospheric composition, or even begin to cool? Will global temperatures, as many reputed scientists believe, continue to rise or even accelerate during the coming centuries? What impacts will occur?

PREDICTING FUTURE TRENDS IN CO_2 AND TEMPERATURE

One of the strongest lines of evidence for those who argue that recent climate change is anthropogenic in origin is the strong **correlation** between CO_2 and temperature over the past 60 years. However, even if two variables, let us refer to them as *A* and *B*, do correlate (even perfectly), does it imply a causal relationship (i.e., that *A* causes *B*)? The answer in simple terms is, no. A strong correlation between two variables does not imply that there is a simple cause-and-effect relationship between the two, even though it is often taken for granted that *A* is causing *B* even when no evidence supports this. This is what is known as a **logical fallacy** because there are several other possibilities:

- *B* may actually be the cause of *A*
- Some unknown third factor is actually the cause of the relationship between *A* and *B* (a factor called a **lurking variable**)
- The relationship is so complex that it can be labeled coincidental—that is, *A* and *B* may have no simple relationship to each other besides the fact that they are occurring at the same time.

In other words, we cannot simply conclude the existence of a cause and effect relationship only from the fact that *A* is correlated with *B*. For example, if researchers found a correlation between individuals' college grades and their income later in life, they might wonder whether doing well in school caused the increased income. It might, but good grades and high income could both be caused by a third (lurking or hidden variable) such as tendency to work hard. I think there is a message here! Unfortunately, we cannot run experiments to determine causation in the context of climate change. We cannot rewind the past 200 years and replay events after making a controlled change to the one important variable—namely CO_2—preferably keeping levels at 280 ppm. What this means is that causation or attribution can only be inferred within some margin of error and never exactly known.

Well, if CO_2 is not the primary driver of global temperature change, then what is? We know that Earth's climate changes naturally, and based on decades of research, we can identify factors that over geologic timescales (i.e., hundreds of thousands to millions of years) have naturally driven changes in the climate, such as tectonic activity, changes in the orbit of Earth about the sun, solar variations, and volcanoes. With regard to the warming during the 20th century, we can rule out drivers like tectonics and orbital variations since they occur too slowly to account for warming over mere decades. Also, we can rule out volcanoes since they affect climate for only a few years, then return to pre-existing conditions. We can also rule out solar variability because our measurements simply have not shown an increase in solar output significant enough to explain Earth's recent temperature increase. Certainly, with a complicated climate system, there will be internal variability (such as the **El Niño/Southern Oscillation**), during which certain parts of Earth are much warmer than normal, but there is no evidence (and no data) supporting this sort of internal variability as a *driver* of climate change. The truth is that over timescales of hundreds of thousands of years, climate scientists do not look at CO_2 as a driving or trigger mechanism so much as a **feedback mechanism** (i.e., something that reinforces the effect). What most scientists think has happened in the past is that small variations in Earth's orbit cause a small initial warming that leads to the release of CO_2, which, in turn, leads to further warming. Since each of these forcing mechanisms of past climate changes can be ruled out to explain 20th century warming, scientists now agree that anthropogenic CO_2 is the most likely cause of modern climate change.

Based off the likelihood of this scenario, the next questions we must address are: (1) Will CO_2 continue to rise during the 21st century and, if so, at what rate? (2) How will Earth's climate respond, more specifically, how will global temperatures and rainfall react? These questions are even more difficult to address because now we move into the world of prediction and ultimately climate models.

No one knows for certain what CO_2 levels will be by the end of the 21st century. The rate of rise of CO_2 will depend on a number of uncertain factors, particularly economic changes and technological innovation. The **Intergovernmental Panel on Climate Change** (IPCC), established in 1988 by the World Meteorological Organization and the United Nations Environmental Program, is charged with evaluating the state of climate science as a basis for informed policy action. Led by government and the top academic scientists and researchers in climate science, the IPCC has published a wide range of future CO_2 scenarios, from 540 ppm to almost 1,000 ppm by the year 2100.[11] So, you probably want to know which is it? Surely, we must narrow that range and reduce

[11] Source: www.climatechange2013.org

the uncertainty if we are to predict how CO_2 will affect temperature, but unfortunately, we cannot. Uncertainty is just part human inquiry (as discussed in Chapter 1), and in a complex system such as the Earth's ocean-atmosphere system, uncertainty is unavoidable. Pretending otherwise would be irresponsible science at best and scientific misconduct at worst. However, what we can say is that future CO_2 levels will (most) likely continue to rise based on the rate of industrial growth in developing countries like China and India and our ongoing dependence on fossil fuels, particularly in developed countries like the United States and ones in Europe. The Energy Information Administration estimates that world CO_2 emissions will increase by 1.9% annually until 2025, and we seem to be right on track for that. As we discussed in Chapter 3, much of the increase in these emissions will occur in the developing world where emerging economies fuel economic development with fossil-based energy. In any case, it does seem prudent to err on the side of caution when predicting emission scenarios. Accordingly, predictions of future temperatures are now based largely on the assumption that global CO_2 levels will double from pre-industrial levels (i.e., from 280 ppm) to between 560 and 600 ppm by 2100.

So, if CO_2 continues its rise and doubles by the end of this century, what will global temperatures do? Well, given what we have said thus far in relation to uncertainty, you can imagine the dilemma I faced when beginning this section on climate prediction! Ultimately, the answer to this question depends on **climate sensitivity,** which is the measure of the climate's response to radiative forcing resulting from increased GHGs along with other anthropogenic and natural causes. Climate sensitivity is defined as the change in average surface temperature due to a doubling of the CO_2 concentration ($\Delta T2_x$) and is estimated to lie between 1.0 and 6.0 °C (about 2 and 11 °F, Figure 6.11), a wide range with very different consequences expected at each end of the spectrum. However, there is now a strong consensus among climate scientists that future temperature change is most likely to be on the order of about another 2 °C (3.6 °F) rise by the end of 2100.[12]

Temperature predictions are made with climate models that are mathematical representations of the interactions among the atmosphere, oceans, land surface, and ice with the sun. These models are far from perfect, and naysayers in the climate change debate most frequently use the "uncertainty of the model projections" to justify their position. To be sure, predicting climate is a very complex task. So models are built to estimate *trends* rather than events. The models also have to be tested to find out if they work. Of course, we cannot wait for 30 years to see if a model is any good or not. Models are therefore tested against the past, against what we know happened. If a model can correctly predict trends from a starting point somewhere in the past, we could expect it to predict with reasonable certainty what might happen in the future. So all models are first tested in a process called **hindcasting**, and the results to date have been quite startling. Researchers have reproduced, with very high confidence, observed continental scale surface temperature patterns and trends over many decades, including more rapid warming since the mid-20th century. Surely if climate models can map past climate changes and get them right, there is very good reason to think their predictions are going to be right too. Of course, all models have limits—uncertainties—for they are modeling very inter-connected systems. No one knows how climate change, natural or human

[12] Source: www.climatechange2013.org

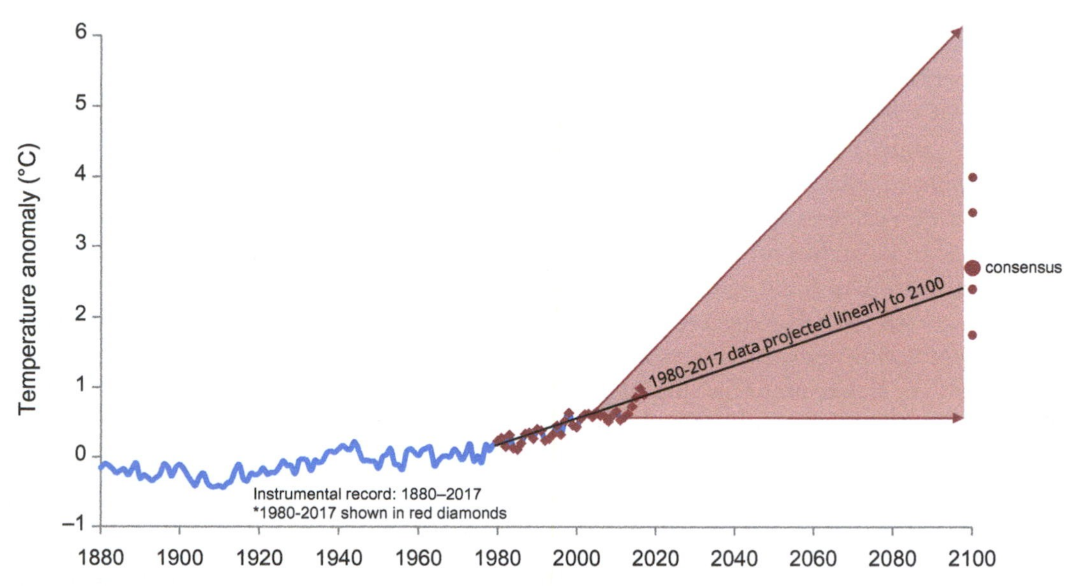

Figure 6.11 Global warming projections using computer models from several climate research centers showing the wide range of possible outcomes through to 2100. The surface record from 1980–2017 are highlighted and used as one projection through to 2100. The red circles at 2100 indicate the best estimate and the likely range assessed for six carbon dioxide scenarios, with the large circle where the consensus is strongest (Source: Intergovernmental Panel on Climate Change—www.ipcc.ch/).

induced, will play out in the real world as opposed to how it plays out in highly sophisticated, yet imperfect, computer models. Nevertheless, as the models have improved, along with increasing sources of high resolution, real-world information from satellites, so too has the confidence in their predictions. In the latest IPCC report,[13] scientists concluded with unprecedented confidence—at least 95% certainty—that the climate will continue to warm and that humans have "been the dominant cause of the observed warming since the mid-20th century." The increase in confidence stems not only from improved models but also from multiple lines of evidence of a warming planet, from increasing ocean heat content to melting Arctic sea ice, which are all consistent with warming due largely to rising amounts of GHGs. It is these lines of evidence that we examine next.

POSSIBLE EFFECTS OF GLOBAL WARMING

If we are already committed to living in a warmer world, then what are the likely impacts of such warming? The possible effects of global warming generally fall into two categories: the impact on the oceans, specifically rising sea level, and changes in the amount and pattern of precipitation. These impacts will operate at local, regional, continental, and global scales. Let us look at rising sea level first.

[13] www.climatechange2013.org

Changes in Sea Level

The physics behind rising sea levels is well understood. In a warmer world, the oceans themselves will expand, raising the sea level. It has been estimated that a warming of the entire world ocean by 1 °C would produce a sea level rise of 1.6 ft just by **thermal expansion**. However, a uniform warming of the entire ocean within a short time is unlikely because the deep ocean warms up much more slowly than the upper layers. Water exchange between these two regions is reduced as the warming happens, thereby slowing down the whole process of sea level rise. A 1.6 foot rise should therefore just be taken as an indication of the order of magnitude of the change possible through thermal expansion. It will most likely be quite a bit lower.

Sea level is also expected to rise due to the addition of fresh water from the melting of land-based glaciers. This is one area where we are on much firmer footing with respect to level of certainty. There is now overwhelming evidence that glaciers around the world are melting and retreating and, furthermore, that the rate of melting is increasing. We all know that a picture is worth a thousand words, and photographs and satellite images showing disappearing ice sheets are now commonplace in the media and scientific literature (see Figures 6.12 and 6.13). The very existence of many of the world's glaciers is now threatened. The snow cap that has covered the top of Mount Kilimanjaro for the past 11,000 years since the last ice age has almost disappeared. Melting appears to be increasing in the Andes, Alps, Himalayas, and the Rockies. An analysis of glaciers worldwide has shown that the volume of glacial ice decreased substantially during the second half of the 20th century, on the order of about 90 km^3 per year.[14] These observations of glacial recession are entirely consistent with the more rapid rise in global temperatures since the mid-20th century, but glacial melt is not only a scientific issue. There is also serious concern about local water resources in these areas. Glaciers retain water on mountains during wet years because snow cover accumulating on the glaciers protects the ice from further melting. In warmer and drier years, glaciers offset reduced rainfall by releasing higher meltwater output. Of particular importance is melting in the Himalayas that produce most of the dry-season water to many of the major rivers of Southeast Asia. In these areas that are so heavily dependent on glacial meltwater, an acceleration of the current rates of retreat will eventually deplete the glacial ice and substantially reduce or eliminate runoff water. A reduction in runoff water will affect people's ability to irrigate their crops and affect their lives in many other ways.

The amount of freshwater added to the oceans from the melting of temperate and alpine glaciers will pale in comparison to the volumes potentially added from melting in the Greenland and West Antarctic Ice Sheets. We know that about 99% of all freshwater ice is in the great ice sheets of polar and subpolar Antarctica and Greenland. In Greenland, several very large glaciers that were stable for centuries/millennia began to retreat in 2000. Satellite images and aerial photographs from the 1950s and the 1970s show glaciers in Greenland were once stable. Now, more sophisticated surveying from both the ground and air shows several glaciers retreating rapidly, some in excess of 100 ft/day. The extent of the Greenland ice melt has been steadily increasing over the past 30 years (Figure 6.14).

---

[14] http://instaar.colorado.edu/other/download/OP55_glaciers.pdf

Figure 6.12 Photograph of the Athabasca Glacier, one of six glaciers that spill down the Canadian Rockies from the Columbia Ice field in Western Canada. Visitors who return to the glacier after their first visit will notice the change brought about by warming temperatures. In the past 125 years, the Athabasca Glacier has lost half of its volume and receded almost a mile, leaving hills of rock in its place. Its retreat is visible in this photo, where the glaciers front edge looms several meters behind the tombstone-like marker that indicates the edge of the ice in 1992.

Antarctica provides the most dramatic example of glacier retreat, where large sections of the Larsen Ice Shelf have been lost. The collapse of the ice shelf has been caused by warmer melt season temperatures. From 1995 to 2001, the Larsen Ice Shelf lost 965 square miles of its area, about two-thirds the size of Rhode Island. In a 35-day period beginning on January 31, 2002, about 1,254 square miles of shelf area disintegrated. The ice sheet is now 40% the size of its previous minimum stable extent.[15]

The result is that sea level has risen by about 0.65 ft (20 cm) over the past 100 years, and predictions from the IPCC suggest that it will continue to rise a further 0.8–2.6 ft (25–80 cm in the next century Figure 6.15).[16] Such a rise would inundate about 7,000 square miles of dry land in the United States (an area the size of Massachusetts) and a similar amount of coastal wetlands, erode recreational beaches, exacerbate coastal flooding, and increase the salinity of coastal aquifers and

[15] National Snow and Ice Data Center (www.nsidc.org/iceshelves/larsenb2002/).

[16] www.climatechange2013.org

Figure 6.13 The Jökulsárlón glacial lake in southeast Iceland, on the edge of Vatnajökull National Park. The lake formed about 70 years ago when the glacier started receding from the edge of the Atlantic Ocean. It has increased in size fourfold since the 1970s with estimates of glacial retreat now about 500 meters per year (Photo: Mike Slattery).

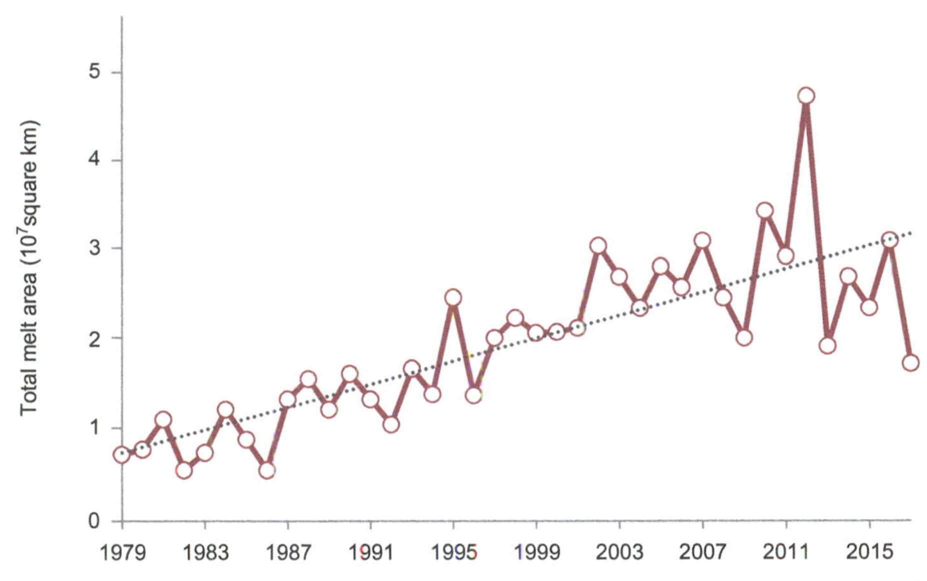

Figure 6.14 Annual estimates of surface melt extent of the Greenland Ice-Sheet from satellite sensors (Source: Mote, T. (2007), *Geophysical Research Letters*, Vol. 34, L22507, updated 2017 through personal communication with author.

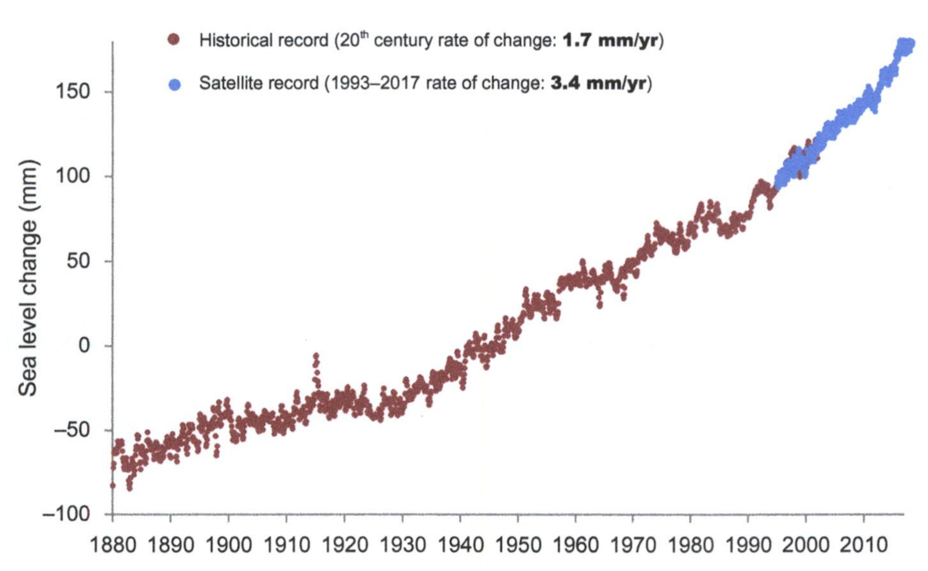

Figure 6.15 Sea level data derived from coastal tide gauge records (1870–1993, red circles) and average sea level since 1993 derived from global satellite measurements, updated here monthly (blue triangles). Sea level rise is associated with the thermal expansion of seawater due to climate warming and widespread melting of land ice (Source: www.climate.nasa.gov/).

estuaries. Several small island nations are already at great risk; some of them will be inundated. Refugees from highly populated deltas such as Bangladesh will likely become critical humanitarian issues. Studies have also predicted much higher sea level rise for the 21st century than the IPCC, exceeding one meter, again depending on the rate at which GHG emissions escalate.[17]

Interestingly, there is some speculation that global warming could lead to cooling, or lesser warming, in the North Atlantic, via a slowing or even shutdown of the **thermohaline circulation**. This circulation is called the ocean conveyor belt or the global conveyor belt and transports warm water to the North Atlantic (Figure 6.16). A melting and influx of fresh water could potentially "turn off" the conveyor. This would affect areas like Ireland, Britain, and Scandinavia that are warmed by the North Atlantic Drift, leaving them much colder than they are today.

A report by the Arctic Climate Impact Assessment—a consortium of eight countries, including Russia and the United States—now confirms climate model predictions of major changes taking place in the Arctic, which are affecting both human and nonhuman communities. The amount of ice in the Arctic decreased by almost 15% between 1979 and 2017 (Figure 6.17). While the reduction of summer ice in the Arctic may be good news for shipping (particularly if the Northwest and Northeast Passage fully opens up in summer), this same phenomenon threatens the Arctic

[17] Source: Rahmstorf, S. (2010). *Nature Reports Climate Change*. (http://www.nature.com/climate/2010/1004/full/climate.2010.29.html)

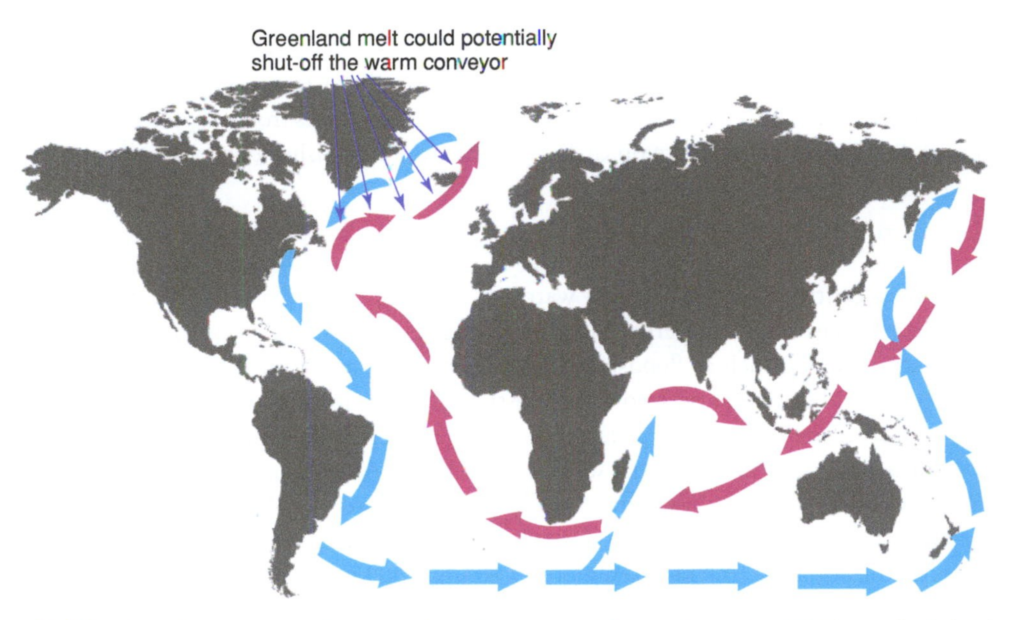

Figure 6.16 Often called a conveyor belt because of its northward transport at the surface, and southward return flow in the abyss in the Atlantic, the ocean circulation system is a slow, three-dimensional pattern of flow involving the surface and deep oceans around the world (Source: www.ncdc.noaa.gov).

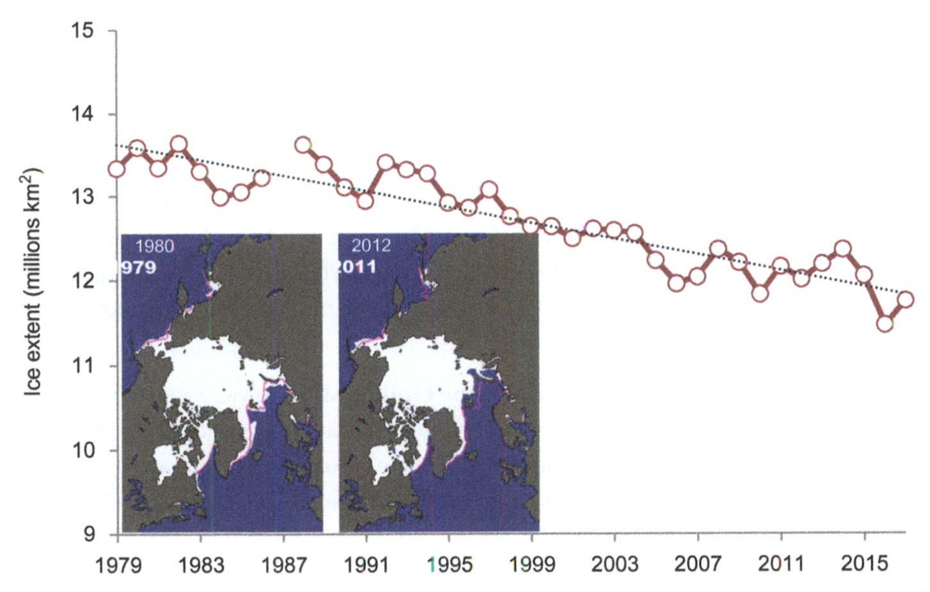

Figure 6.17 Average monthly (December) Arctic sea ice extent, 1979–2017. Sea ice extent over the 30-plus-year period shows a decline of 3.3% per decade. Insets show 1980 (left) and 2012 (right) sea ice extent with the purple line the median ice extent over the entire monitoring period (Source: http//nsidc.org/).

ecosystem, most notably polar bears that depend on ice floes. Subsistence hunters such as the Inuit peoples have found their livelihoods and cultures increasingly threatened as the Arctic ecosystem changes. However, an ice-free Arctic would cut 5,000 nautical miles from shipping routes between Europe and Asia and allow offshore oil drilling and maybe new fisheries access.

Changes in Precipitation

Arguably, more controversial than global warming's impact on sea level are the possible changes to the planet's precipitation patterns. Many have suggested that the frequency and intensity of extreme weather events, such as floods, droughts, hurricanes, and tornadoes, will increase. At the end of 2005—the hottest year on record—the Atlantic basin had just wrapped up its most active hurricane season in recorded history. Extreme weather events like hurricanes Rita and Katrina inevitably raised the question: Is global warming to blame? After Hurricane Katrina impacted the Gulf Coast and became the costliest natural disaster in U.S. history, some Americans began viewing monster hurricanes as the greatest threat posed by a warming world. A host of reputable magazines ran articles on the potential link between global warming and hurricanes. In *Time* magazine, a rather prophetic piece entitled "Is Global Warming Making Hurricanes Worse?" was published just three weeks before Katrina. Yet Max Mayfield, Director of the National Hurricane Center during the 2005 season, noted that science simply does not support a link between global warming and recent hurricane activity. According to Mayfield, Katrina and Rita are part of a natural cycle.[18] Certainly, having two very large storms in one season is hardly unprecedented: two major hurricanes hit the Gulf Coast only six weeks apart in 1915, mimicking the double whammy of Katrina and Rita. However, a year later, scientists from the National Center for Atmospheric Research (NCAR) in Boulder, Colorado, published a report noting that global warming accounted for around half of the warmth in the waters of the tropical North Atlantic in 2005, while natural cycles were only a minor factor.[19]

These conflicting views on the link between hurricanes and temperature change raise the question whether scientists really know if hurricanes like Katrina are a direct result of climate change? Can human-made GHGs really be blamed for the intensity of these headline-grabbing storms, or are there, as other experts insist, too many additional variables to say one way or the other? We do know that 2005's activity was related to very favorable upper-level winds and the extremely warm sea surface temperatures in the Gulf of Mexico. But climate change should exacerbate the problem of hurricanes because warmer air can easily translate into warmer oceans, and warm oceans are the fuel that drives the hurricane's turbine.

When Katrina hit at the end of August, 2005, the Gulf of Mexico was a veritable hurricane refueling station, with water up to 5 °F higher than normal. Rita too drew its killer strength from the Gulf, making its way past southern Florida as a Category 1 storm, then exploding into a Category 5 as it moved westward. However, recent stormy years in the North Atlantic have also been preceded by many very quiet ones, all occurring at the same time that global temperatures were marching upward. An analysis of the global hurricane record by scientists at the Geophysical Fluid Dynamics

[18] Article appeared in *USA Today*, September 25, 2005.

[19] National Center for Atmospheric Research (www.ucar.edu/news/releases/2006/hurricanes.shtml).

Laboratory in collaboration with Princeton University suggests that it is premature to conclude that human activity—and particularly greenhouse warming—has already had a detectable impact on Atlantic hurricane activity.[20] Such human activity may have already caused substantial changes that simply cannot be detected at the present time. However, the study concludes that anthropogenic warming over the next century will *likely* increase the frequency of the strongest hurricanes in the Atlantic roughly by a factor of two by the end of the century.[21] The potential impact on just our oil production alone should be cause for concern (Figure 6.18).

The potential impacts of climate change extend well beyond more frequent and more intense storms. Climate change is modifying the circulation of water on, above, and below the surface of the Earth. Consequently, droughts and floods will likely become more frequent and widespread. Higher temperatures increase the amount of moisture that evaporates from land and water, leading to drought in many areas. If temperatures continue to rise globally, a highly likely scenario, droughts will undoubtedly intensify, with potentially devastating consequences for agriculture and water supply. This phenomenon has already been observed in some parts of Asia and Africa, where droughts have become longer and more intense. Hot temperatures and dry conditions also increase the likelihood of forest fires. In the conifer forests of the western United States, earlier snowmelts, longer summers, and an increase in spring and summer temperatures have increased fire frequency (Figure 6.19). A recent study by scientists at the University of California concluded that rising temperatures associated with climate change could result in many more severe forest fires in the coming decades.[22] The research team found that by 2050, forest fires would likely cause a major shift in the Greater Yellowstone Ecosystem and affect the region's wildlife, hydrology, and aesthetics.

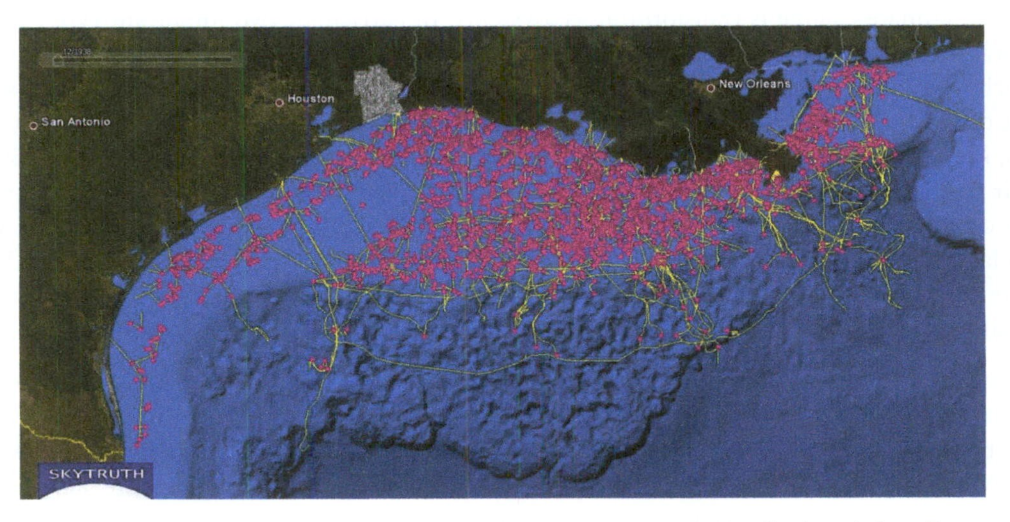

Figure 6.18 Map on the northern part of the Gulf of Mexico showing the more than 4,000 active oil and gas platforms (Sources: www.noaa.gov and www.skytruth.org).

[20] http://www.gfdl.noaa.gov/global-warming-and-hurricanes

[21] Source: Bender, M.A. et al. (2010). *Science*, 327: 454–458.

[22] Source: Westerling, A.L. (2011). *Proceedings of the National Academy of Sciences.*

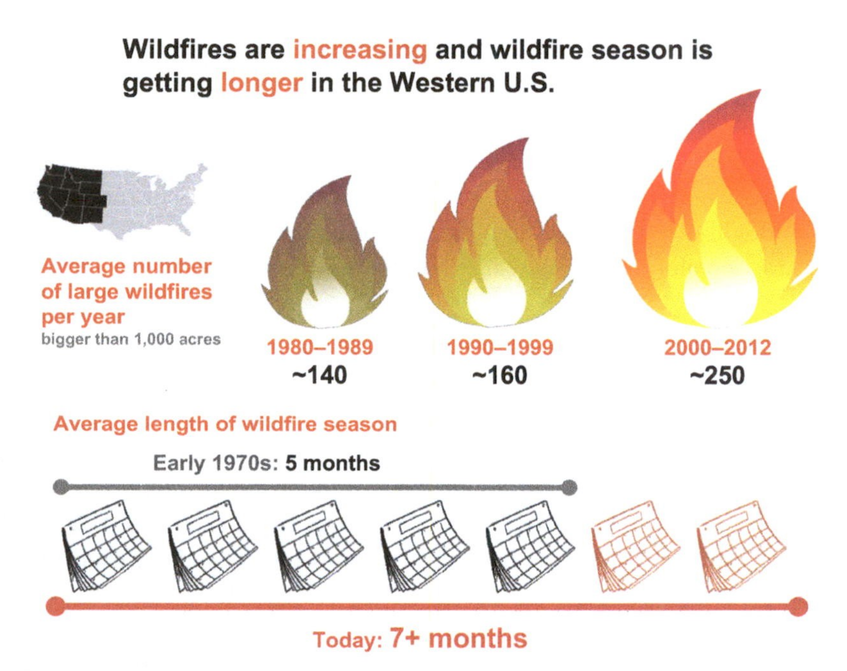

Figure 6.19 The number of large wildfires, defined as those covering more than 1,000 acres—is increasing throughout the Western U.S. Over the past 12 years, every state in the Western U.S. has experienced an increase in the average number of large wildfires per year compared to the annual average from 1980 to 2000. The wildfire season has also lengthened over the past 40 years (Source: Union of Concerned Scientists, https://www.ucsusa.org/global-warming/science-and-impacts/impacts).

Finally, climate change is transforming ecosystems. In particular, two important types of ecological impacts of climate change have been observed: shifts in species' *ranges* (the locations in which they can survive and reproduce) and shifts in *phenology* (the timing of biological activities that take place seasonally). Examples of these types of impacts have been observed in many species, in many regions, and over long periods of time. As Earth warms, many species are shifting their ranges to areas with more tolerable climate conditions. However, some organisms—those that cannot move fast enough or those whose ranges are actually shrinking—are being left with no place to go. For example, a 2012 study of changing mountain vegetation in Europe found that some alpine meadows could disappear within the next few decades.[23] Cold-loving plants traditionally found in alpine regions are being pushed out of many habitats by warm-loving plants.

Climate change is also driving changes in the timing of seasonal biological activities. Many biological events, especially those in the spring and fall, are based on seasonal cues. Studies have found that the seasonal behaviors of many species now happen 15–20 days earlier than several decades ago. Migrant birds are arriving earlier, butterflies are emerging sooner, and plants are budding and blooming earlier. It is worth emphasizing that, while species have responded to climatic changes throughout their evolutionary history, a primary concern for species and their ecosystems is this rapid rate of change.

[23] Source: Gottfried, M. (2012). *Nature Climate Change*, 2, 111–115.

MOVING FORWARD

I have used the phrase "scientific consensus" at several points in this chapter, but what does that really mean? How many scientists does it take to make the term "consensus" a valid one? Multiple studies published in peer-reviewed scientific journals show that 97% or more of actively publishing climate scientists agree that climate-warming trends over the past century are likely due to human activities.[24] In addition, most of the leading scientific organizations worldwide have issued public statements endorsing this position.[25]

And while there is some scientific uncertainty, including the exact amount of climate change expected in the future and, especially, how changes will vary from region to region across the globe, we should not use this as justification for complacency. Yet key questions remain to be resolved: What (if anything) should be done? What could be done cost effectively to reduce or reverse future warming? How will we deal with the expected consequences?

The key challenge with climate change has been our inability to reach a legally binding global agreement on how to reduce our CO_2 emissions. As discussed in the previous chapter, we were successful in doing so for CFCs, and we seem to be reversing stratospheric ozone depletion as a result. We cannot say the same for carbon. The Kyoto Protocol was the original global treaty aimed at taking actions to reduce the extent or likelihood of climate change. It was adopted in December 1997 and came into force on February 16, 2005 with 160 countries responsible for 55% of global GHG emissions) committing to reduce their emissions of CO_2. Between 2008 and 2012, developed countries which signed the protocol were slated to reduce their GHG emissions by an average of 5% below their 1990 levels whereas developing economies had no binding GHG restrictions. Australian Prime Minister John Howard refused to ratify the Agreement during his tenure from 1996 to 2007, arguing that the protocol would cost Australians their jobs, since countries with booming economies and massive populations such as China and India had no reduction obligations under the Protocol.[26]As noted in the introduction to this chapter, President George W. Bush also refused to submit the treaty for ratification, not because he did not support the Kyoto principles but principally because of the exemption granted to China. He also opposed the treaty because of the strain he believed that it would put on the U.S. economy and emphasized, on several occasions, the uncertainties which he asserted are present in the climate change issue.

On 12 December 2015, the Kyoto Protocol was effectively replaced by the Paris Agreement. The goal of this agreement is to keep global temperature rise this century below 2 degrees Celsius, the degree change usually agreed upon as being the tipping point to preventing massive effects of climate change. As of August 2018, 196 nations had signed the agreement, including the U.S., China, and India, the three largest global emitters of CO_2 (Figure 6.20). Under the agreement,

[24] While a "consensus" is a general agreement of opinion, the scientific method steers us away from this to an objective framework. Remember, in science, facts or observations are explained by a hypothesis (a statement of a possible explanation for some natural phenomenon), which can then be tested and retested until it is refuted (or disproved). Eventually, a group of hypotheses are integrated and generalized into a scientific theory.

[25] Source: https://climate.nasa.gov/scientific-consensus/#*

[26] Australia ratified the agreement in December 2007 with the election of Mr. Kevin Rudd.

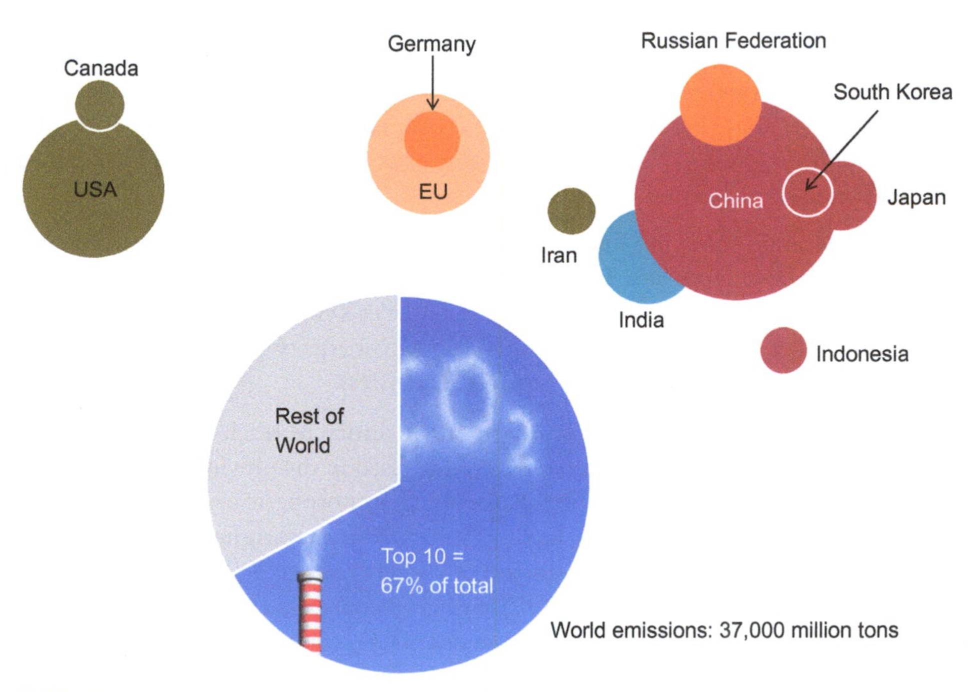

Figure 6.20 Top ten carbon dioxide emitting countries as of December 2017. The data for each country (and the European Union) are sized relative to the global output pie chart (37 Gt of carbon dioxide per year). China (10,433 million tons) has sped ahead of the U.S. (5,011 million tons) with India (2,534 million tons) now the third largest emitter globally (Source: www.eia.gov).

each country sets its own goals, a "bottom-up approach", but there is no mechanism to force a country to set a specific target by a specific date. It is, in effect, voluntary. That said, commitments are rolling in: for example, Norway will ban the sale of petrol- and diesel-powered cars by 2025; the Netherlands will do the same by 2030.

In June 2017, President Trump announced his intention to withdraw the U.S. from the agreement, a move many environmental groups see as shortsighted, reducing the ability of the U.S. government to provide global leadership on climate change. Since the announcement by the President, more than 1,200 universities, colleges, investors, businesses, mayors, and governors from around the country have declared in unison that they are still part of the Paris Agreement and have sent a letter to the United Nations to underline their commitment to continue to address carbon emissions. Despite not being bound to any international agreement, U.S. CO_2 emissions are actually down 14% from their peak in 2005 (Figure 6.21). In fact, emissions are now at their lowest level since 1994, primarily the result of reduced transportation sector emissions and more widespread use of lower-priced natural gas in place of coal. Overall, it is a significant improvement but, as some suggest, a drop in the bucket by global emission standards.

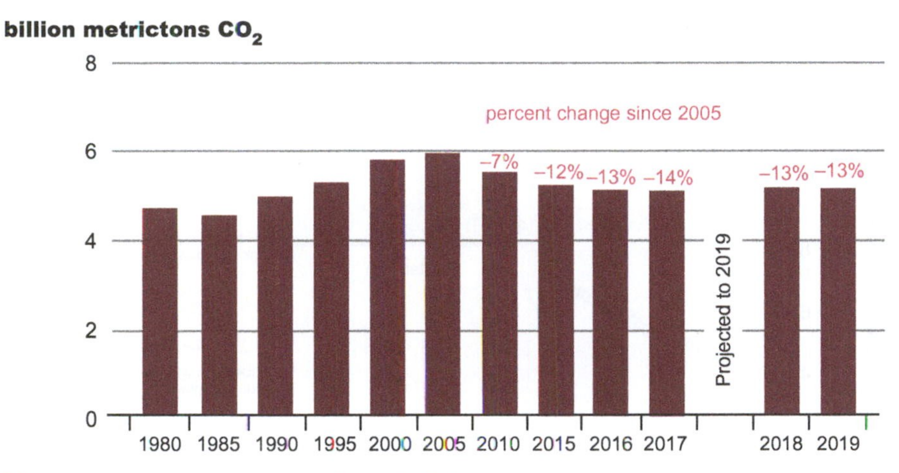

Figure 6.21 Energy-related carbon dioxide emissions, 1990–2019 (Source: www.eia.gov).

Ultimately, there is no single solution to mitigate the worst effects of climate change. Alternative power (like solar, wind, and nuclear) is not going to be sufficient to replace all coal and oil use. Efficiency will not improve fast enough, and we cannot (yet) effectively and economically capture and store carbon in the ground (a strategy called **carbon capture and sequestration** or CCS). These are all true but only in isolation. A solution that will work must come from a combination of multiple, varied efforts. Professor Robert Socolow from Princeton University has captured this complexity elegantly in a concept he calls **stabilization wedges** (Figure 6.22). The concept is a simple tool for conveying the emission cuts that will need to be made to avoid the worst impacts of future climate change. Two futures are considered: allowing emissions to double to 16 GtC per year versus keeping emissions at current levels for the next 50 years. The emissions-doubling path approximately extends the climb for the past 50 years. Emissions could well be higher or lower over that period, but this path appears to be a reasonable reference scenario. The argument put forward by Socolow and his colleagues is that we can prevent a doubling of CO_2 if we can keep emissions flat for the next 50 years and then work to reduce emissions in the second half of the century. The authors point out that this path will likely allow us to dodge the worst predicted consequences of climate change. Keeping emissions flat will require cutting projected carbon output by about 8 billion tons per year by 2060, keeping a total of about 200 billion tons of carbon from entering the atmosphere. This carbon savings is what is called the **stabilization triangle**.

A widely-held assumption is that only revolutionary technologies, like nuclear fusion, could enable such large emissions cuts. However, there is no reason why one technology should have to solve the whole problem. Thus, to make the problem more workable, the stabilization triangle is actually divided into eight wedges. A wedge represents a carbon-cutting strategy that has the potential to

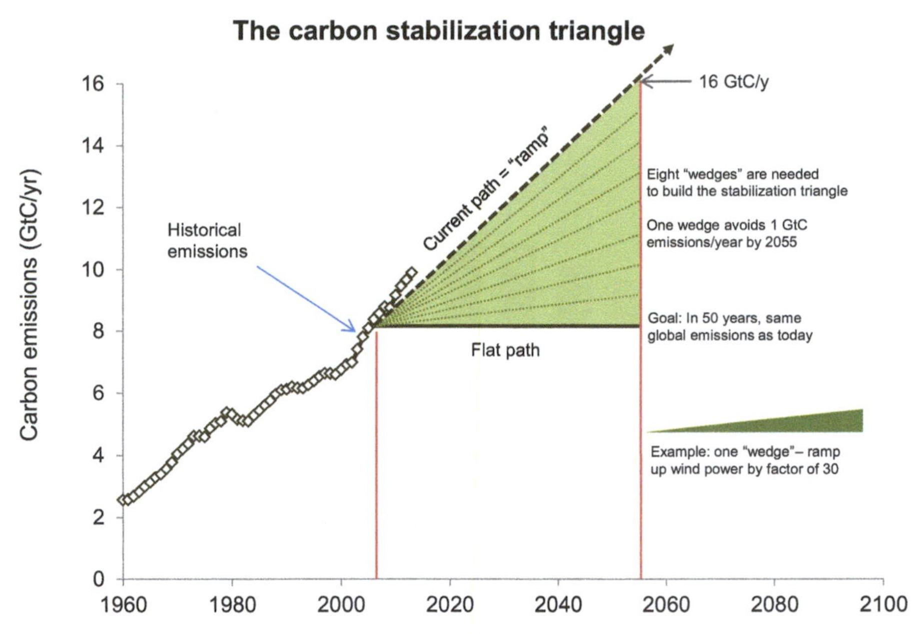

Figure 6.22 Carbon dioxide stabilization wedges. The diamonds represent the historical emissions path for global carbon as CO_2 from fossil fuel and cement manufacturing. The curve rises 1.5% per year starting from 7 GtC per year in 2004. The stabilization triangle shows avoided emissions with actual global emissions fixed at eight GtC per year. The stabilization triangle is divided into eight wedges, each of which reaches 1 GtC per year in 2060 (Source: Pacala, S. and Socolow, R. (2004), *Science*, Vol. 305: p. 968–972).

avoid 1 billion tons of carbon emissions per year by 2060 or one-eighth of the stabilization triangle. Keeping emissions flat will require the world's societies to "fill in" the eight wedges of the stabilization triangle. In the analysis, at least 15 strategies are available now that, with scaling up, could each take care of at least one wedge of emissions reduction.[27] No one strategy can take care of the whole triangle. New strategies will be needed to address both fuel and electricity needs, and some wedge strategies compete with others to replace emissions from the same source. But there is already a more than adequate portfolio of tools available to control carbon emissions for the next 50 years. For example, a wedge of emissions savings would be achieved if the fuel efficiency of all the cars projected for 2060 were doubled from 30 mpg to 60 mpg. Efficiency improvements could come from using hybrid and diesel engine technologies as well as making vehicles out of strong but lighter materials. Cutting carbon emissions from trucks and planes by making these engines more efficient can also help with this wedge. Adding new nuclear electric plants to triple the world's current nuclear capacity would cut emissions by one wedge if coal plants were displaced. We would also gain a wedge of emissions savings from wind displacing coal-based electricity, with current wind capacity being scaled up by a factor of 30. Each wedge is difficult but achievable. As Professor Scolow himself has said, this approach "decomposes a heroic challenge (eliminating the emissions in the stabilization triangle) into a limited set of merely monumental tasks."[28]

[27] http://cmi.princeton.edu/wedges/intro.php

[28] Source: Pacala, S. and Socolow, R. (2004), Science, Vol. 305: p. 968–972, my parenthesis.

Many like the "we can do this" spirit of the stabilization wedge model as opposed to the "we are doomed" scenarios often portrayed by scientists and environmentalists. The strategies put forth in the wedge approach will require cooperation at a global scale and as we have discussed in this chapter, this has so far proven difficult. Some have argued that the reason for this is due to a global governance gap, where politicians think in terms of the next election cycle and corporations in terms of the next quarter's results. However, things are being done at smaller scales. For example, the Western Climate Initiative[29] (WCI) is a collaboration of independent jurisdictions (e.g., California and British Columbia) working together to identify, evaluate, and implement emissions trading policies to tackle climate change at a regional level. This is a comprehensive effort to reduce GHG pollution, spur investment in clean-energy technologies that create green jobs and reduce dependence on imported oil. The Regional Greenhouse Gas Initiative[30] (RGGI) is a cooperative effort among the states of Connecticut, Delaware, Maine, Maryland, Massachusetts, New Hampshire, New York, Rhode Island, and Vermont to cap and then reduce CO_2 emissions by 2.5% each year through 2020.

As important as these regional efforts are, reducing carbon emissions is going to require action from all of us. The old adage "think globally, act locally" is very relevant in the context of climate change because we can help curb further warming of the planet by taking sensible steps. For example, buying local produce and ditching bottled water are two simple actions that will reduce your carbon footprint because anything shipped long distances requires carbon-intensive fuel for transport. However, as discussed in Chapter 3, the biggest single step we can take is to require cars and trucks to go farther on a gallon of gas. According to the National Academy of Sciences, currently available technology can make cars and trucks nearly double their gas mileage to an average of 40 mpg within a decade without reducing the size, power, or variety of cars available to consumers. This will also save Americans billions of dollars and reduce pollution and further GHG emissions. It is vital that we all play a part in identifying cost-effective steps such as these that we can take *now* to contribute to the long-term reduction in net global GHG emissions. Action taken now to reduce the build-up of GHGs in the atmosphere will lessen the magnitude and rate of climate change. As the United Nations Framework Convention on Climate Change (UNFCCC) recognizes, a lack of full scientific certainty about some aspects of climate change is not a reason for delaying an immediate response that will, at a reasonable cost, prevent dangerous anthropogenic interference with the climate system.

CONCLUDING THOUGHT

In the debate about climate change, you frequently hear that the science is too uncertain, the impacts are too far in the future, that it is all part of some natural cycle, and that there is no readily identifiable "villain." This leads skeptics to argue that we would be foolish to make it a major policy issue and that doing so would, in effect, be tantamount to rolling the dice with our future.

29 www.westernclimateinitiative.org

30 www.rggi.org

The difficulty is that climate change is an exceedingly complex phenomenon and is not really open to elegant, simple and consensual solutions. Even when scientists, politicians, and the public agree on the basic principles and the most robust findings of climate science, there is still plenty of room for disagreement about what the implications of that science are for action. Science thrives on disagreement and can only progress through disagreement and challenge. However, disagreements presented as disputes about scientific evidence, theory, or prediction may often be rooted more in fundamental differences between the protagonists. "Climate change as scientific controversy" is a compelling discourse to which the media and other social actors are readily attracted.

Climate scientists have constructed and presented a powerful consensus about the physical transformation of the world's climates. I agree, and I believe that the risks posed to society by the physical attributes of climate change are now clearly tangible and serious. There is no doubt in my mind that the global climate is changing and that human activity is largely to blame. The scientific community has made it abundantly clear: we are potentially in deep trouble. This global issue does not care about race, color, or creed, nor political affiliation, although, ironically, the people who produce the least emissions will probably be the ones to suffer the most. We need to stop thinking about it as being "a problem waiting for a solution" or as a "fact" waiting to be discovered, proved, or disproved, using the methods of science. It is an environmental, cultural, and political phenomenon that is reshaping the way we think about ourselves, our societies, and humanity's place on Earth.

7

Deforestation

Contributed by Makayla Klein. © Kendall Hunt Publishing Company

INTRODUCTION

Each year, I take a group of students from my university to Costa Rica for a three-week environmental stewardship course. One of the most overpowering sights of the trip occurs along the drive up to the world-renowned Monteverde Cloud Forest (Figure 7.1). During most of the drive, all that can be seen is deforested, eroded pastures. Unbridled tree cutting and cattle grazing reduces the land to dust during the dry season and mudslides during the wet season. This would be Costa Rica's future without conservation measures in place. However, within the mist at Monteverde, you see the other possibility: productive land for farmers alongside large patches of untouched and protected forest. The contrast between the two areas is astounding. For many of the students, it is their first real glimpse of environmental destruction, and the impact is noticeable. The following excerpt is from the field journal of a graphic design major:

> *I had one of those surreal moments today. One of those moments that make you stand still—where time stops. I stood on a platform raised just as high as the tree tops. I could see to all ends of the world. A grey mist covered the trees, and just the peaks were able to make their way out of the fog. No ground to be seen, we were just floating. If I jumped, I still may not have found the ground, like in a dream. It was a*

Figure 7.1 Deforested and overgrazed hillsides on the road to Monteverde, Costa Rica (Photo: Mike Slattery).

gloomy day, but not even the darkest blanket could hide the beauty that was underneath. I knew, though I could not see it, that there was something extravagant under those clouds, but the Earth was hiding it from us that day, as if it knew we were not capable of beholding what was underneath. Some day that beauty will be revealed to us, and on that day, no one will ever think about bringing harm to it again.

WHAT IS DEFORESTATION?

Clearing forests to other land uses such as agriculture, grazing, and new settlements has been ongoing for many centuries. This process, known as **deforestation**, has accelerated in recent decades and is now widely recognized as one of the most important environmental problems facing the world. From news and media, we frequently hear that if current rates of deforestation continue, the world's rainforests will vanish within two centuries, perhaps much faster than that. Vivid images of burning trees in the Amazon are everywhere (Figure 7.2). However, the loss of tropical forest is far more profound than the destruction and burning of merely beautiful areas. Tropical forests are the most diverse ecosystems on Earth and are home to more than half of the world's known plant and animal species (a link we will make explicitly in Chapter 8). According to a report by the U.S. National Academy of Sciences, a four square mile (1,000 hectare) patch of rainforest contains up to 1,500 species of flowering plants, 750 species of trees, 125 species of mammals, 400 species of birds, 150 species of butterflies, 100 species of reptiles, and 60 species of amphibians.[1]

Figure 7.2 Destruction and fire in the Amazon (Photo: iStockphoto.com/Brazil2).

[1] See The Nature Conservancy (www.nature.org).

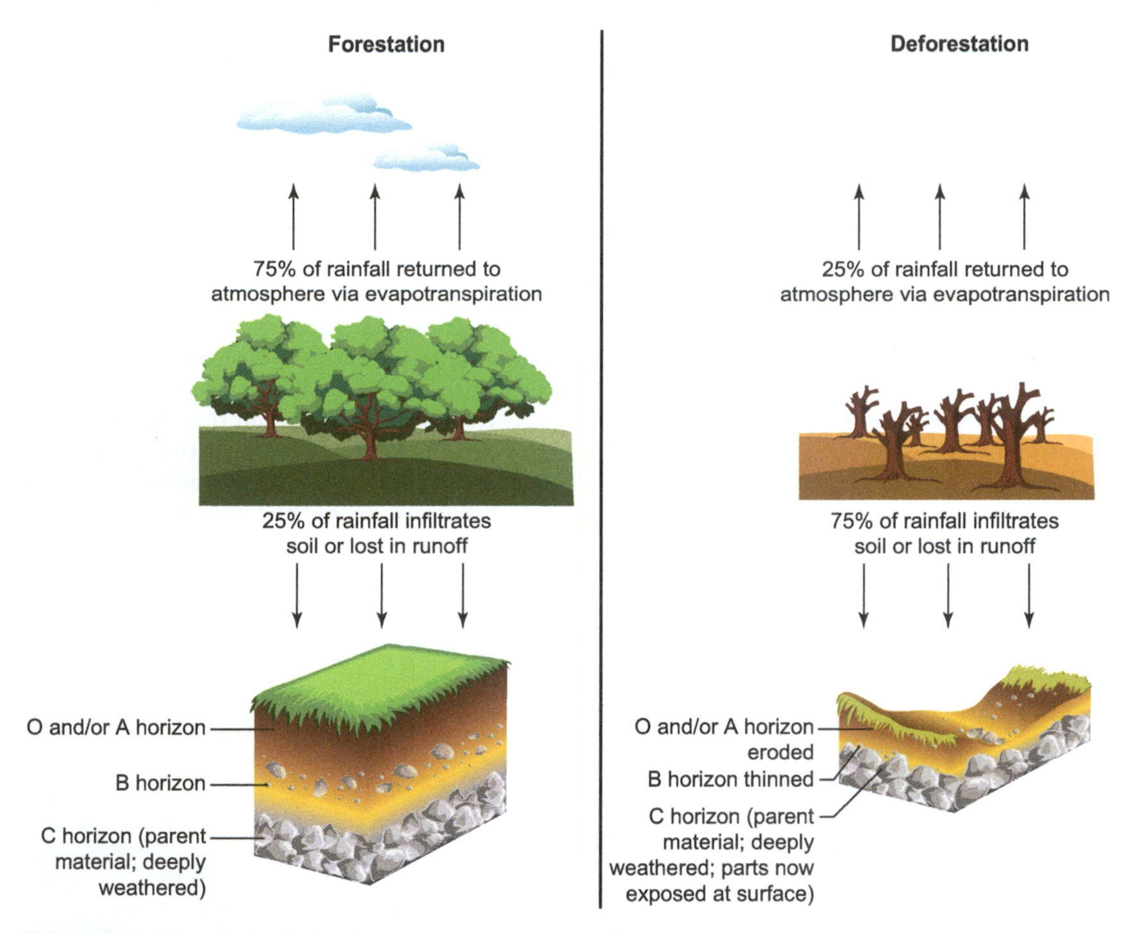

Figure 7.3 Schematic showing how forests act as system regulators. Deforestation results in less evapotranspiration, more runoff, and increased soil erosion.

Tropical forests act not only as habitat but also as **system regulators** (Figure 7.3). Rainforests moderate air temperatures, maintain atmospheric humidity levels through **evapotranspiration**, and regulate stream flows by allowing rainfall to enter streams more slowly. Forest canopy also protects underlying soil from erosion. The Amazon Rainforest, which alone comprises 30% of the world's rainforests, plays a critical role as Earth's "lung," absorbing carbon dioxide and generating oxygen. Tropical forests also provide us with a wide range of industrial wood products that account for about 25% of a $400 billion global market each year.[2] They are also important sources of new pharmaceuticals. As one example, the periwinkle plant from the Madagascar rainforest provides a drug that has proven very successful in treating lymphocytic leukemia.

While everybody in the world benefits from the rainforests, some 150 million native or indigenous peoples rely solely on the forests for their ways of life. The forests provide food and shelter and

[2] The World Commission on Forests and Sustainable Development (www.iisd.org/wcfsd/).

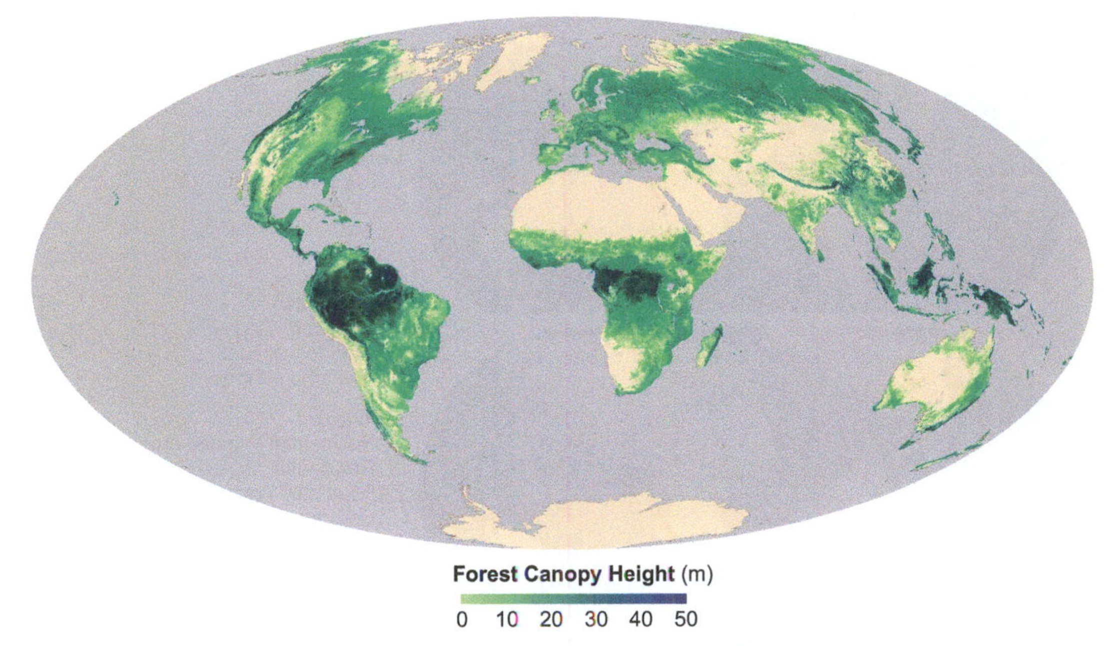

Figure 7.4 Global forest canopy heights. The map shows that, in general, forest canopy heights are highest near the equator and decrease the closer forests are to the poles. The tallest forests, shown in dark green on the map, tower higher than 40 meters (130 feet) and are found in a band in the tropics that includes the rainforests of the Amazon, central Africa, and Indonesia (Source: www.earthobservatory.nasa.gov and published in Simard, M. (2011), *Journal of Geophysical Research*, Vol. 116).

play a major role in their religious and cultural traditions. Most of the rainforest timber sold on the international market is exported to wealthy countries where it is sold for hundreds of times greater than the price paid to the indigenous people whose forests have been plundered.

It is not within the scope of this book to undertake a global examination of deforestation. In this chapter, we focus primarily on tropical forests (Figure 7.4) for the reasons noted above but acknowledge that many other forest ecosystems are under threat, such as temperate and northern old-growth forests that are destroyed for timber, paper, and other agricultural commodities.

RATES OF TROPICAL DEFORESTATION

Humans have profoundly transformed the surface of the Earth. Starting with practices like hunting and gathering and moving toward the increasingly permanent use of land for agriculture and settlement, humans have transformed ecosystem form and process locally, regionally, and globally. Research conducted by scientists at the Laboratory for Anthropogenic Landscape Ecology[3] has shown that 40% of Earth's total ice-free surface has been transformed into agricultural land and settlements since 1700 (Figure 7.5).

---

[3] www.ecotope.org

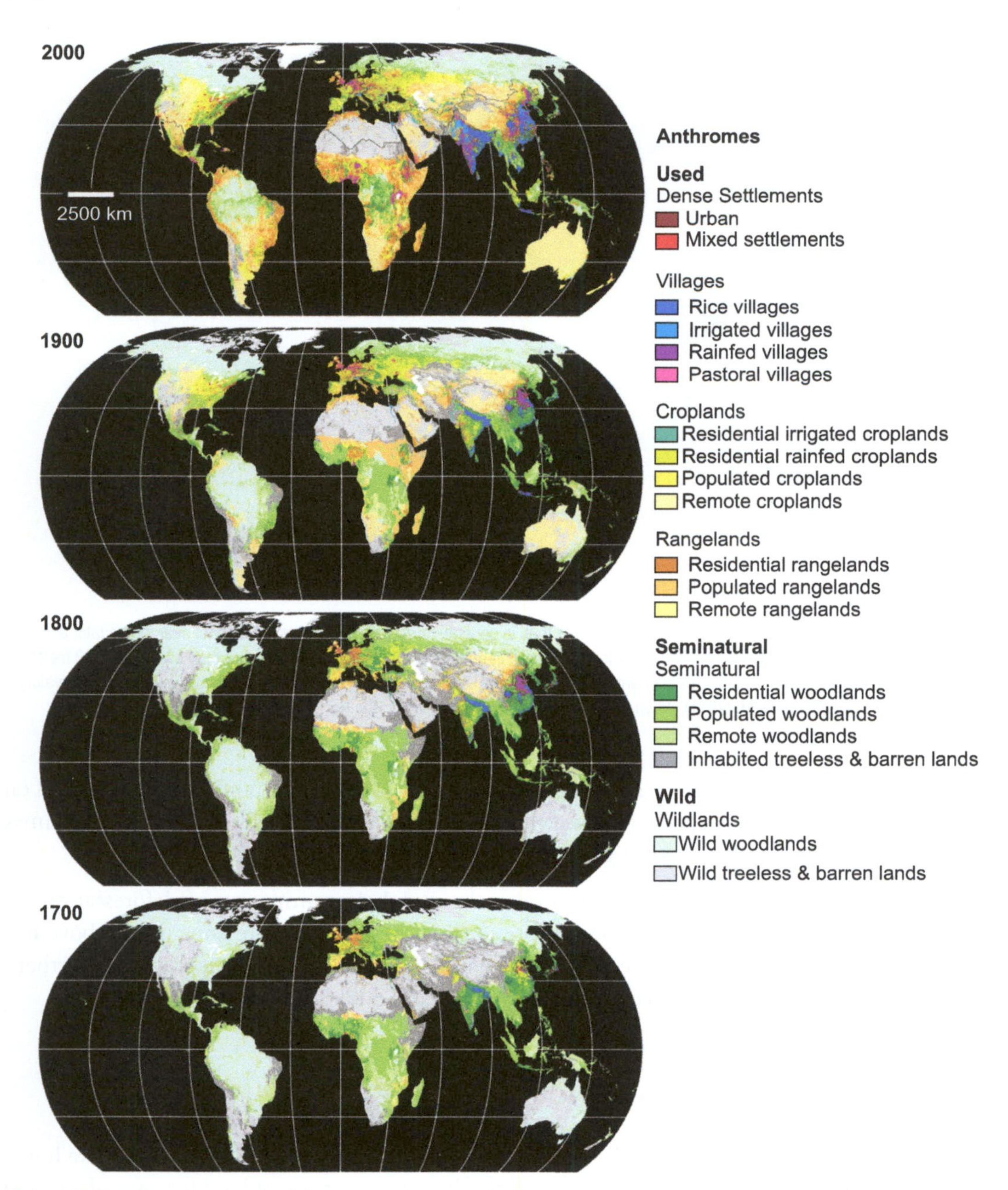

Figure 7.5 Changes to the Earth's surface, 1700–2000. In 1700, nearly half of the terrestrial biosphere was wild, without human settlements or substantial land use. Most of the remainder was in a semi-natural state (45%) having only minor use for agriculture and settlements. By 2000, the opposite was true, with the majority of the biosphere in agricultural and settled anthropogenic biomes (or anthromes), less than 20% semi-natural and only a quarter left wild (Source: Ellis et al. (2010), *Global Ecology and Biogeography*, Vol. 19: p. 589–606).

Although the world's total area of tropical forest has continuously declined for centuries, we seem to have put that process on speed dial over the past 60 or so years. That said, it is difficult to know the exact rates of deforestation in many regions of the world, primarily because the majority of tropical forests reside in developing countries where detailed monitoring is often too expensive to fund. Nevertheless, in the early 1980s, the Food and Agriculture Organization of the United Nations (FAO) estimated that about 14 million hectares[4] of tropical forests (rainforest and other forest types, such as seasonally dry tropical forest) were being destroyed each year, an area the size of North Carolina. Of this, 5.4 million hectares (about the size of West Virginia) were cleared annually in South America alone, primarily from the Amazon Basin. These data are alarming because, until the late 1970s, deforestation in Brazil was considered only a minor problem with a limited, locally contained impact. However, from the mid-1980s, the situation changed dramatically (Figure 7.6). Approximately 40 million hectares of forest were cleared (that's more land than the entire state of California), accounting for nearly 14% of the Brazilian Amazon. This was deforestation at an unprecedented scale.

Deforestation rates have often proven controversial, particularly in the Brazilian Amazon. It is common to see headlines equating forest loss to large geographical areas, such as "The world lost 40 football fields of tropical trees every minute in 2017."[5] Expressing deforestation in this way gives the public a more tangible sense of the *scale* of the issue. According to the FAO, Brazil's average deforestation rate from 1978 to 1988 was 2.15 million hectares per year (21,500 km^2,

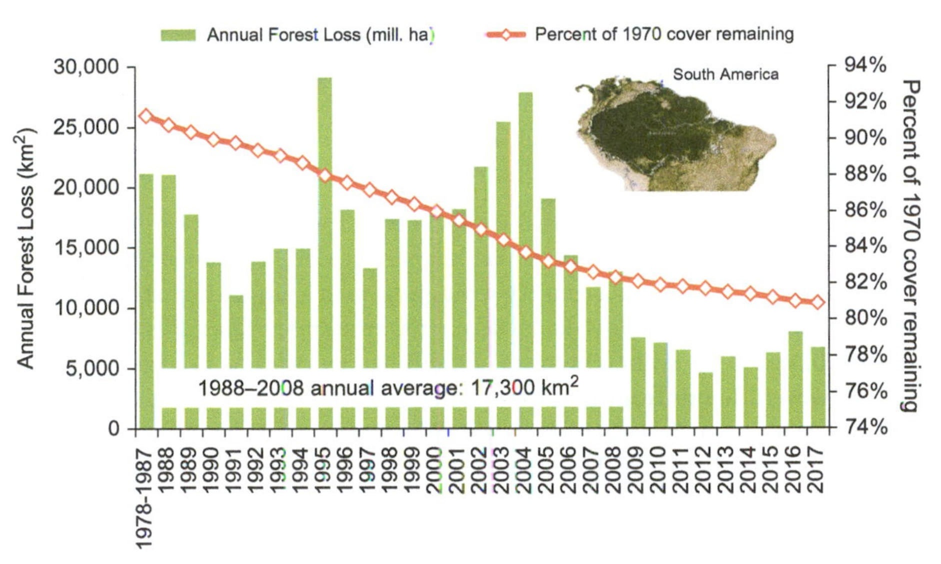

Figure 7.6 Deforestation rates for the Amazon (Source: Brazilian National Institute of Space Research (INPE) and the United Nations Food and Agriculture Organization.

[4] Forest loss is usually reported in square miles, square kilometers, or hectares. As noted in Chapter 2 (footnote 12), a hectare is 100 m by 100 m, or the size of two football fields.

[5] Source: e360.yale.edu, June 27, 2018.

see Figure 7.6). Rates decreased to an average of 1.5 million hectares per year (about 15,000 km^2) during the 1990s (except for the 1995 peak), leading the Environmental Minister, Jose Sarney Filho, to claim in 2000 that, "The tendency of an increase in deforestation has been controlled." However, rates then increased during the first half of the 2000s, peaking in 2004 at 2.8 million hectares (28,000 km^2). Since then, rates declined reaching a minimum of around 4,700 km^2 in 2012. The most recent figures from the Brazilian government and the FAO show Amazon deforestation down more than 75% from the 2004 peak (Figure 7.6). In 2017, the Amazon lost 660,000 hectares (6,600 km^2), the fourth lowest figure recorded since detailed monitoring of the forest began.[6] Falling commodity prices, increased enforcement efforts, and government conservation initiatives are all credited for the drop in rates throughout the 2000s. Furthermore, the Brazilian government have set a goal of reducing deforestation rates in the Amazon region by another 70% over the next 10 years. They established an Amazon fund, where foreign nations are being encouraged by Brazil to contribute financially to the conservation of the vast Amazon region. Norway, for example, committed to give $1billion over the next five years.

While it is important to recognize the efforts made by the federal and state governments, as well as Brazilian society in general, it is clear that deforestation remains an urgent issue. Despite the declines noted above, Brazil had the dubious honor of leading the world's countries in total tree cover loss in 2017, mostly due to fires set to clear land. This appears partly due to recent political instability and a government that has rolled back environmental protections, according to Global Forest Watch, but it is not just the tropical rainforest that is being decimated in Brazil.[7] The most critical situation is found in the *cerrado*, a vast tropical savanna ecoregion (see Figure 7.7 for location). The *cerrado* is the most biologically diverse savanna in the world and is being cleared for large-scale farming at rates that are twice those in the Amazon. Unlike the heralded rainforest it borders, the loss of the *cerrado* and its rich savanna so far has failed to attract much notice, despite the fact that more than 60% of the savannah's former 200 million hectares has disappeared under the plow, mostly within the past two decades.

Do these rates of deforestation in Brazil really tell the full story? The answer, in short, is no. One potential pitfall is that the Brazilian government has been criticized for under-reporting deforestation rates because it is bad press for a country struggling with a large international debt. Furthermore, these absolute rates of forest loss do not account for the area of forest *affected* by clearance. For example, by 1990, nearly 24 million hectares of the Brazilian Amazon had been cut down, about 7% in total. However, due to **fragmentation** (areas that are isolated after deforestation), plants and animals are cut off from the larger forest area. When this is accounted for, a total of 16.5% of the forest (an area nearly the size of Texas) had actually been affected by deforestation, more than twice the reported rate.[8] This is evident in Figure 7.7, two satellite images of deforestation in the Brazilian state of Rondônia, one of the most deforested parts of the Amazon. Typically, deforestation follows a fairly predictable pattern in this region. The first clearings that appear in the forest are in a **fishbone pattern**, arrayed along the edges of roads. Over time, the fishbones collapse into

[6] This loss in 2011 is equivalent of the size of Delaware.

[7] https://www.globalforestwatch.org/

[8] Source: Skole, D.L. and Tucker, C.J. (1993), Science, Vol. 260: p. 1905–1910.

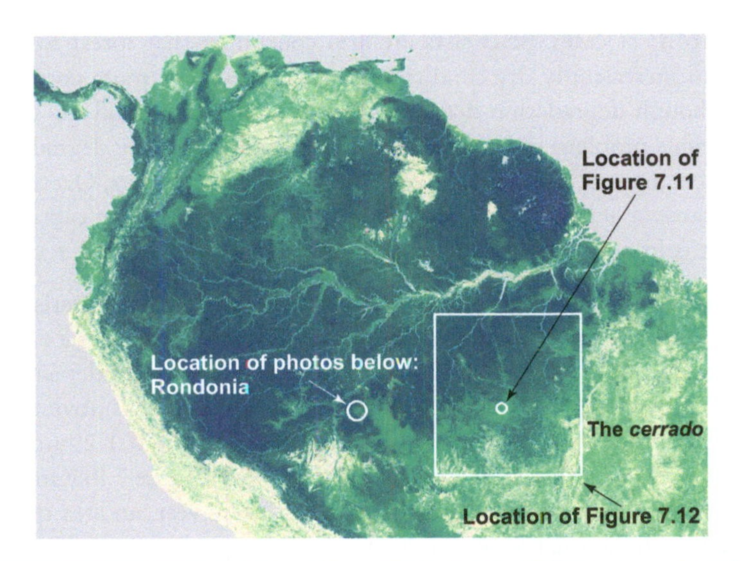

Figure 7.7 Deforestation in the remote northwest corner of Rondônia, Brazil pictured in two NASA Terra satellite images. Intact forest is deep green, while cleared areas are tan (bare ground) or light green (crops, pasture, or occasionally, second-growth forest). The two photographs span a period of ten years, 2000–2010. The top map shows the location of these images, along with the location of the images in Figures 7.11 and 7.12. Note that the tropical savannah *cerrado* is located south and east of the main body of the Amazon.

a mixture of forest remnants, cleared areas, and settlements. This pattern follows a common trajectory in the Amazon. Legal and illegal roads penetrate a remote part of the forest, and small-scale farmers migrate to the area. They claim land along the road and clear a portion for fast growing crops, generally about one to two hectares. Within a few years, heavy rains and erosion deplete the soil, and crop yields fall. Farmers then convert the degraded land to cattle pasture and clear more forest for crops. Eventually, having cleared much of their land, the small landholders sell it or abandon it to large cattle holders who consolidate the plots into large areas of pasture.

A report published in the prestigious scientific journal *Nature* questioned the official Brazilian government's figures of rainforest destruction. The report presented field surveys of wood mills and forest burning across the Amazon and shows that logging crews severely damaged 1.5 million hectares (15,000 km^2) of forest that had not been included in deforestation mapping programs. The work also found that fires burn additional large areas of standing forest, which was also not documented in most cases. Scientists are also concerned that forest loss could escalate in the Amazon due to increasingly dry conditions. Such impacts are more correctly termed **forest degradation**. Although degradation does not involve a change in land use (e.g., from forest to pasture), it is a serious problem in the tropics. Millions of hectares are degraded each year by the action of exploitative loggers, firewood collectors, and livestock herders. Overall, the report found that present estimates of annual deforestation for the Brazilian Amazon capture less than half of the forest area that is impoverished each year and even less during years of severe drought.

Globally, the news on deforestation shows some positive signs. Official statistics released by the FAO in its 2015 *Global Forest Resources Assessment*[9] suggests that, while the extent of the world's forests continues to decline as human populations continue to grow and demand for food and land increases, the *rate* of net forest loss has been cut by over 50%. That is obviously excellent news. However, we cannot forget that, since 1990, the world has lost about 1.29 million km^2 of forest, equivalent to an area covering three times the size of California (Figure 7.8). On average, we are still losing about 3.3 million hectares (33,000 km^2) of forest every year, an area the size of Maryland. A study published in the journal *Science* in 2013 provides arguably the most accurate quantification of global forest change to date using Earth observation satellite data at a spatial resolution of 30 m.[10] Several important findings have come out of this study. First, while Brazil's total rainforest loss over the past decade was confirmed, it also exhibited the largest decline in annual forest loss globally. This lends support to the Brazilian government's reports on reduction in rainforest clearing

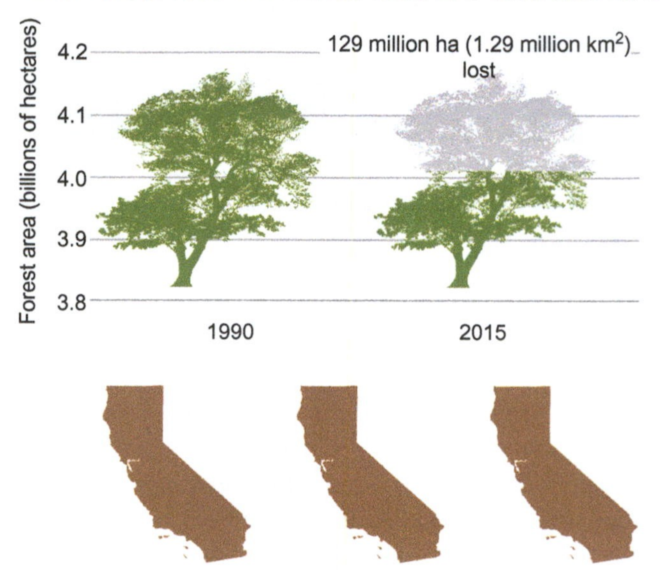

Figure 7.8 Total forest loss worldwide between 1990 and 2015. The total area of forest lost is equal to an area the size of three California's. The infographic is derived from data provided by the World Resources Institute (www. wri.org) and the Global Forest Assessment of the FAO. Images from Shutterstock.com

[9] http://www.fao.org/publications/e-book-collection/en/

[10] Source: Hansen, M.C. et al. (2013), *Science*, Vol. 342: p. 850–853.

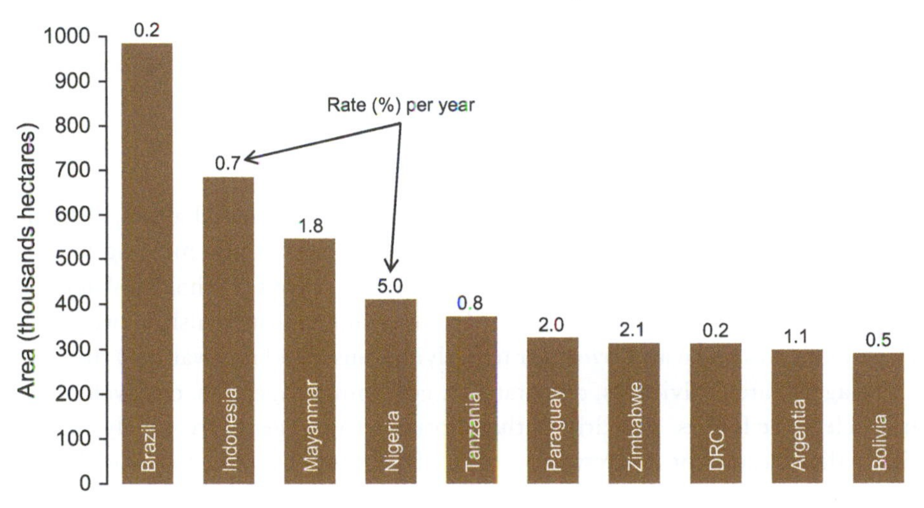

Figure 7.9 Top ten countries reporting the greatest annual net loss of forest area, 2010-2015 (Source: FAO Global Forest Assessment, 2015).

and is, again, very good news. However, increased annual loss of Eurasian, African, and other South American tropical forests have more than offset the slowing of Brazilian deforestation (Figure 7.9). Africa is facing large-scale deforestation with Nigeria leading the African nations in absolute forest loss (4,100 km^2 annually) due largely to logging, subsistence agriculture, and the collection of fuelwood. Even more telling is the percentage of a country's forests cleared over time (Figure 7.10).

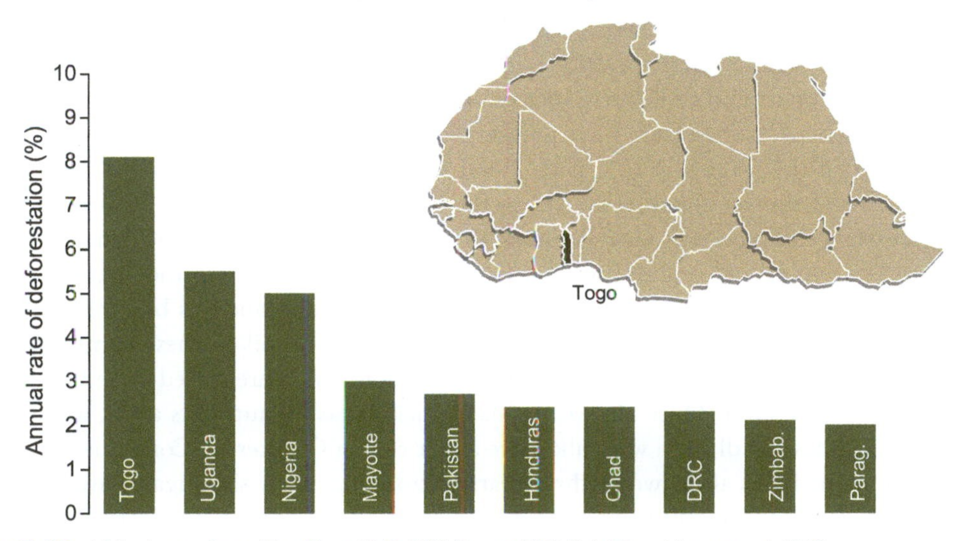

Figure 7.10 Highest percentage of forest loss, 2010–2015 (Source: FAO Global Forest Assessment, 2015).

By this metric, the tiny country of Togo in West Africa (just slightly smaller in area than West Virginia) fares worst in the world, having cleared 8% of its forest each year between 2010 and 2015, mostly due to slash-and-burn agriculture, fuelwood cutting, and illegal logging. Uganda and Nigeria rank second and third, respectively, highlighting the significance of deforestation in Africa.

AGENTS OF DEFORESTATION

People have been deforesting the Earth for thousands of years. As noted above, although tropical forests reside predominantly in developing countries, they are not just meeting local or national needs. Economic globalization means that the needs and wants of the whole global population are bearing down on these forests as well. Thus, it is important to distinguish between the *agents* (or direct causes) of deforestation and *drivers* (or underlying causes and motivations) that lead to deforestation. The agents are individuals, corporations, government agencies, or development projects that actually clear the forests. The drivers that motivate these agents are widely varied. Rarely is there a single direct cause for deforestation. Most often, multiple processes work simultaneously or sequentially to cause deforestation.

The single biggest direct cause of tropical deforestation is conversion to cropland and pasture. Agriculture occurs on two scales: large plantation (i.e., commercial) farming and small subsistence farming. In plantation farming, large areas of forest are destroyed to grow crops such as rubber, sugar, palm oil, soy, coffee, or tropical fruits (Figure 7.11). Plantations that produce only one species of tree or one type of food (known as **monoculture plantations**) on rainforest soil are referred to as cash crops because the intention of planting them is to make money quickly, with little concern about the environmental damage caused by deforestation. Much of the produce grown from these plantations are exported to rich, industrialized countries rather than feeding the local populace. Large cattle pastures also often replace rainforest to grow beef for the world market.

In all geographic areas, **subsistence farmers** rank high as important agents of deforestation. These are impoverished, landless people who follow roads into already damaged rainforest areas and establish small-scale farming operations. Unlike commercial farmers, subsistence farmers only grow enough to support and feed their families. They chop down small areas (typically less than two hectares) and burn the tree trunks in a process called **slash-and-burn agriculture** (also known as swidden agriculture, Figure 7.12). Their important crops are corn, beans, cassava (a starchy tuber), plantains (a shorter starchier cousin to the banana), and upland rice (depending on the region). Subsistence farmers also raise livestock to meet their daily needs. The damage from this type of agriculture is extensive: the FAO estimates that small farming families account for more than 60% of tropical forest loss. As populations have grown and land has become scarcer, this type of farming has become more intensive. However, the type of soil on these farms is generally not suited for sustainable farming. Nutrients in these tropical soils are cycled close to the surface within the vegetation layer. Once cleared, rainfall quickly washes nutrients away, and the soil's productivity declines rapidly (we will talk more about this in Chapter 9). Consequently, farmers must abandon their fields after two or three years of cropping. They shift again, usually moving further into the rainforest and burning more of it. The World Resources Institute notes that one of the primary forces pushing landless migrants into the forests is the inequitable distribution of

Figure 7.11 A NASA satellite captured this image of deforestation from Mato Grosso, an inland state of central Brazil, deep in the Amazon interior, in July 2006 (see Figure 7.7 for location). Widespread forest clearing, visible as rectangles of gray-beige, in this region of Brazil is primarily due to large agricultural clearings, such as for soy plantations. The area shown here is approximately 160 km^2. The Earth Observatory website contains high quality imagery from many different satellites with detailed descriptions and articles (see http://earthobservatory.nasa.gov/).

agricultural land. For example, in Brazil, approximately 42% of cultivated land is owned by a mere 1% of the population, and landless people make up half of Brazil's population.[11]

Subsistence farmers do not generally move into pristine areas of undisturbed rainforests. They follow roads made principally for logging operations. In the Amazon, nearly every road is unauthorized except for a few federal and state highways, including the east-west Trans-Amazonian Highway and the soybean corridor. Scientists estimate that there are more than 100,000 miles of roads in Brazil made illegally by loggers. The Brazilian government has certainly played its part, initially encouraging loggers and farmers to colonize the Amazon by cutting down **primary forest** to build the Trans-Amazonian Highway. This established a pattern, where colonists cut down forest, used the soil until it lost its productivity, and then abandoned it or developed it into pastures. As a result, tropical rainforests disappeared gradually from the roadside. Today, 80% of deforested land is within 30 miles of a road.[12]

---

[11] Source: www.wri.org

[12] Source: www.rainforestportal.org

Figure 7.12 This NASA satellite image reveals the "slash-and-burn" deforestation by which people clear farm and pastureland out of the rainforest. The name describes the process of cutting down the trees and setting fire to what is left, using the fire to return soil-fertilizing nutrients back to the soil from the vegetation. The scene shows a section of southern Pará state of Brazil, just north of the border with Mato Grosso state on July 29, 2004 (see Figure 7.7 for location). Right of image center, the Xingu River flows in from the south, but its obvious course disappears in a large clearing around the expanding airport town of Sao Felix. Left of center, a road cuts up through the forest, its ragged appearance the result of clearing that is occurring on either side. Fire activity is shown in red (Source: http://visiblearth.nasa.gov/)

Logging roads lead to an estimated 90% of the destruction caused by the slash-and-burn farmers. For this reason, the World Resources Institute and other groups actually rank commercial logging as the biggest cause of tropical deforestation. Commercial logging is exactly what it sounds like: cutting trees for sale as timber or pulp. Logging can occur selectively, where only the economically valuable species are cut, or by **clear cutting**, where all the trees are cut. Heavy machinery, such as bulldozers, road graders, and log skidders, are used to remove cut trees and build roads. This equipment ends up damaging the forests almost as much as the chainsaws damage individual trees (Figure 7.13).

Selective logging is frequently portrayed as an environmentally sensitive alternative to clear cutting. According to the timber trade, selective logging ensures that the forest re-grows naturally and in time will be ready again for safe logging practices. In most cases, this simply does not add

Figure 7.13 Deforestation in the Amazon due to clearcutting. © iStockphoto/Luoman.

up because removing only a few logs damages and/or destroys large areas of rainforest. Unselected trees are felled to get to those of interest, and soil is compacted by heavy machinery that leads to runoff and erosion in heavy rain. The felling of one selected tree tears down other vegetation such as climbers, vines, and epiphytes (air plants that live on the tree). Large holes are also left in the canopy causing damage to the lower portions of the forest. Overall, the system is so damaged that complete regeneration would take hundreds of years.

There is considerable variation from region to region and country to country as to which groups are the most important agents of deforestation. In many tropical countries, the majority of deforestation results from the actions of poor subsistence cultivators. However, in Brazil, only about one-third of recent deforestation can be linked to these shifted cultivators. Historically, a large portion of deforestation in Brazil can be attributed to land clearing for pastureland by commercial interests. Overall, cattle ranching is the leading cause of deforestation in the Brazilian Amazon, and this has been the case since at least the 1970s. Currently, cattle ranching is thought to account for 65–70% of deforestation in the Amazon.[13] In contrast, in Southeast Asia, commercial farming, logging, and plantations play a more significant role. For example, in Borneo, the conversion of tropical forest to commercial palm tree plantations to produce palm oil for export is the major

[13] Source: www.mongabay.com

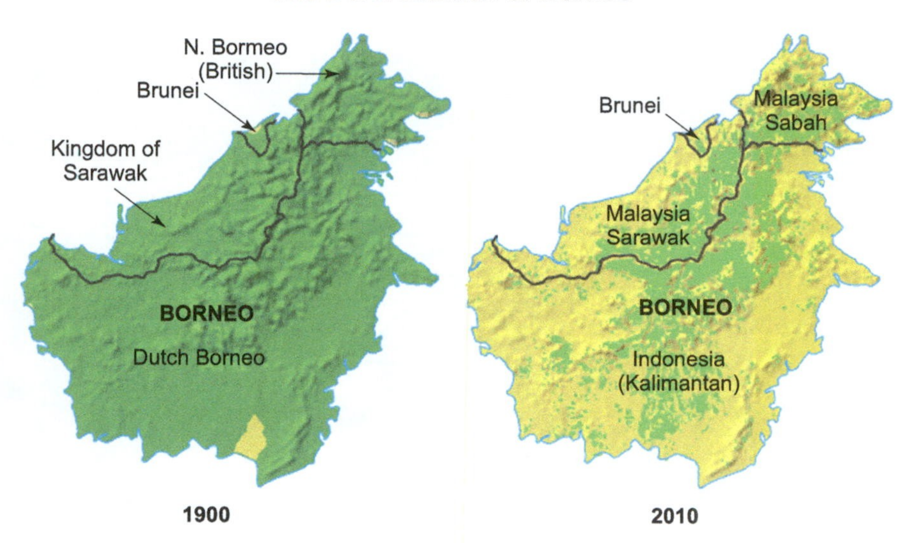

Figure 7.14 Figure 7.14. Extent of deforestation in Borneo from 1990 to 2010. According to World Wildlife Fund estimates, more than 170,000 km^2 of forest has been lost to oil palm plantations (Reproduced by permission. © 2008 Scientific American, inc. All rights reserved).

cause of deforestation, and more than two-thirds of the islands' forests could be destroyed by 2020 (Figure 7.14). The situation in Africa is a complex mixture of overgrazing in the dry forest zones and slash-and-burn farming along with high-grade logging in the moist forests of West and Central Africa.

DRIVERS OF DEFORESTATION

The motivations behind the agents of deforestation are not easy to define. Although poverty and overpopulation are often cited as the underlying causes of tropical deforestation, analyses of multiple scientific studies indicate that explanation is an oversimplification. Poverty does drive people to migrate to forest frontiers, where they engage in slash-and-burn forest clearing for subsistence. International agencies, such as the FAO and intergovernmental bodies, believe that they can solve the problem by encouraging economic development and trying to reduce population growth. The World Rainforest Movement[14] and many other non-governmental organizations (NGOs), believe that unrestrained development and the excessive consumption habits of rich, industrialized countries are directly responsible for the most forest loss, but rarely does one factor alone bear the sole responsibility for tropical deforestation.

[14] http://www.wrm.org.uy/

The connection between population growth and deforestation is inherently unclear. In some African and Asian countries, overpopulation may be an important cause, but generally, countries with the greatest amount of tropical rainforest are those with the lowest human population densities. Research indicates that deforestation most often involves non-demographic mechanisms resulting from credit and capital market failures, securing property rights, uneven land distribution, consumption patterns in developed countries, and profit-driven multinational companies.

In particular, the inequitable distribution of land ownership creates the most pressure on tropical forests. As noted earlier, the government and a very small percentage of people own the majority of the land in many developing countries. This has resulted in the expulsion of poor peasants to the forest frontier areas, leading to the slash-and-burn subsistence farming. Most tropical countries are very poor by U.S. standards, and farming is a basic way of life for a large part of the population. For example, the mean GDP per capita in Brazil in 2016 was $15,838 compared to the mean in the U.S. of $54,630. In Bolivia, which holds part of the Amazon rainforest, the mean GDP per capita was $6,629. However, in Brazil, there is an enormous gap between the mean GDP of $15,838 and what subsistence farmers actually earn. The reality is that 5% of the population owns 85% of the wealth and there are a quarter of Brazilians living on less than $1 a day.[15] Government land policies do encourage farmers to settle on forested land with each squatter acquiring the right (known as a usufruct right) to continue using a piece of land by living on a plot of unclaimed public land (no matter how marginal the land) and "using" it for at least one year and a day. This use, as discussed earlier, is unsustainable in the long run.

Government policies encouraging economic development, such as road and railway expansion projects, have caused significant, unintentional deforestation in many parts of the world, including in the Amazon and Central America. Agricultural subsidies and tax breaks as well as timber concessions have encouraged forest clearing as well. Global economic factors such as a country's foreign debt and expanding global markets for rainforest timber and pulpwood can also encourage deforestation. A highly competitive global economy drives the need for money in economically-challenged tropical countries. At the national level, governments frequently sell logging concessions (i.e., permits to clear forest land) to raise money for projects, pay international debt, or develop industry. In the 1970s and 1980s, the governments of poorer developing countries borrowed vast sums of money from development agencies in industrialized countries in order to improve their own economies. For example, Brazil had an international debt of $550 billion as of 31 December 2017, on which it must make payments each year.[16] Most countries in similar situations to Brazil are still battling to make repayments partially resulting from escalating interest rates.

Since 1987, debt repayments from developing countries have exceeded the amount of aid money received by those countries. What this means is that poor countries are often forced to exploit their natural resources, including their forests, mainly to earn foreign exchange for servicing their debts. NGOs in developing countries have pointed out for several years that without a solution

[15] Source: UK Department for International Development (DFID).

[16] Source: https://www.indexmundi.com/brazil/debt_external.html

to the debt crisis, no chance really exists to stop impoverishment and environmental destruction. In some Southeast Asian countries, the construction of roads for logging operations was funded by Japanese aid. Later, the forests were exploited by Japanese timber companies. The timber companies carried home the profit, and in the end, the Southeast Asian countries owed Japan money for the road construction. It seems that deforestation is the inevitable result of current social and economic policies being carried out in the name of development. It is this push for development that gives rise to commercial logging, cash crops, cattle ranching, colonization schemes, and the dispossession of landless people and indigenous people.

CASE STUDY: DEFORESTATION AND FOREST RECOVERY IN COSTA RICA

While deforestation continues at a high rate in many countries, afforestation and natural expansion of forests in some countries and regions have reduced the net loss of forest area significantly at the global level. One country that has experienced both deforestation and afforestation is Costa Rica, one of my favorite places in the world!

Located north of Panama and south of Nicaragua and measuring only 185 miles across at its widest point, Costa Rica is one of the smallest countries in the Americas, but it boasts one of the most diverse selection of flora and fauna found anywhere on the planet. In fact, 5% of all known species on Earth occur here, even though the country comprises only 0.01% of the global landmass (Figure 7.15). More than twice the number of bird species exists in Costa Rica than in all of the United States! This biodiversity is in part due to its position as a transition zone between South and North America, where migrating animals and plants meet. Costa Rica's latitude (10ºN) contributes to its steady temperatures year-round, and abundant precipitation creates hospitable conditions for many forms of life. There is also a complex vertical distribution of microclimates created by differences in altitude: a chain of volcanoes and mountains divide the country like a backbone, and topography ranges from the bleak, treeless *paramo*, 12,000 ft above sea level, to rainforests on the coasts just 50 miles away.

Costa Rica's rich array of flora and fauna has been under threat for decades from widespread habitat destruction. The primary cause is deforestation.

Before the 1950s, Costa Rica's forests slowly declined as the country's agricultural society began to emerge, with large coffee-producing landowners as the dominant agent. The country then experienced a staggering four-fold increase in total population, from less than 800,000 to more than 3 million people, in less than two generations following World War II (remember that Great Acceleration we talked about in Chapter 1?). In the same period, about 50% of the forest was cleared. During the 1970s and 1980s, Costa Rica had some of the highest rates of deforestation in the world, so much so that, by 1983, only 26% of the country retained forest cover. The deforestation rate had risen to 50,000 hectares (or 1%) per year (Figure 7.16).

Multiple factors contributed to Costa Rica's forest loss. Initially, the government favored laws that allowed people to move into the forest to clear areas for cattle raising. As strange as it may sound,

Figure 7.15 Costa Rica's forests house enormous biodiversity, about 5% of all Earth's identified species (Photos: Mike Slattery).

land titling laws were passed that *rewarded* deforestation. As the demand for beef in the United States increased, dollar-strapped nations such as Costa Rica chose clear cutting and deforestation as a formula to earn money. The government established specialized exchange rates and credit instruments in order to help cattle ranchers expand beef exports or attract new investors. Beef exports within Costa Rica increased nearly 500% from the 1960s to the early 1980s. During this time, cattle pasture land increased from 27% of the land mass to 54%. Added to this was the increased demand for other agricultural products such as bananas and palms.

Rampant and unchecked logging also played a major role in forest loss. Under Costa Rican law, clear cutting was actually considered land improvement. The standing forest did not have any value for the land owner, so they began selling land cheaply for wood production. Loggers paid for valuable species only (i.e., selective logging), and once those trees were removed, the remaining forest was clear cut and burned by the land owner to establish pastures or plots for crops. All of these practices ensured that the nation's forest resources were transformed into cash profits.

Then something amazing started to happen. Forest protection laws began to be passed in the early 1970s, and protected areas increased dramatically during a 15-year period, covering almost 1 million hectares by 1986. By 1989, the annual deforestation rate had dropped to 22,000 hectares per year. The figure dropped even lower to 4,000 hectares per year by 1994 and, in 1998, the deforestation rate had dropped to zero.

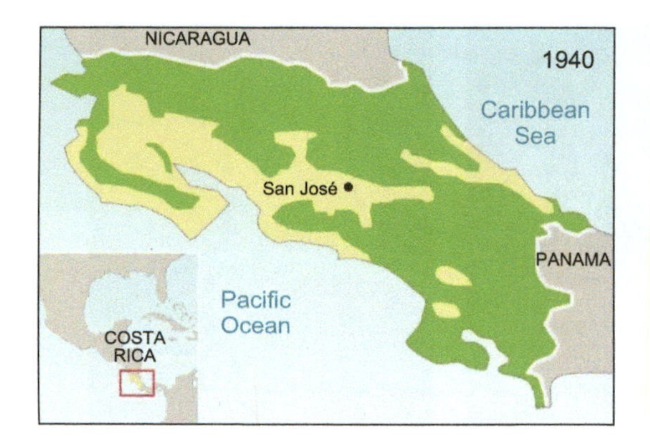

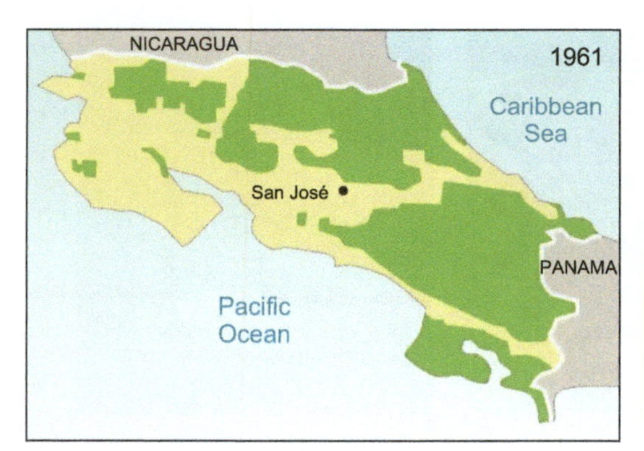

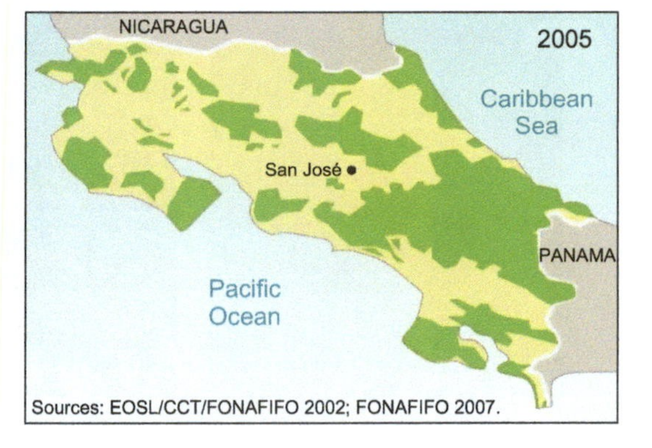

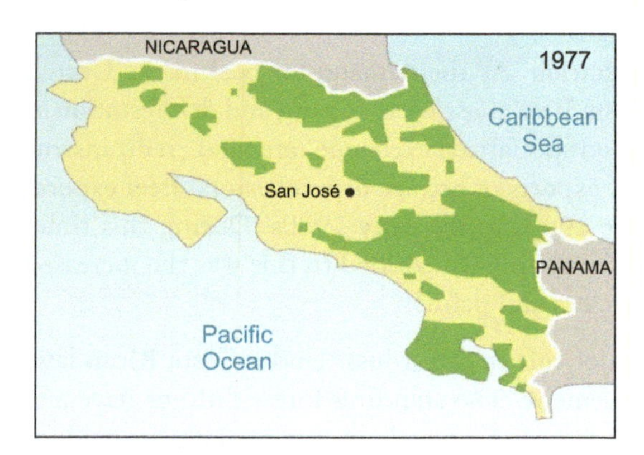

Figure 7.16 Change in forest cover in Costa Rica, 1940–2005. Source: EOSL/CCT/FONAFIFO 2002; FONAFIFO 2007 (http://maps.grida.no/go/graphic/change-forest-cover-costa-rica). Cartographer/designer/author credit *Philippe Rekacewics assisted by Cecile Marin, Agnes Stienne, Guillo Frigieri, Riccardo Pravettoni, Laura Margueritte and Marion Lecoquierre*

Laws that favored deforestation were followed by a series of laws intended to halt deforestation. Several important steps were put in place, including an income tax deduction for locals of $200/ reforested hectare.[17] The current program is the Payment for Environmental Services (PES) initiative, which gives economic retribution to forest owners or land owners that want to establish forest plantations, due to environmental services they provide to society. These services include carbon sequestration, biodiversity protection, water regulation, and landscape beauty. The program makes direct cash transfers to private landowners of $64 per hectare preserved through five-year contracts. The aim is to protect primary forest and allow secondary forest to flourish, and it has worked! Today, forest cover has increased to 54% (double 1983 levels), and the government has set the ambitious goal of further increasing this figure to 70% by 2021.

While the PES scheme has been a success, there are a number of obstacles to the program. Because all funds for the PES program are funneled through the central government and administered by the Ministry of Finance, beaurocracy is causing financial bottlenecks, with claims that money is being diverted away from the PES program to other infrastructural developments, such as roads. As with many developing countries, monitoring is difficult because the program is understaffed and the workers themselves are overloaded. Nevertheless, the PES initiative has been quite successful. As part of a much broader conservation initiative, that includes foreign investment in private reserves, about 25% of Costa Rica's land is devoted to national parks or reserves and is protected from further development.

An emerging threat to Costa Rica's landscape is the rapid expansion of pineapple farms. By 2005, exports of fresh pineapple surpassed coffee to become the country's second-largest agricultural export, after bananas. By 2008, pineapple plantations covered more than 40,000 hectares, an increase in 185% from 2004. Pineapple is a particularly damaging crop: it is highly mechanized, requires extensive application of herbicides and insecticides, and has a superficial root zone of between 10 and 15 inches which makes the soils susceptible to erosion. Large areas of the landscape remain exposed to the intense tropical rainfall during the growing season, increasing soil erosion and clogging rivers with sediment and pollutants. Although these issues relate more to soil degradation and are covered in greater detail in Chapter 9, I mention them here because the revenue associated with pineapple has begun to lure landowners to abandon forest protection and the PES program for more lucrative land uses such as pineapple.

CONCLUDING THOUGHTS

The FAO has been coordinating global forest resources assessments since 1946. Much of the data in this chapter comes from the 2015 assessment, supplemented by analysis of over 650,000 satellite images.[18] It is clear from both reports that progress toward sustainable forest management can be described as mixed, at best. While many trends remain alarming, particularly in Southeast Asia and Africa, there have also been many positive developments over the past two decades.

[17] The average monthly household income in Costa Rica is about U.S. $965.

[18] Hansen, M.C. et al. (2013), *Science*, Vol. 342: p. 850–853.

Overall, rates of deforestation are showing signs of slowing down at the global level and some countries have made significant progress to reduce their rates of forest loss, especially in the past decade. Most of the net loss of forest still happens in the tropics while most of the net gain takes place in the temperate and boreal zones and in some emerging economies, such as India. Brazil, which had the highest net loss of forest in the 1990s, has significantly reduced their *rate* of loss, but vast areas of the Amazon continue to be cleared. Indonesia and Malaysia appear to be particularly vulnerable to further forest loss to due expansion of palm oil, and several countires in Africa are losing their forest at rates in excess of 2% per year.

Overall, however, the area of forest designated for conservation of biological diversity and the protection of soil and water has increased globally. The area of planted forest has also increased and, although only accounting for 7% of the total forest area, planted forests supply an increasing share of the demand for wood. A large number of policies and laws have been enacted or updated and countries like Costa Rica, among many others, have begun to make major gains in forested cover. This is all very good news, but the loss of the equivalent of 60,000 football fields of forest every day is still enormous.

As we have discussed, the causes of deforestation are many: illegal logging, an expansion of agriculture, cattle ranching, overconsumption in industrialized countries, and an inequity of land distribution, where poor farmers are forced to clear and work lands on steep hillsides and other marginal areas. The consequences are potentially devastating: loss of genetic reserves of incalculable value, soil erosion, flooding, sedimentation in canals and rivers, water pollution, and loss of natural beauty. As we discussed in Chapter 2, population is very likely to increase substantially over the next several decades in these tropical regions, and it is not hard to imagine that this population boom will increasingly threaten land still covered with forest.

What can we do to save the world's remaining forests? First, there is a pressing need to preserve intact sections of tropical forest, as much as we can and as quickly as we can. Many people and conservation organizations are working toward this end.[19] Equally, there is an urgent need to address the economic needs of the lesser-developed nations in which almost all of the tropical forests reside. While broad-scale commercial and conservation strategies need to be developed, these must consider the economic and environmental constraints of the particular country. As an example, Costa Rica has one of the most enlightened and dedicated approaches to conservation in the world, and the country has made an impressive effort to preserve its natural resources. Yet are they really on a level playing field when it comes to global (and fair) trade? I believe that it is incumbent upon the world's wealthy and developed nations to help create a more stable global economic climate so that the pressures of debt and poor terms of trade will not cause ecology and economy to collide at a local scale, in this context, the world's great tropical forests.

[19] For example, see www.nature.org

8

Biodiversity: The Sixth Mass Extinction

INTRODUCTION

I remember my first walk through the rainforests of Costa Rica. I was in the Monteverde Cloud Forest Reserve,[1] a 10,500-hectare reserve of immense beauty. I literally could not see the forest for the trees. There were plants growing on trees (called epiphytes, or "air plants") and trees battling trees for every inch of canopy space and sunlight (Figure 8.1). The constant sound of bird and animal calls was both deafening and soothing, a symbiotic symphony of sorts. I remember thinking, "Why would anyone want to destroy this? How could anyone not see the value in protecting and preserving this truly wondrous natural resource?" Since that walk, I have come to understand that the answer to this seemingly simple question is quite complex.

The most diverse ecosystems on Earth are the tropical forests, most of which are in developing countries. As outlined in Chapter 7, many of these countries are burdened by poverty while at the same time trying to develop, making it difficult to finance forest management programs. Can we really expect developing countries to forfeit income from economic activities that destruct forests so that more-developed countries benefit from forest services, such as carbon storage, water filtration, biodiversity preservation, and climate regulation? The Coalition for Rainforest Nations believe that the answer

Figure 8.1 The Monteverde Cloud Forest, Costa Rica (Photo: Mike Slattery).

[1] http://reservamonteverde.com/

is *no.*[2] This ambitious project involves 52 countries collaborating to reconcile forest stewardship and species preservation with economic development. "We are simultaneously struggling to defeat poverty while challenged with responsibility over a majority of the world's biodiversity," reads a statement on the website of the organization. They are partnering with industrialized nations, such as the United States, to support fair trade and improved market access for developing countries. Their ultimate goal is to create a more stable, equitable economic climate by reducing debt pressure and poor terms of trade. This lessens the chance of conflict between the economy and ecology. In this chapter, we explore biological diversity, or **biodiversity**, a term used to describe the variety of life on Earth and the natural distribution and patterns of organisms. The biodiversity we see today is the result of billions of years of evolution, shaped by natural processes, and, increasingly, by the influence of humans. However, even with our influence, we must remember that as the species *homo sapiens*, we only compose one part of the intricate web of life, and we are fully dependent on that web to help sustain our species.

HOW MANY SPECIES EXIST?

Before we can truly measure biodiversity in any habitat around the world, we need to ask a fundamental question: how many species are there on Earth? For decades, scientists have been asking and trying to answer this question. And the simple answer is, we do not really know. Scientists actually have a better understanding of how many stars exist in the galaxy than how many species live on Earth! Guesses, estimates, and seemingly sophisticated calculations of global species diversity vary from 2 to 100 million species, with a best estimate of somewhere between 2 and 8 million. A 2011 study estimated approximately 8.7 million species globally.[3] However, it is important to stress the word *estimate*, for as we discuss below, continuing studies reveal how research is only scratching the surface of the possible numbers of species living on Earth.

According to the *Catalog of Life*,[4] the most comprehensive and authoritative global index of species, about 1.9 million species have now been catalogued and named. Groups of organisms, such as flowering plants, vertebrate animals, and butterflies, are relatively well known (about 17,500 species of butterfly alone have been identified, Figure 8.2). Still, scientists are constantly discovering new species. For example, researchers at the California Academy of Sciences, along with several dozen international collaborators, added 85 new plant and animal species to Earth's tree of life in 2017, including 16 flowering plants, three scorpions, 10 sharks, 22 fish, a lizard, an elephant-shrew, and a slew of species with unique names, including 'bat-wing' sea slug named after Dumbo the Flying Elephant![5] Each year, the SUNY College of Environmental Science and Forestry releases a list of the top ten new animal discoveries, and this year's is a great one, including a beetle that lives exclusively among one species of army ant and a species of bacterium that colonized a newly-formed underwater volcanic dome off the Canary Islands.[6] In total, between 15,000 and 18,000 new species are identified each year, with about half of those being insects.

[2] http://www.rainforestcoalition.org
[3] More, C. et al. (2011), *PLoS Biology*, 9: e1001127, 1–8.
[4] http://catalogueoflife.org/
[5] Source: http://earthsky.org/earth/85-new-species
[6] Source: http://www.esf.edu/top10/

Figure 8.2 According to the World Butterfly Index, 17,457 species have been identified with estimates that another 1,500 await discovery. We will never know the true total because many will have become extinct even before they are discovered. If moths are included, the total increases tenfold to 174,240 currently known species (Photo: iStockphoto.com).

There are groups of organisms that are barely known at all. The group known as fungi (mushrooms), which makes up an entire **taxonomic kingdom**, is very poorly known. It is speculated that as little as 5% of fungal species are known at present. The same goes for organisms such as nematodes (roundworms) and mites. Edward O. Wilson, the renowned Harvard biologist I referred to earlier in the book, calls such creatures "the black hole of taxonomy."

Many species live in areas that are not often studied. Scientists believe that there could be as many as a million undiscovered species just on the ocean floor. Other parts of the world are simply difficult to get to, such as the desolate Kerguelen Islands in the southern Indian Ocean (you can only get there with a six-day boat ride from an island off the coast of Madagascar). Political instability also precludes researchers from venturing into certain areas. In the next few years, for example, scientists expect to see a whole batch of new species emerging from Cuba, where U.S.-based researchers have long been barred from entering the country. With the loosening of travel restrictions, a new crop of scientists are lining up to visit. Other parts of the world are simply a no-go for most researchers, such as Northern Mali, Afghanistan, and the Eastern Congo.

We have only accumulated a fraction of the scientific knowledge needed to fully comprehend the vast biological diversity that is characteristic of our planet. Knowing the total number of species in the world, or at least having a good approximation, is important. Without baseline knowledge about how many species are out there, it becomes difficult to know exactly what we are managing and estimate, for example, how many species are being lost due to human activity. With current rates of rainforest destruction, we are most likely losing countless numbers of species that we never even knew existed.

BIODIVERSITY PRINCIPLES

Even with the uncertainty of not knowing how many species exist, ecologists are compelled to describe and compare the biodiversity within different types of habitats. Obviously, many more species exist within the rainforests than in more simple habitats like meadows or small woodlands, but it is important to draw out patterns and concepts that can give us an idea of the extent of the biodiversity. This section presents concepts that help us quantify biodiversity and make it easier to compare different habitats to each other.

The first and most fundamental concept of species diversity is called **species richness.** This involves simply counting or estimating the number of species that exist naturally within a habitat. The word ***naturally*** infers that we should include only resident species, not accidental (i.e., non-native) species or temporary-immigrant species (a difficult task!). The one problem with species richness is that it only counts the numbers of species and does not address what species are common throughout the habitat and which ones are rare. For example, take a community with two species divided into two extreme ways, as follows:

	Community 1	Community 2
Species A	99	50
Species B	1	50

In Community 1, species B would be considered rare (only 1%), while in Community 2, species B is as common as Species A. This difference in relative abundance draws out another important concept in defining biodiversity known as **equitability** or evenness. Species richness and evenness can then be combined into a single index of species diversity, referred to as heterogeneity. Heterogeneity will be higher within a habitat when there are more species and when the species have equal abundance.

A more quantitative measure of biodiversity is known as **Simpson's Diversity Index. Simpson's index** (D) measures the probability that two individuals randomly selected from an area will belong to the same species. To calculate this, we use the following equation:

$$D = \sum \left(\frac{n}{N} \right)^2 \tag{8.1}$$

where n is the number of individuals of a particular species and N the total number of individuals of all species present. For example, let us assume that we have five species of plants in a small area of grassland:

Species	Number (n)
Big blue stem	3
Buffalo grass	11
Milkweed	1
Purple coneflower	1
Stinging nettle	4
Total (N)	20

Putting the figures into the formula for Simpson's index:

$$D = \sum\left(\frac{3}{20}\right)^2 + \left(\frac{11}{20}\right)^2 + \left(\frac{1}{20}\right)^2 + \left(\frac{1}{20}\right)^2 + \left(\frac{4}{20}\right)^2 \tag{8.2}$$

$$D = 0.35$$

This D-value means that within this grassland there is a 35% chance of randomly picking two samples that belong to the same species. The value of D can range between 0 and 1. A value of 0 represents infinite diversity (i.e., a 0% chance that you will randomly select two individuals from the same species), and a value of 1 represents no diversity (or a 100% chance that you will randomly pick the same species every time). Simply stated, the closer the value of D is to 1, the lower the diversity.

At first glance, this may not make a lot of sense because, typically, when you have a higher number, you like to think that you have more of something. Therefore, scientists subtract the diversity index, D, from 1, or:

Simpson's index of diversity

$$(D) = 1 - D \tag{8.3}$$

The range of Simpson's index of diversity is still between 0 and 1, but now the index represents the probability that two individuals randomly selected from a sample will belong to *different* species. This also means that the greater the D value, the higher the diversity. From the grassland example above, $D = 0.65$, which means that there is a 65% chance that two, randomly selected individuals will belong to different species.

So far, we have been referring to biodiversity in terms of **species diversity**. However, biodiversity can describe other scales of diversity. For example, **genetic diversity** includes looking at the genetic differences within an individual species, such as between breeds of livestock. Chromosomes, genes, and DNA—the building blocks of life—determine the uniqueness of each individual and each species. **Ecosystem diversity** is another type of biodiversity determined by the species composition, physical structure (i.e., the physical location or type of environment in which an organism or biological population occurs), and processes within an ecosystem.

GLOBAL DIVERSITY AND HOTSPOTS

The tropics are the world's mega-diverse zones (Figure 8.3).[7] Many areas in the tropics have abundant rainfall and warm temperatures year-round so that ecosystems are highly productive. The year-round dependability of food, moisture, and warmth supports a great exuberance of life and allows a high degree of specialization in the physical shape and behavior of species. The general decrease in diversity from the tropics toward the poles is known as the **latitudinal diversity gradient** (LDG) and is one of the most widely recognized patterns in ecology. However, there are also zones outside the tropics

[7] The decrease in species richness or biodiversity that occurs from the tropics to the poles, often referred to as the latitudinal diversity gradient (LDG), is one of the most widely recognized patterns in ecology.

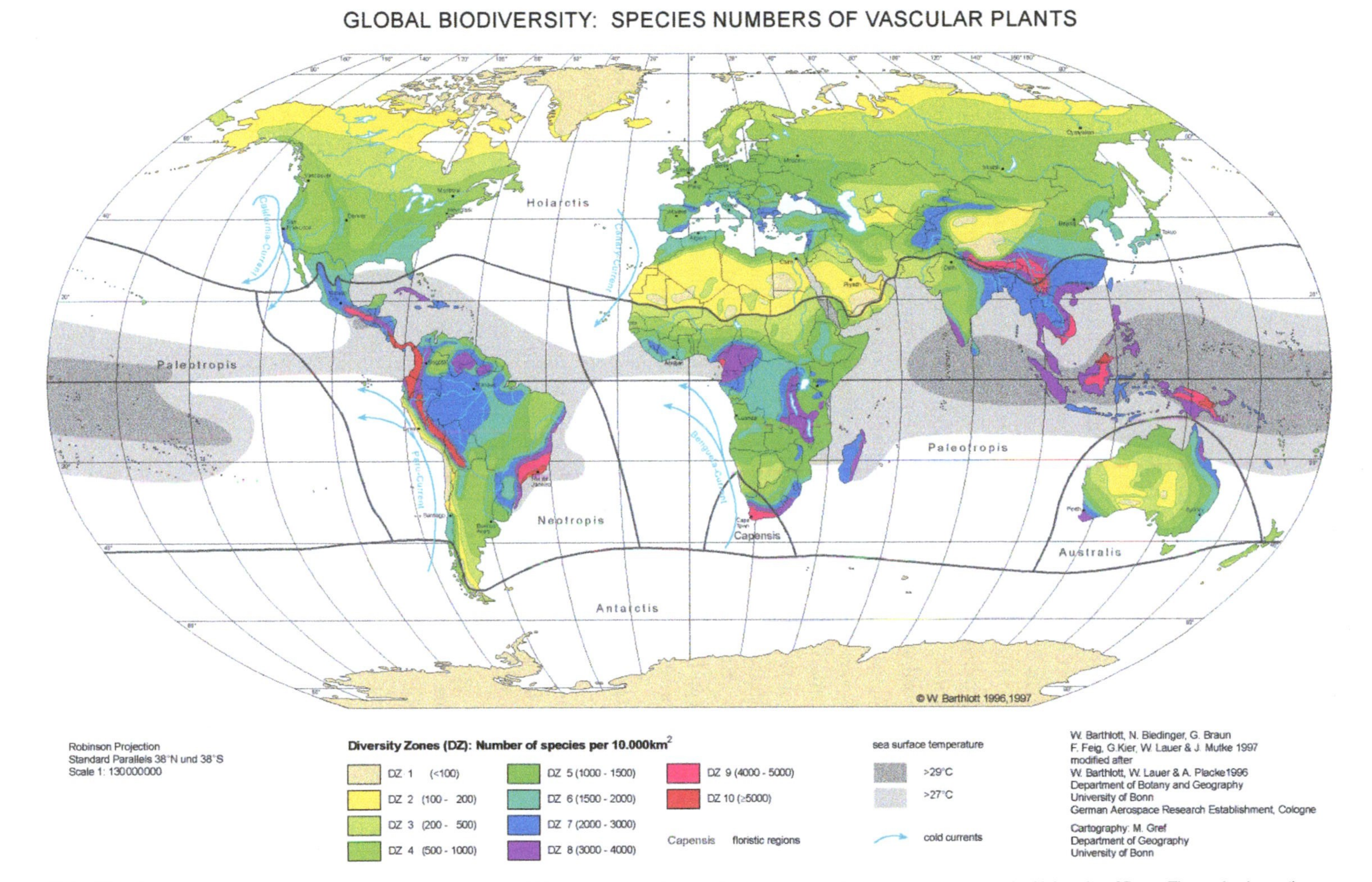

Figure 8.3 World map of species richness of vascular plants after a study by Professor Dr. Wilhelm Barthlott and his research team at the University of Bonn. The scale shows the number of species per 10,000 km^2 (Source: http://www.nees.uni-bonn.de/index_en.html and published in Barthlott, W. et al. (2005), *Nova Acta Leopoldina*, Vol. 92: p. 61–83).

where plant diversity is highly concentrated, for example mountainous areas such as the Andes in South America, where elevation changes allow for different climate zones and hence greater diversity of habitat. This is known as the elevational diversity gradient (EDG). These areas are increasingly under direct threat from human activity.

In 1988, British ecologist Norman Myers introduced a biodiversity concept called a **hotspot**, to address the dilemma that conservationists face, namely, what are the most important places in the world to conserve in terms of biodiversity? To qualify as a hotspot, a region must contain a large number of species of plants as **endemics** (i.e., the degree to which species are found only in a given place), and it has to have lost at least 70% of its original habitat. Currently, there are 35 hotspots covering slightly more than 2% of the Earth's land area (Figure 8.4). I encourage you to visit the web site for Conservational International[8] and the Critical Ecosystem Partnership Fund[9] where you will find a wealth of information on each important region. For example, Costa Rica is part of the Mesoamerica Hotspot, whose ecosystems are a complex mosaic of dry forests, montane forests, and rainforests, as well as coastal swamps and mangrove forests along the Pacific coast from Mexico to Panama. Forests in the Mesoamerica hotspot are the third largest among the world's hotspots, and their spectacular endemic species include quetzals, howler monkeys, and over 17,000 plant species (see Figure 7.15 and Figure 8.5). The region is also a corridor for many neotropical migrant bird species. The hotspot's montane forests are also important for amphibians of which many endemic species are in dramatic decline due to an interaction between habitat loss, fungal disease, and climate change. Collectively, the world's hotspots hold at least 150,000 endemic plant species (about 50% of the world's total) and about 12,000 endemic terrestrial vertebrate species (about 42% of the world's total), meaning once the hotspot is destroyed, all of these species will also become extinct.

Obviously, the hotspots shown in Figure 8.4 contain high biodiversity, but which of those are *the* most important to save? If we think about this in economic terms, in what areas would a given dollar amount slow the extinction rate the most? We also need to decide which species we should consider saving. Intuitively, we want to conserve the most threatened species first. Ironically, that is not always what happens. Humans inevitably tend to focus on charismatic vertebrates, such as polar bears, gorillas, elephants, and rhinos (see Box 8.1 later in this chapter). To a certain degree, this is practical, because these are the species for which we currently have the best data. It also somehow "feels right" to preserve these glamorous species first. However, some of the most **endangered species** are very small species that are critical in terms of ecosystem functioning. These are referred to as **keystone species.**[10] The more than 1 million identified invertebrate species have a far greater impact on the planet than do nonhuman vertebrates, partly because they have a much larger total body mass **(biomass)**. In 1987, Edward O. Wilson,[11] the ecologist whom I referred to earlier, described insects as "the little things that run the world." Yet, in determining the International Union for Conservation

[8] http://www.conservation.org/how/pages/hotspots.aspx

[9] http://www.cepf.net/resources/hotspots/Pages/default.aspx

[10] There are species whose impacts are disproportionately large relative to their abundance, termed **keystone species**. Keystone species, because of their proportionately large influence on species diversity and community structure, have become a popular target for conservation efforts. The reasoning is sound: protect one, key species, and in doing so stabilize an entire community.

[11] http://eowilsonfoundation.org/

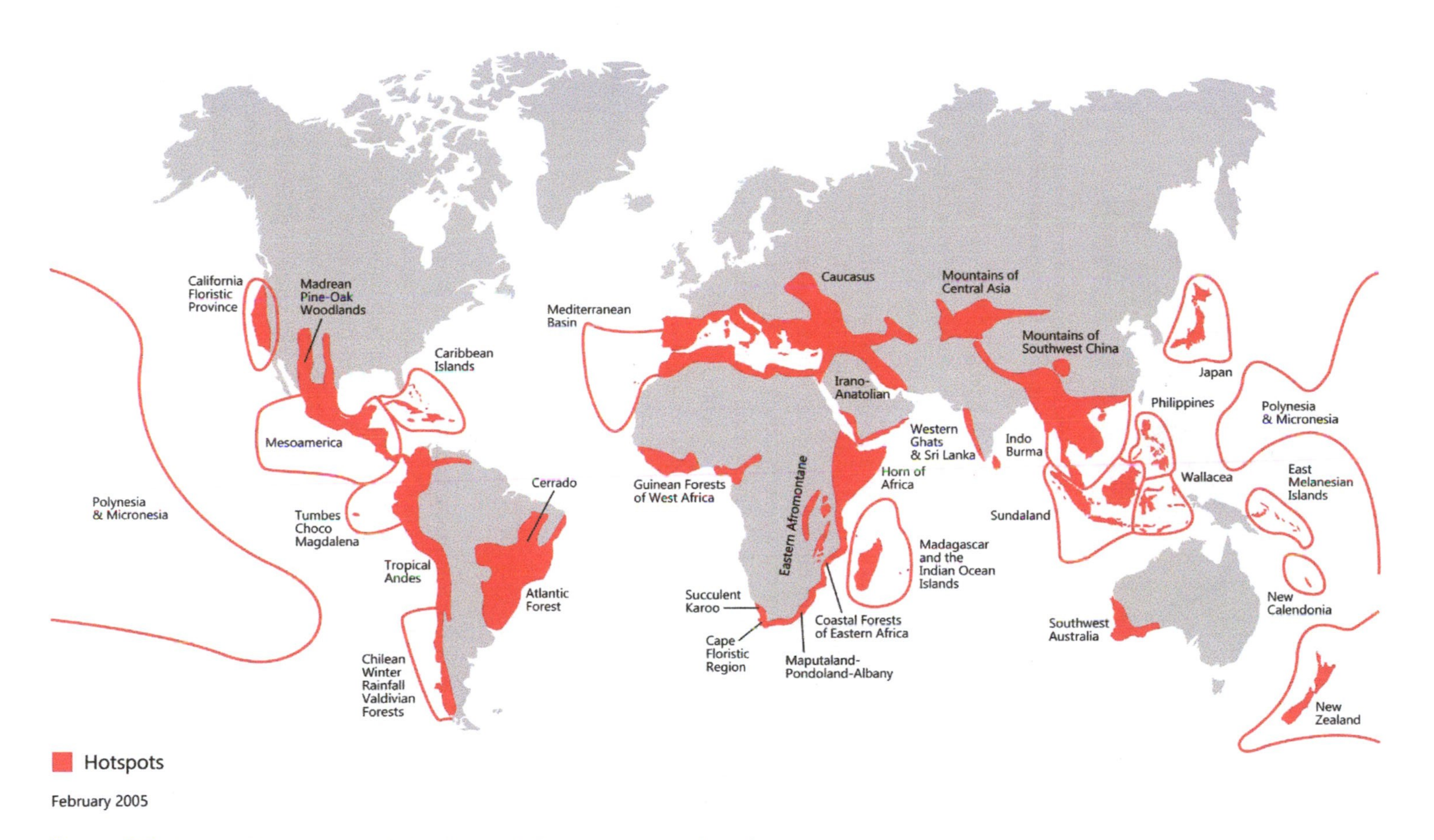

Figure 8.4 Location of the world's major biodiversity hotspots (Source: www.conservation.org).

Figure 8.5 The strikingly colored Resplendent Quetzal was considered divine, associated with the "snake god", by Pre-Colombian Mesoamerican civilizations. It is one of Central America's most sought after birds among birdwatchers and tourists (Photo: Mike Slattery).

of Nature's (IUCN) so-called Red List of Threatened Species,[12] 90% of the world's mammals have been evaluated compared to only 0.3% of invertebrates. According to scientists like Wilson, the world could get on very well if all vertebrates, including humans, were to disappear!

Hotspots are not the only system devised for assessing global conservation priorities. BirdLife International,[13] a global alliance of conservation groups, has identified 218 "Endemic Bird Areas" (EBAs), each of which hold two or more bird species found nowhere else. The World Wildlife Fund in the United States has derived a system called the "Global 200 Ecoregions," the aim of which is to select priority Ecoregions for conservation within each of 14 terrestrial, 3 freshwater, and 4 marine habitat types.[14] The Ecoregions are chosen for their species richness, endemism, taxonomic uniqueness, unusual ecological or evolutionary phenomena, and global rarity. All classified hotspots contain at least one Global 200 Ecoregion and all but three contain at least one EBA. Overall, 60% of Global 200 terrestrial Ecoregions and 78% of EBAs overlap with hotspots. These are clearly important areas to focus our conservation efforts.

SPECIES LOSS AND EXTINCTION

Extinction is the gravest aspect of a looming biodiversity crisis as it is irreversible. According to BirdLife International, 1,253 bird species (12.5% of the total, or one in eight) are globally threatened with extinction because of their small and declining populations or ranges. Of these, 189 species face an extremely high risk of extinction in the immediate future,[15] but modern extinction rates are actually very difficult to quantify. As with carbon dioxide and temperature, discussed in Chapter 6, we need a longer-term frame of reference to make sense of potential human impacts. We can get that from the fossil record by determining the **background extinction** rate, which is the constant rate of extinction throughout geologic time. Scientists estimate that individual species live about 10 million years *on average* before going extinct. If we use this estimate and assume that there are approximately 10 million species on earth today, then we would expect to lose somewhere

[12] www.iucnredlist.org
[13] www.birdlife.org
[14] www.worldwildlife.org
[15] http://www.birdlife.org/datazone/sowb/state/STATE2

between 1 and 10 species per year due to natural causes outside of human impact. This comes out to between 0.0001% and 0.00001% of our total number of species per year or 0.01% to 0.001% per century. Under natural conditions, we would expect that this would be balanced by the evolution of new species, leading to little net loss.

While extinction is a natural process, it appears that human impacts have elevated the extinction rate by at least a hundred (and possibly several thousand) times the natural rate.[16] Some ecologists have suggested that we are in the midst of a **mass extinction**, the magnitude of which has occurred only five times in the history of our planet. The last mass extinction brought the end of the dinosaur age about 65 million years ago (Figure 8.6). Today, many species are being extinguished even before they are discovered, and no one knows how many.

Declines in the numbers of animals such as tigers, elephants, rhinos, and various species of birds have drawn the world's attention to species at risk. An estimated 34,000 plant and 5,200 animal species, including one in eight of the world's bird species, as noted above, face extinction.[17] The gravest threat to biological diversity is the fragmentation, degradation, or the complete loss of

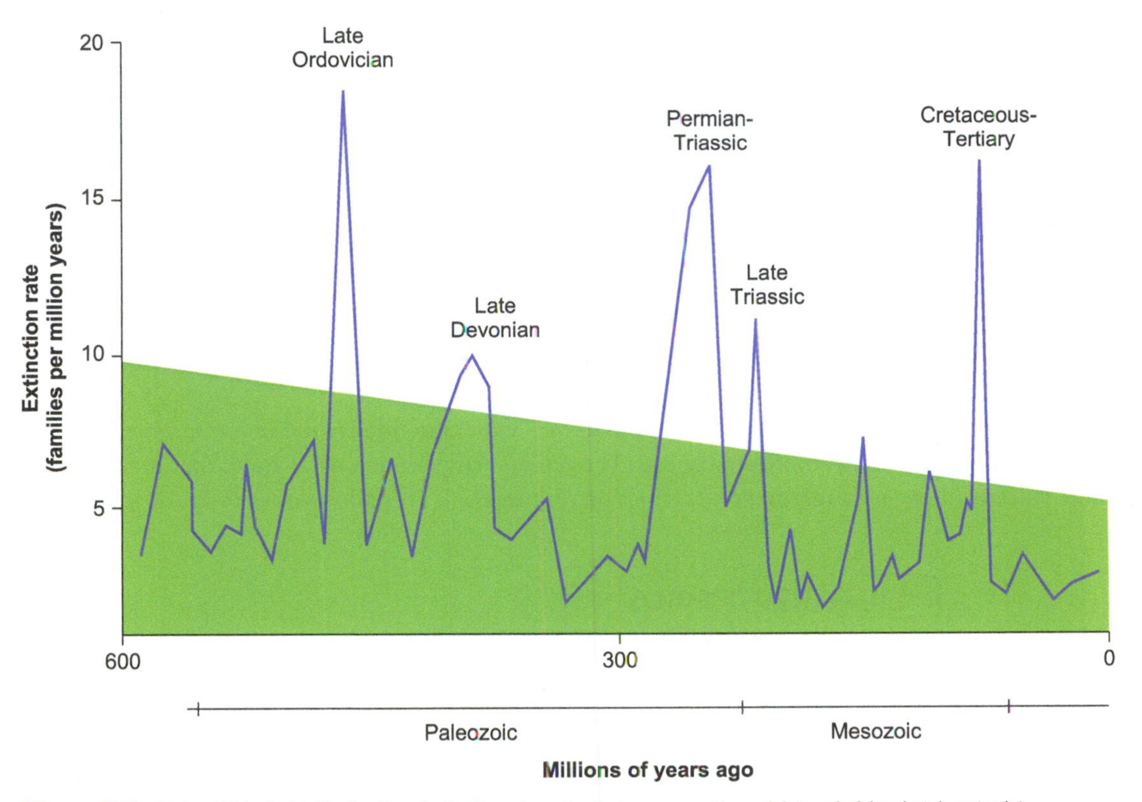

Figure 8.6 Rate of biological extinction (i.e., the fraction of species that are present in each interval of time but do not exist in the following interval) over the last 600 million years. Note that the background rate is punctuated by five major mass extinctions (Source: Rohde and Muller (2005) published in *Nature*, Vol 434 and Raup and Sepkoski (1982) published in *Science*, Vol. 215).

[16] Source: Ceballos et al. (2015), *Science Advances*, Vol. 1, DOI: 10.1126/sciadv.1400253

[17] United Nations Convention of Biological Diversity (www.cbd.int/default.shtml).

ecosystems like forests, wetlands, and coral reefs. In Chapter 7, we noted that forests are home to much of the known terrestrial biodiversity, but almost half of Earth's original forests are gone, cleared mostly during the 20th century. Up to 10% of coral reefs, which are one of the planet's richest yet most fragile ecosystems, have been destroyed, and one-third of the remaining coral reefs face possible collapse over the next 10–20 years. Coastal mangroves, a vital nursery habitat for countless species, are also vulnerable, sensitive ecosystems, with half already gone.

Why is this loss of biodiversity so important to our future? For starters, it reduces the productivity of ecosystems and shrinks the planet's basket of goods and services that we constantly draw from. It destabilizes ecosystems, weakening their ability to deal with natural disasters and stresses, such as floods, droughts, and hurricanes, and human-caused stresses, such as pollution and climate change. Already, we are spending huge sums in response to flood and storm damage which is exacerbated by habitat loss. In February 2005, *National Geographic* magazine reported on the ongoing loss of wetlands along Louisiana's Gulf Coast, a linchpin in the U.S. oil and gas infrastructure. Warnings that the disappearance of the marshes made the coastal region vulnerable to frequent flooding and severe hurricane damage went unheeded. Unfortunately, just a month later, Hurricane Katrina provided the tragic postscript to that forecast, and Louisiana's wetlands are continually being washed away with an area the size of a football field disappears every 35 minutes! At the present net rate of wetlands loss, Louisiana will have lost this crucial habitat in about 200 years, a very short period of time, even in human terms.

Can we save the world's ecosystems and, along with them, the millions of species we have yet to discover, some of which may produce the foods and medicines of tomorrow? The answers to these questions lie in our ability to bring our demands into line with nature's ability to produce what we need and to safely absorb what we throw away.

CHARACTERISTICS OF THREATENED SPECIES AND CONSERVATION STRATEGIES

Understanding the ultimate causes of species decline and extinction is vital in our quest to stop the accelerated rate of current biodiversity loss. We can identify characteristic traits that make species more susceptible to endangerment or extinction (Figure 8.7). This allows us to monitor their status

Figure 8.7 A light-hearted look at endangered species (Copyright: Sidney Harris, ScienceCartoonsPlus.com with permission).

more carefully and to promote stronger protection and preservation of their habitat. Generally speaking, many endangered species possess share one or several of the following characteristics:

1. **Small localized range:** Species that are restricted to a relatively small geographical area are inherently vulnerable to extinction. Populations on islands are especially vulnerable to extinction. They have incurred the greatest number of extinctions in the past 400 years. Examples include the Dodo (Figure 8.8) and the Moa, a giant flightless bird that inhabited the islands of New Zealand about 1,000 years ago.

2. **Specialized habitat and/or diet:** Species that depend on a certain type of habitat or food source do not adapt well to either natural or human-caused changes and are more prone to extinction.

3. **Low reproductive rates and low natural mortality:** Slow-reproducing species that have few young at longer intervals and low natural mortality rates tend to be less resilient to population losses than those species that reproduce at frequent intervals.

4. **Slowmoving animals:** These species are helpless in the face of hunting and predation by humans and/or introduced predators. Tortoises and sea turtles are killed for trade or by vandals for sport. Other large animals are vulnerable to over-hunting, many times killed merely because they make large targets or trophies. Animals of large size also tend to require considerable amounts of habitat and, therefore, are naturally rarer than species with smaller habitat requirements.

5. **Perceived value:** Wild animals and plants that have a value as food, pets, ceremonial objects, or marketable products to humans are prime candidates for extinction (Box 8.1). For example, the once-abundant sturgeon of the Caspian Sea (i.e., sources of Beluga and other expensive caviar) are now critically endangered as a result of unrestricted fishing and poaching for the luxury gourmet market.

Expanding on the topic of range, extinctions of island species make up 80% of all extinctions recorded. The most famous example is the Dodo bird, first discovered by Portuguese sailors in

Figure 8.8 Vintage engraving of the Dodo, a flightless bird endemic to the Indian Ocean island of Mauritius. Related to pigeons and doves, it stood about a meter tall, weighing about 20 kilograms, living on fruit and nesting on the ground. The Dodo has been extinct since the mid-to-late 17th century.

1598 on the island of Mauritius. It became extinct in 1681 after sailors finished them off for dinner (Figure 8.8). The Dodo had lost its ability to fly as it did not ever need to in its native habitat. It lived and nested on the ground and ate fruits that had fallen from trees. There were no mammals on the island and a high diversity of bird species lived in the dense forests. Never had these birds seen predators the likes of man on the island before. Interestingly, the Dodo is just one of the bird species driven to extinction on Mauritius. Many others were lost in the 19th century when the dense Mauritian forests were converted into tea and sugar plantations.

As shown in Figure 7.5, humans have appropriated or transformed vast areas of our planet's surface through agriculture, urbanization, and resource extraction, among others. Through our appropriation of land, we have fragmented it, thereby leaving islands of pristine habitat in a sea of human dominated landscape (much like the Dodo on Mauritius). For example in Chapter 7, we saw satellite images depicting the fragmentation of forests in Amazonia. If you were to sum all of the patches of forest together, the total area of forest remaining might be quite large. However, each patch is often too small to support viable populations of species.[18] **Fragmentation** not only reduces the total area covered by a forest but also exposes the organisms remaining in the fragment to the conditions of a different surrounding ecosystem; these are defined as, **edge effects**. This concept in ecology describes the juxtaposition of contrasting environments, such as the boundary between natural habitats, especially forests, and disturbed or developed land. The net result of fragmentation is that remaining forest fragments tend to lose native biodiversity.

Our strategy as environmental scientists should be to conserve as many species as possible in their natural, *unfragmented* habitat, but what is the real-world situation in which conservationists work? There is not much land that people are willing to devote to conservation, given the pressures of supporting 7 billion-plus people, and land that can be appropriated for conservation tends to be highly fragmented anyway. So what is the best way to conserve lots of species, given that their habitat is fragmented into islands at the outset? To answer this question, we must know what controls species richness on islands and apply it to conservation areas.

Species richness on islands is determined by the balance between immigration (i.e., species entering the island) and extinction rates. Graphically, the number of species for any given island is where the immigration and extinction curves intersect (Figure 8.9). Small islands that are far from the mainland are predicted to have the least number of species while large islands near the mainland are predicted to have the highest island species richness. The conservation implications of this simple theory, known as **island biogeography theory**, are profound. If we consider conservation areas where we want to have maximum species richness as islands, island biogeography theory tells us that they should be large and near to the mainland. What is the implication of the term *near* in this context? Near islands have higher richness because of their increased connectivity with the mainland. It is easy to migrate from the mainland to the island. Conservation areas should mimic this, by increasing the connectivity between several reserves. If the population of any species experiences a decline for any reason, it can be bolstered by immigrating species from other reserves. This is the idea behind **biological corridors**. They are passes through which species can move from reserve to reserve without crossing

[18] In biology, the concept of Minimum Viable Population states that there is a minimum number of individuals needed to maintain the population. If the population declines below this minimum number, the remaining population no longer has the genetic variability to successfully breed. In a matter of a few generations, the species will most likely become extinct.

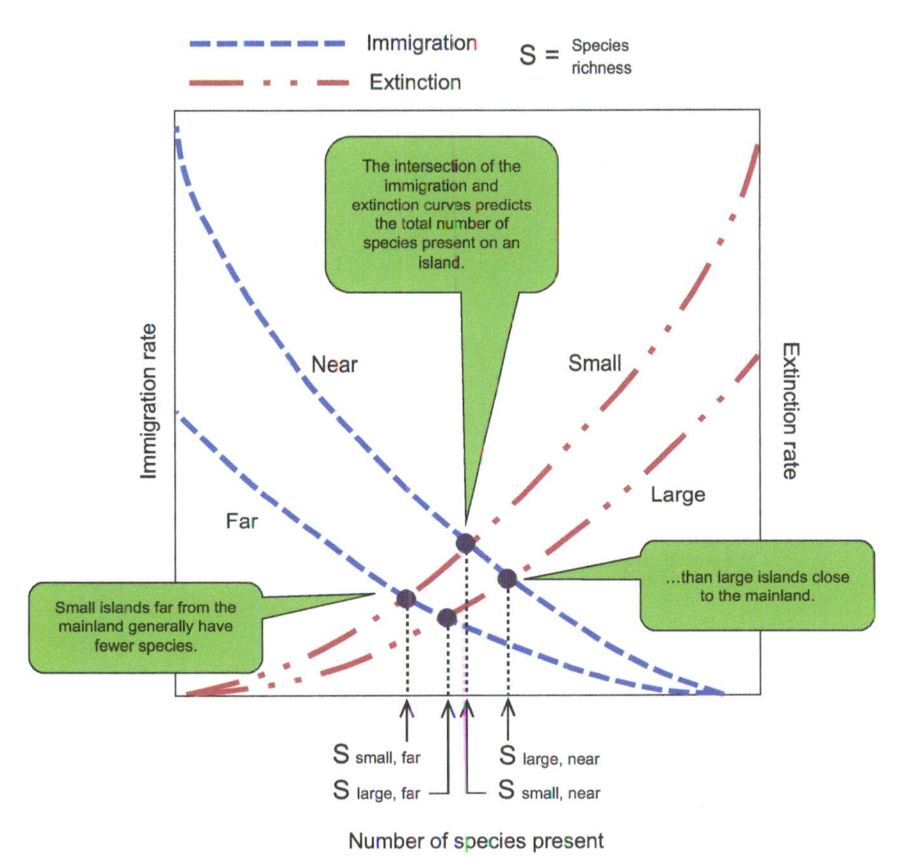

Figure 8.9 Graphical depiction of island biogeography theory. Note that species richness is greater on islands that are large and near to the mainland (Redrawn from the original paper by MacArthur and Wilson (2001), *The Theory of Island Biogeography*).

an excessively human-dominated landscape. An example of this is the Mesoamerican Biological Corridor (MBC) shown in Figure 8.10. Around 11% of Mesoamerica is currently under some category of protection for biodiversity conservation, ranging from <1% in El Salvador to 25% in Costa Rica. Protected areas are regionally integrated into a single functional conservation area by these corridors.

Now, let us consider the term *large* in terms of islands. We know from our earlier discussion that the larger the reserve, the better for maintaining species richness. This is easier said than done in the real world where land available for conservation is neither cheap nor abundant. Human activities on the land often stop just at the boundary of the reserve. One moment you are in lush forest, the next you are in the middle of roads with shops, cars, and people buzzing around. Normal human business (noise, air pollution, etc.) can affect protected organisms and ecosystems, even deep within a reserve. When the boundary between man and nature—city and reserve—is very marked, it effectively reduces the size of the reserve where species can maintain populations for survival.

How do we effectively expand the area of conservation reserves while considering our need for land? This question has been answered most effectively by using **buffer zones** to surround conservation reserves (Figure 8.11). In building a reserve, a protected **core area** is surrounded by a buffer zone

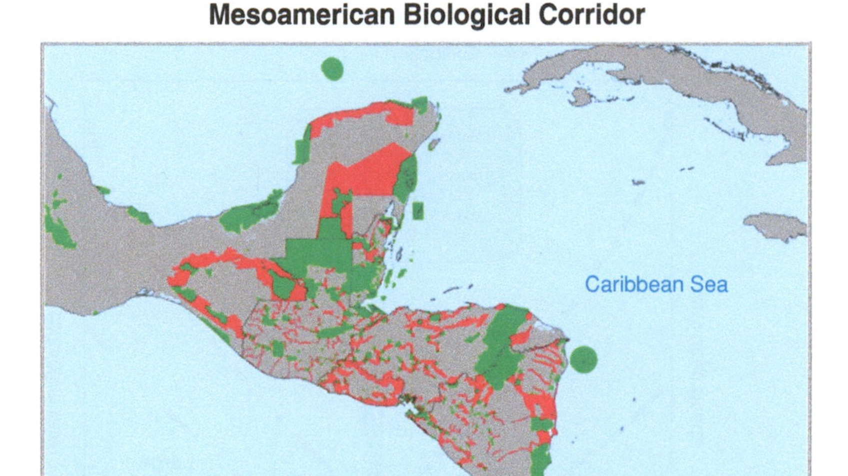

Figure 8.10 The Mesoamerican Biological Corridor (MBC). Since the initiation of the MBC, the government of Costa Rica has officially recognized biological corridors and adopted them as one of its principal conservation strategies, although these areas are not legally defined as conservation areas. Rather, corridors owe their existence to grassroots initiatives and are managed by local councils. To date, 47 biological corridors have been proposed in Costa Rica, covering some 1,753,822 hectares, representing 35% of the country's land area (Source: Ibrahim, M. et al. (2011): http://policymix.nina.no/Casestudies/CostaRica.aspx#2).

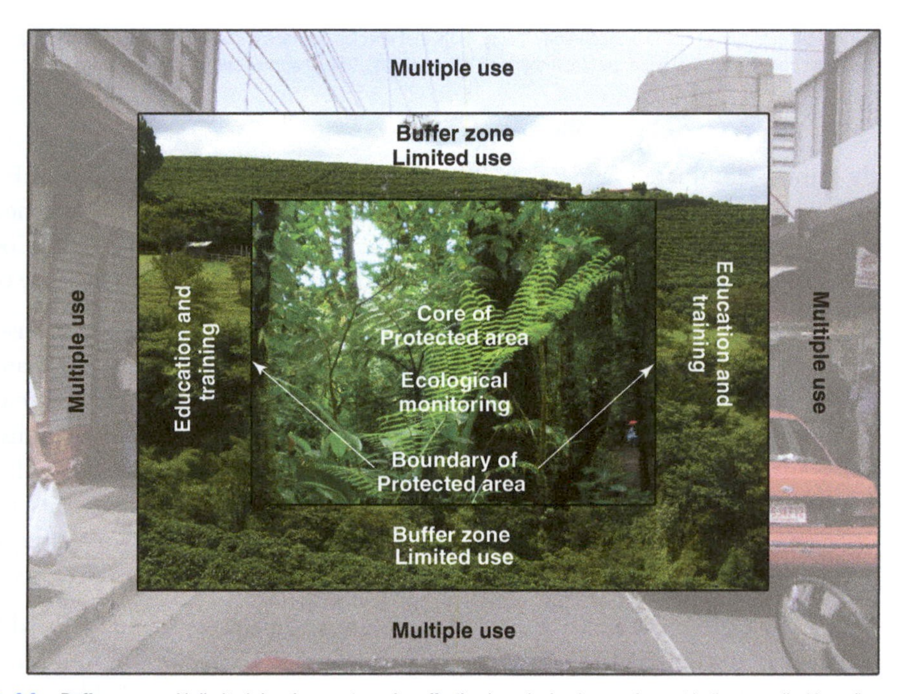

Figure 8.11 Buffer zones with limited development can be effective in reducing human impact in the so-called "core" area of a reserve.

and then a transition area. The core area and buffer zone should each have a definite boundary, and the transition zone as a whole is usually not strictly delineated. The core area excludes all human use except for strictly-controlled scientific research. Only activities, such as certain research, education, training, recreation, and tourism, which do not conflict with the protection of the core zone, are allowed within the buffer zone, while development activities are permitted in the transition area. Establishing buffer zones has been shown to protect core areas more effectively in biological conservation initiatives in many parts of the world.

The Rhino Crisis in South Africa

The rhinoceros is one of the most charismatic megafauna we have left in the wild. Once abundant throughout Africa and Asia, with population estimates approaching 500,000 in the early 20th century, rhinos are now being poached to the brink of extinction. Nowhere is the situation worse than in South Africa, home to three-quarters of all wild rhinos worldwide. The killing spree started in 2008 and is now a major crisis. In the last five years alone, South Africa has lost over 5,500 rhinos. What this translates into is one rhino brutally killed every 8 hours. Conservationists agree that rhino deaths have now reached a "tipping point" overtaking births. Simply put, this means that rhinos could go extinct in the very near future.

The poaching is driven by the illegal trade in rhino horn, itself fuelled by a growing demand in Asian countries, mainly China and Vietnam, where horn is believed to have medicinal properties. Modern science has proven that these coveted horns carry no medicinal value whatsoever as they are nothing more than compressed hair or keratin. In Vietnam, the primary gateway for rhino horn, it has become such a status symbol that consumers readily (though illegally) pay upward of $6,000 per 100 g (0.1 kg or 3.5 oz)*. It is seen as a cocaine-like party drug, virility enhancer, and luxury item—the "alcoholic drink of millionaires," as one Vietnamese news site called it (ground rhino horn is mixed in a cocktail). There are also reports of people giving rhino horn to family and loved ones who are terminally ill as a sign of respect and worth. At these prices, horn is more expensive than gold and cocaine, which has attracted the involvement of ruthless criminal syndicates who use high-tech equipment to track down and kill the rhinos. Part of the tragedy lies in the method: poachers hack off the rhinos' horns while they are still alive (shooting them would create noise and attention) and leave them to bleed to death.

(Continued)

*The average rhino horn weighs 3–4 kg, meaning that each horn is worth between $180,000 and $240,000! This is a conservative estimate as horns have been bought for up to $10,000 per 100 g. Dr. Fowlds himself has weighed a horn at 12 kg, which would be valued at about $720,000.

(Continued)

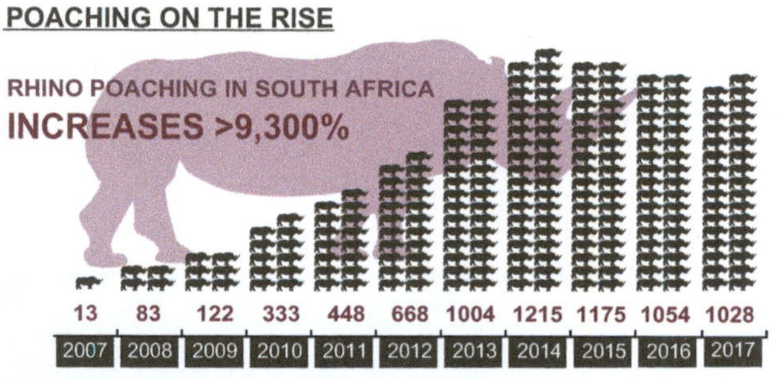

One rhino is **POACHED** every **8 HOURS**

Source: Mike Slattery; images © Shutterstock.com

While law enforcement is critical in deterring poachers, there appears to be no single answer to combat the current poaching crisis. It seems that we need a multi-faceted approach, including ongoing anti-poaching and monitoring patrols, community-based conservation and environmental education schemes, rhino translocations to more protected areas, and demand-reduction projects in Asia. One man leading the fight to save the rhino is Dr. William Fowlds, a South African veterinary surgeon who works, quite literally, at the face of the crisis conducting life-saving surgery on poached rhinos. I am proud to work alongside Dr. Fowlds to help in this fight. It is an enormous task because the poachers have sophisticated equipment and can get the horn off an animal and into the black markets in Asia within 48 hours.

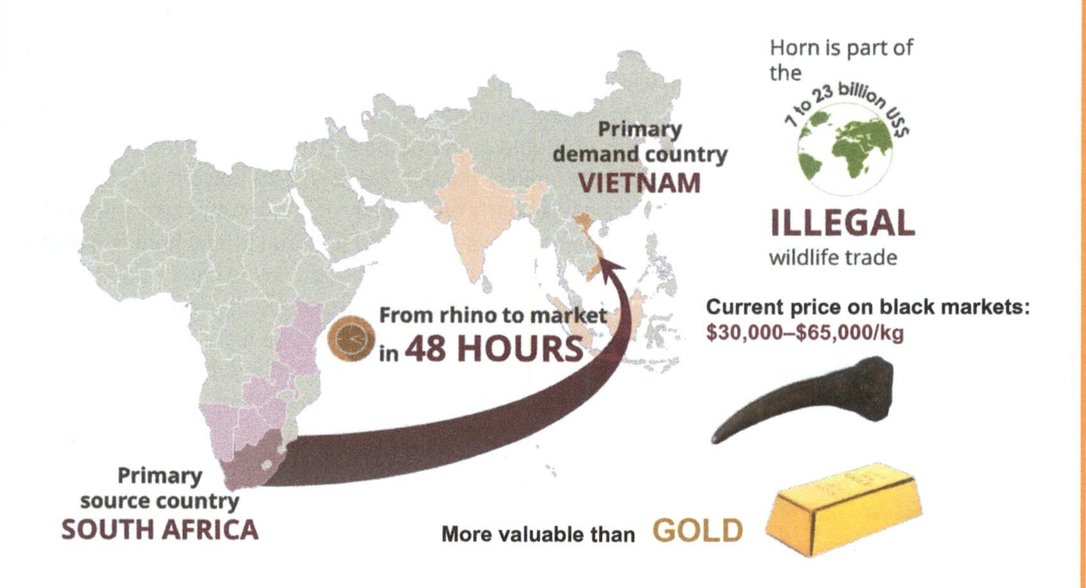

If you are interested in learning more about this crisis and how you can help, please visit our website, www.planetrhino.com. I know from personal experience what it is like to work on a rhino that has been poached, the pain the animal experiences, and the long road ahead for rhinos in rehabilitation. While the situation often feels hopeless, there are many people doing wonderful work, and you can get involved and make a difference.

Photo: Mike Slattery

PROTECTING BIODIVERSITY IN THE U.S.: THE ENDANGERED SPECIES ACT

Passed in 1973, the Endangered Species Act (ESA) is a safety net for wildlife, plants, and fish that are on the brink of extinction. Because of the strength of this landmark law, the ESA's history has been filled with controversy and grabs more headlines than nearly any other federal environmental law. Over the past decade, supporters like the Endangered Species Coalition[19] have trumpeted its virtues and necessity, while opponents of the Act (for example, the National Endangered Species Act Reform Coalition)[20] decry its regulatory heavy hand. Therefore, it is little wonder that the ESA elicits strong opinions from all sides of the debate.

[19] http://www.stopextinction.org

[20] http://www.nesarc.org

The ESA prohibits any person from killing or even harming an endangered species or significantly altering the habitat that the species requires for survival, and it imposes civil and criminal penalties to enforce these prohibitions. It is based on three key elements: listing species as threatened or endangered, designating habitat essential for their survival and recovery, and ultimately restoring healthy populations of the species so they can be removed from the list. The protection afforded by the ESA currently extends to over 2,400 species, and many of them have completely recovered, partly recovered, had their habitat protected, or had their populations stabilized or increased as a result.[21] Equally as important, millions of acres of forests, beaches, and wetlands (those species' essential habitats) have been protected from further degradation and development. The ESA works with citizen involvement to preserve not only large and charismatic species (e.g., grizzly bears and bald eagles), but also those that are small, equally unique, and (to some) beautiful. The farseeing vision of the Endangered Species Act is that all these species will not merely survive in the sterile confines of zoos but thrive in their natural, wild environments.

How then does a species get listed? A declining species has to be added to the official list of endangered and threatened species before it receives any federal protection. The authority to list species as threatened or endangered is shared by the National Marine Fisheries Service (NMFS), which is responsible for listing most marine species, and the Fish and Wildlife Service (FWS), which administers the listing of all other plants and animals. Getting on the list though can be the hardest part.

Any person may petition the government to list a species as either endangered or threatened. By definition, an endangered species is any species that is "in danger of extinction through all or a significant portion of its range." A **threatened species** is any species "which is likely to become an endangered species within the foreseeable future." The listing process is designed to take no more than 27 months. In some limited circumstances, an expedited or emergency listing may be given temporarily.

The ESA requires the designation of critical habitat for all endangered and threatened species. **Critical habitat** is an area "essential to the conservation of the species," or, in other words, enough space to accommodate the endangered or threatened species as it hopefully increases its numbers and recovers. Habitat loss is the most prevalent cause of endangerment, affecting more than 95% of all listed species according to one study. Therefore, we must protect critical habitat if we hope to conserve endangered and threatened species.

Recovery plans, as part of the Fish and Wildlife Service's Recovery Program, are designed to reverse the decline of a threatened or endangered species and eventually bring the population to a self-sustaining level. When a species is delisted, it does not necessarily mean that it is not afforded other federal and state protection. In fact, the ESA requires the Fish and Wildlife Service to continue to protect these species and take steps to ensure their continued recovery for at least five years. For example, the American bald eagle was delisted in 2007. Nesting pairs rose nationwide to over 7,000, up from a low of 417 in 1963, when high levels of DDT were damaging their eggs. The bald eagle will continue to be protected by the Bald and Golden Eagle Protection Act (BGEPA) and the Migratory Bird Treaty Act (MBTA). Some environmental groups caution that these laws

[21] http://ecos.fws.gov/tess_public/pub/Boxscore.do

will need to be closely reviewed to protect the eagle's habitat from logging and development, a protection that has benefited not only the bald eagle but also other forest species impacted by human activities.

SAVING THE WORLD'S SPECIES: WHAT'S IN IT FOR US?

We hear tales of woe all the time when it comes to the world's biodiversity. Whether it is tigers, pandas, rhinos, California condors, or coral reefs, much of the world's wildlife is under threat. To most people it is upsetting. Eventually, it seems to just become numbing.

I am often asked about my work with rhinos and whether it is worth worrying about their survival at all (Figure 8.12). Sure, it will be sad if there are not any more rhinos (or pandas, or tigers, etc.) on the planet, but it is not like we depend on them. Besides, surely it is more important to take care of humans? How can we justify spending millions of dollars preserving animals when over a billion people go to bed hungry at night? It just does not seem right.

One study in 2012 estimated that it would cost $76 billion a year to preserve threatened land animals.[22] The cost of saving endangered marine species may be even more. The costs involved are staggering. As extinction is a natural process anyway, why should we bother at all in trying to stop it?

Perhaps the most obvious rationale for protecting and conserving biodiversity is simply that we want to. I love the natural world. I think animals are majestic and fascinating. Nothing lifts my spirits and fills my cup like watching a herd of elephant or a dazzle of zebras drinking from a waterhole as the sun sets over the African savannah. Yes, nature is beautiful, and many conservationists argue that this aesthetic value is reason enough to keep and preserve it. But what about the animals and plants that people are less fond of: the ugly, the dangerous, or the just plain obscure? It seems a bit harsh that if we do not find them appealing, they are out.

So, the fact that some of us find nature beautiful, by itself, will not do. There needs to be a more practical reason to keep species around.

We frequently hear the argument that we should keep ecosystems like tropical rainforests because they may contain very useful things, like medicines. For example, researchers at the National Cancer Institute unlocked the Pacific yew tree's potential to treat ovarian and breast cancer. In 2018, plant scientists unraveled the complex chemistry of the Madagascar periwinkle in a breakthrough that opens up the potential for rapid synthesis of cancer-fighting compounds. Undoubtedly, the natural world still holds many secrets that could inspire a new generation of biologic drugs, but pharmaceutical companies are finding plenty of ways of finding new medicines that do not involve trekking to remote corners of the world trying to find the next miracle drug. And what happens to all the species that do not make useful things like medicines?

[22] Source: McCarthy et al. (2012), *Science*, Vol. 338, p. 946–949.

Perhaps the most compelling argument as to why we should save our biodiversity centers on the ways animals and plants benefit us just by simply being there. Nature provides us with innumerable *ecosystem services* that would be extremely costly or impossible to replace (remember the discussion on putting a price on nature's goods and services in Chapter 1?). How could we replace green plants that provide us with the oxygen we breathe or pollinating insects like bumblebees? Many of our crop plants rely on these insects to produce seeds and would not survive, let alone provide us with food, without them. The scale of these ecosystem services are enormous. The economic and environmental benefits of biodiversity in the U.S. alone are estimated at over $300 billion a year.[23] If you buy into this economic argument, then our very survival depends on the conservation of biodiversity, and we should protect it, if simply to avoid severe economic repercussions.

Many people feel that we should protect other species because they have an intrinsic right to exist on our planet, irrespective of whether they provide economic benefits or are beautiful. This line of thinking, often called "nature for itself," underpins the writings of John Muir, the father of the so-called preservationist ethic and founder of the Sierra Club. Muir traveled extensively through California and the Sierra Nevada. He wrote that these natural areas offer emotional refreshment and even religious and spiritual experiences. In Muir's world, philosophers, poets, artists, spiritual minds—indeed any human—requires the stimuli of natural beauty for growth. Like Muir, many writers have recognized nature's value in and of itself, apart from its value to humanity. In Muir's *My First Summer in the Sierra,* he writes:

> *The snow on the high mountains is melting fast and the streams are singing bankfull, swaying softly through the level meadows and bogs, quivering with sun-spangles, swirling in pot-holes, resting in deep pools, leaping, shouting in wild, exulting energy over rough boulder dams, joyful, beautiful in all their forms. No Sierra landscape that I have seen holds anything truly dead or dull, or any trace of what in manufactories is called rubbish or waste; everything is perfectly clean and pure and full of divine lessons. This quick inevitable interest attaching to everything seems marvelous until the hand of God becomes visible; then it seems reasonable that what interests Him may well interest us. When we try to pick out anything by itself, we find it hitched to everything else in the universe.*

The idea of respecting nature is also the historic, underlying philosophy of the Kuna Indians of Panama, who lead a lifestyle based on subsistence living. They emphasize respect for the land and believe one should maintain a deep, intimate relationship with it:

> *For the Kuna culture, the land is our mother and all living things that we live on are brothers in such a manner that we must take care of her and live in a harmonious manner on her, because the extinction of one thing is also the end of another.*[24]

Whether we put it in economic terms or not, science is telling us that ecosystems provide us with a host of things we cannot do without and that the more diverse each ecosystem is, the better. So,

[23] The economic and environmental benefits of biodiversity, by Pimental *et al.* (1997), *Bioscience*, vol. 47, 247–257.

[24] From a delegate of the Kuna people of Panama to the 4th World Wilderness Congress, Denver, Colorado, 1987.

Figure 8.12 The southern white rhinoceros, a species under great threat for its horn. Sadly, less than 20,000 of this particular species remain, down from an estimated 500,000 at the start of the 20th century (Photo © Victoria Bennett).

for our own good, both in terms of practical things like food and water, and less physical needs like beauty, we should protect them.

CONCLUDING THOUGHTS

At the 1992 Earth Summit in Rio de Janeiro, world leaders agreed on a comprehensive strategy for sustainable development: development that meets our needs while ensuring that we leave a healthy and viable world for future generations. One of the key agreements adopted at Rio was the Convention on Biological Diversity. Signed by 150 government leaders, this historical pact provides guidelines and policies for maintaining the world's ecological foundation as we go about the business of economic development. The Convention establishes three main goals: (1) the conservation of biological diversity, (2) the sustainable use of its components, and (3) the fair and equitable sharing of the benefits from the use of genetic resources. In April 2002, the Parties to the Convention committed themselves to achieve, by 2010, a significant reduction of the current rate of biodiversity loss at the global, regional, and national level as a contribution to poverty alleviation and to the benefit of all life on Earth. This 2010 target was subsequently endorsed by the heads of state and government at the World Summit on Sustainable Development in Johannesburg, South Africa. In 2010, Parties to the convention met in Nagoya, Japan, and adopted a revised strategic plan for biodiversity, including the so-called **Aichi Biodiversity Targets**, for the period

2011–2020.[25] Some examples of these targets are to at least halve and, where feasible, bring close to zero the rate of loss of natural habitats including forests, and to make special efforts to reduce the pressures faced by coral reefs.

It is important to stress that the Convention on Biological Diversity recognizes that the conservation of species is a common concern of humankind and is an integral part of the development process. The Convention recognizes that biological diversity is about more than plants, animals, microorganisms, and their ecosystems—it is about people and our need for food, security, medicines, fresh air and water, shelter, and a clean and healthy environment in which to live. The agreement is legally binding, meaning that countries that joined the convention are obliged to implement its provisions.

Although still in its infancy, the Convention on Biological Diversity is already having positive results. The philosophy of sustainable development, the ecosystem approach, and the emphasis on building partnerships are all helping to shape global action on biodiversity. The data and reports that governments gather and share with each other provide a sound basis for understanding the challenges and collaborating on the solutions. There are many positive signs, with the number and size of protected areas increasing.

However, on the basis of information available, a common message emerges: biodiversity is in decline at all levels and geographical scales, and it appears that this decline is accelerating. Much more needs to be done. With human population on the march toward 9 billion-plus, the environmental problems we currently face will only grow, thereby exacerbating current stresses on biodiversity, unless we take immediate, deliberate action. International agreements such as the Convention offer a comprehensive, global strategy for preventing such a tragedy. If we apply the concepts embodied in the Convention and make the conservation and sustainable use of biological diversity a real priority, we can ensure a new and sustainable relationship between humanity and the natural world for generations to come.

[25] http://www.cbd.int/sp/targets/

9

Soil Degradation

Courtesy of the Jay N. 'Ding' Darling Wildlife Society

INTRODUCTION

Soil, which people frequently think of as worthless "dirt," covers most of the land surface with a fragile, thin, but invaluable layer. Simply put, without soil no plant life could exist. Soil not only serves as the medium for terrestrial plant growth but also functions as a recycling system for nutrients and organic wastes as well as a habitat for soil organisms. A single gram[1] of productive agricultural soil typically contains several million bacteria and tens of thousands of algae, fungi, and other soil life (Figure 9.1). It is remarkable stuff!

While we need soil to be able to grow food, we are losing soil at rates far greater than soil can form. In this chapter, we focus on **soil degradation** and how it affects our ability to produce sustainable food supplies for our growing world population. Soil degradation generally refers to the overall

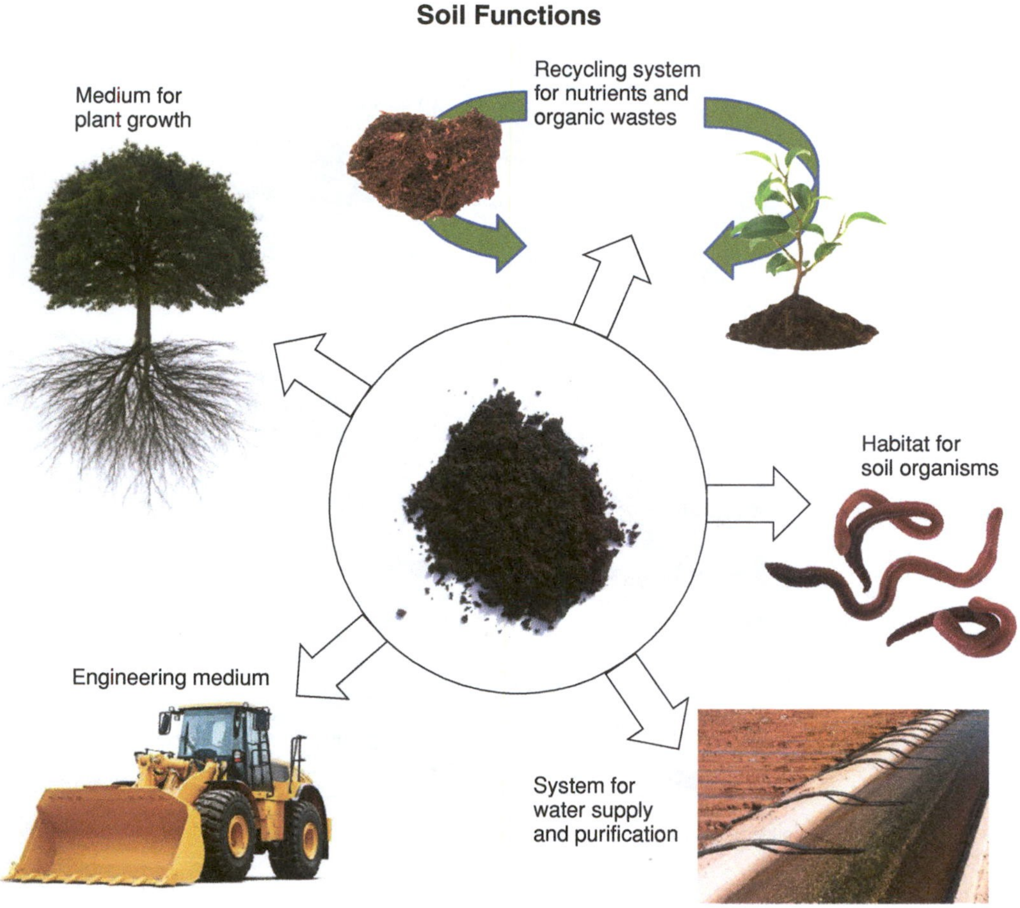

Figure 9.1 The various functions of soil.

---

[1] One gram equals about 1/5 teaspoonful of soil.

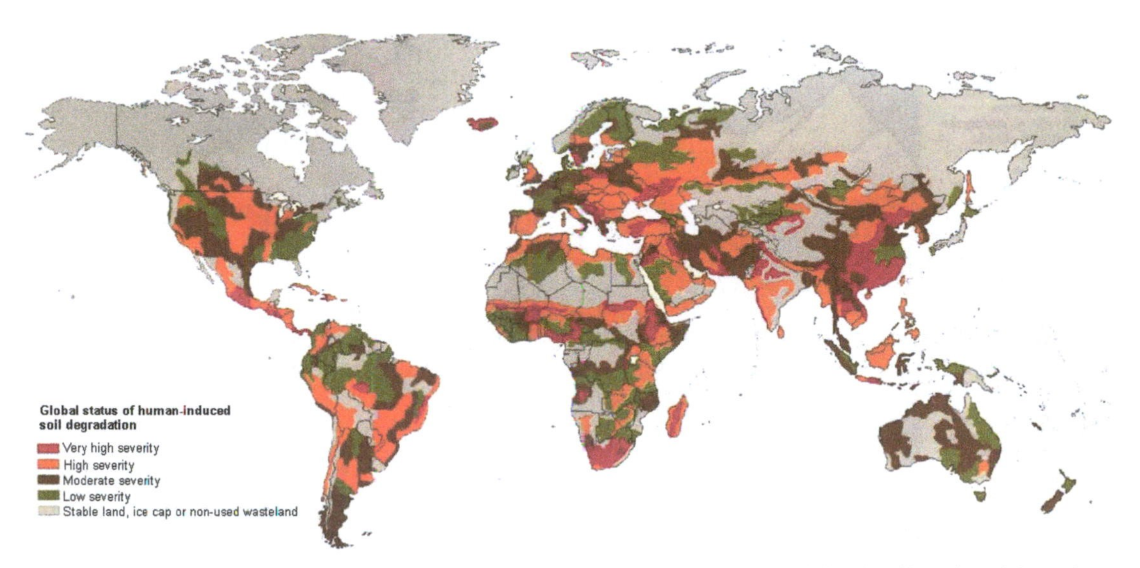

Figure 9.2 Human impact on the world's soil. Highly erodible soils in Southeast Asia are especially vulnerable to degradation and erosion, an acute problem given that regions' expected population growth and economic development (Source: USDA-NRCS, Soil Science Division, World Soil Resources, Washington D.C.)

reduction of the productive potential of the soil. Erosion by wind and water is the primary cause of soil becoming degraded, but chemical degradation (such as a loss of nutrients) and physical degradation (such as compaction) are also significant. The degree of degradation ranges from "light" (where only part of the topsoil is removed and most of the vegetation remains) to "severe" (where all topsoil is lost and less than 30% of the vegetation remains). In general, lightly degraded soils can be improved by farm practices such as **crop rotation** (planting different crops in different years) and minimum tillage techniques (drilling the seed directly into the soil without plowing). However, more severely degraded soils are more difficult to restore and frequently have to be abandoned. Restoring degraded soils to full function turns out to be an expensive and difficult problem.

According to the FAO, most of the world's soils show some level of degradation, largely resulting from human activities (Figure 9.2). Most of the United States has degraded or very degraded soil, particularly within the central United States, the so-called "Breadbasket of America." We have some serious soil degradation problems that will only worsen with time. Unfortunately, these problems are largely being ignored because, let us face it, not many people care about dirt?

THE SCIENCE OF SOIL

Let us take a closer look at this important substance. Soil actually forms very slowly. It can take thousands of years to produce just a few inches of soil. The process begins with the **weathering** (or breakdown) of rock into loose material. **Soil development** (or soil formation) refers to changes within this loose material (called **parent material**) over time into layers or horizons. Soil formation involves a complex suite of processes that add, remove, translocate, and transform material within the profile to produce the horizons (Figure 9.3). The end product, soil, is a dynamic entity having properties derived from the combined effects of climate activity and biotic activity as modified by relief (topography), acting on parent materials over periods of time on the scale of hundreds to thousands of years. These relationships are referred to as the **soil formation factors**, and can be written as

$$\text{Soil formation} = f(c, o, r, p, t\ldots) \tag{9.1}$$

where c = climate, o = organic matter, r = relief, p = parent material, and t = time. For example, soils developed from weathered bedrock on steeper slopes would tend to have thin A horizons

Major Soil Horizons and Soil Formation Processes

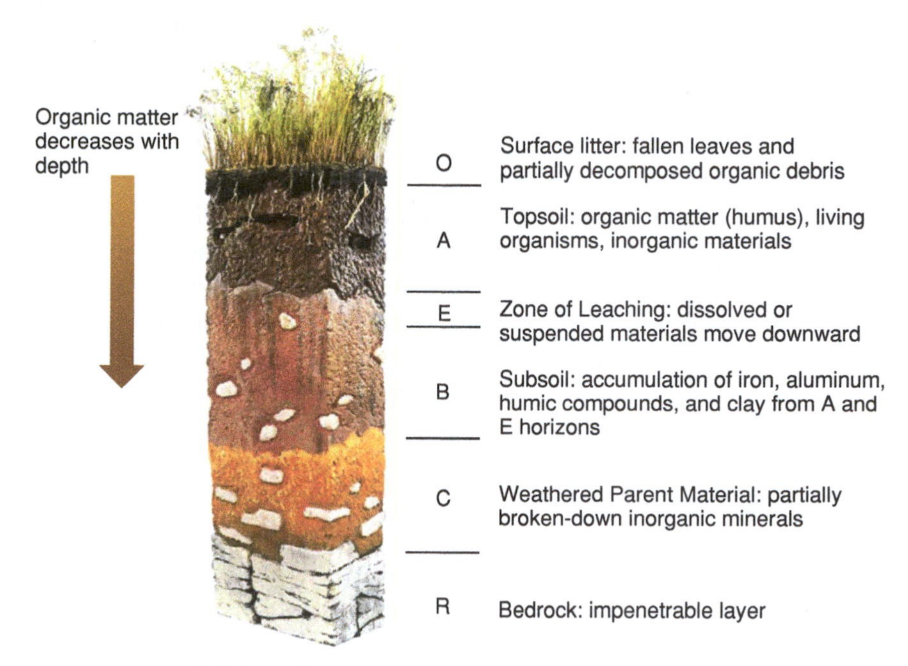

Soil formation processes

Additions *- organic matter is added as organic matter falls to the top layer (leaf* litter, death). Precipitation with dissolved ions and suspended particulates (O and A horizon).

Removal of material *- uptake from plants removes ions and erosion removes* ions, particulates, and organic matter (O and A horizon). Leaching also causes loss of ions from profile.

Translocation *- ions, humus, compounds, and clays get moved down the profile* from O and A through E into B. Ions move up the soil profile through capillary action.

Transformation *- leaf litter and other organic matter is transformed by soil* macro and micro-invertebrates into humus. Minerals are transformed via chemical weathering (carbonation, oxidation).

Figure 9.3 Soil forming processes in relation to the major soil horizons. Note that it would be extremely rare for a soil to have all six horizons in one profile.

Figure 9.4 (Left) Mollisol formed under tall grass prairie found throughout SW Iowa, NW Missouri, NE Kansas, and SE Nebraska. Most of these soils are now used for cultivation of corn, soybeans, small grains, and hay. The Ap designates a plowed A horizon whilst Bt signifies the accumulation of clays in the subsoil. (Right) A Mollisol in South Dakota. The thick, dark A horizon has developed as a result of the proliferation and subsequent decomposition of fine and very fine roots. The subscript k indicates accumulation of carbonates (Source: http://www.cals.uidaho.edu/soilorders/orders.htm).

over weakly developed B horizons, whereas soils developed in floodplain sediment within the same landscape would tend to have thick, clay-rich horizons with higher **organic matter** content.

Soils are the product of an incredibly complex, symbiotic relationship between the mineral world and the living (organic) world. Organic matter is particularly critical because it is this material that gives soils much of their fertility, erosion resistance, and water-holding capacity that supports plant growth. The soil profile shown in Figure 9.4 is a **mollisol** (*molli-* from the Latin word meaning "soft"). Mollisols are very productive agricultural soils underlying the breadbasket of the United States (Figure 9.5). The dark surface horizon is composed of **humus** (partly decomposed organic matter) and mineral grains. This is the "A" horizon and is commonly referred to as **topsoil**. This is the most important soil horizon because it is the zone that supports vegetation and is thus referred to as the root zone. "B" horizons have relatively little organic material but contain materials that are leached down from the topsoil. The leached material is often small clay particles that pass easily through the large empty pore spaces in the topsoil. The "C" horizon, typically found several feet below the land surface, is either slightly broken-up bedrock or material transported in from elsewhere (such as alluvium or river-born sediment). This horizon is the parent material in which the soil forms.

SOIL EROSION: NATURAL VS. ACCELERATED

While the profiles shown in Figure 9.4 can be considered typical soil profiles, many other types exist, depending on climate, local geology, and native biological communities that influence the organic and inorganic matter within soil. Soil profiles can vary greatly in thickness, depending on

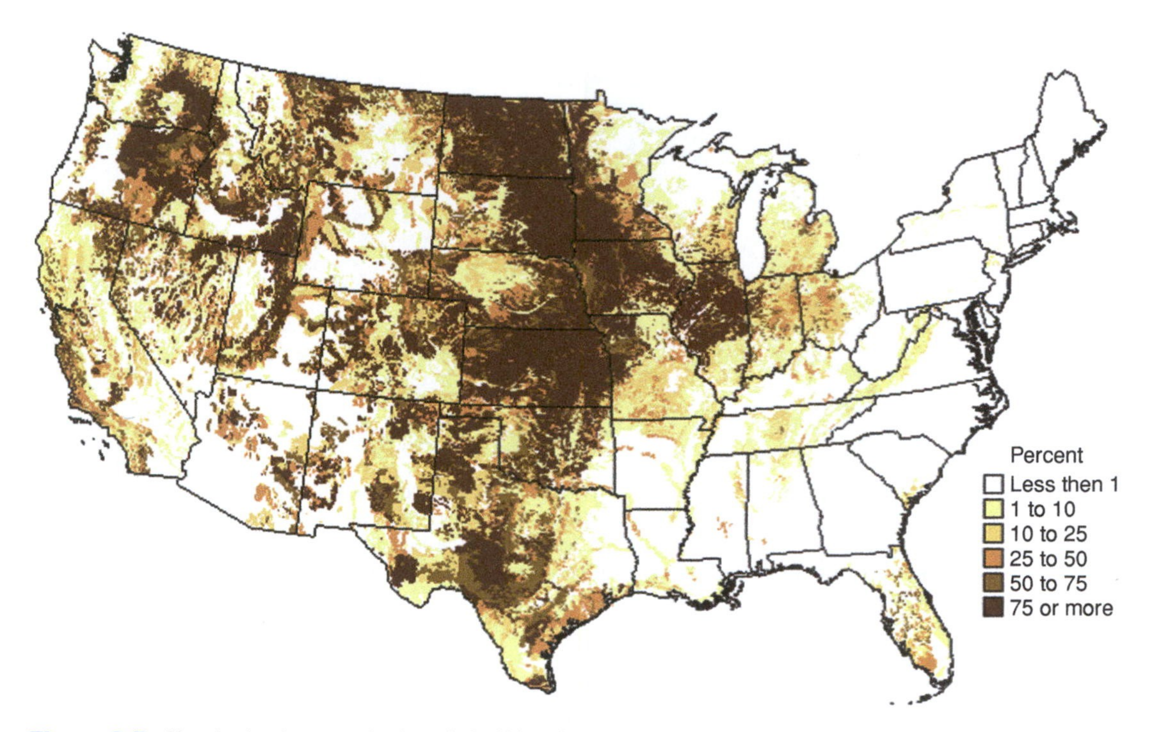

Figure 9.5 Map showing the percent land area in the U.S. underlain by Mollisols (Source: www.nrcs.usda.gov.).

the balance of soil formation to **soil erosion**. Erosion removes soil from an area and is a two-phase process that involves the following: (1) the detachment of particles from the surface and (2) the transport of particles by erosive agents like wind or water. Over geologic time, erosion can lower the elevation of entire landscapes (a process called **denudation**).

The amount of erosion in an area depends on the intensity of the erosional processes and the resistance of the rock or soil in that region. If erosional forces are intense (i.e., huge windstorms or rain events), most likely, a lot of erosion will occur. If the rocks and soil in the region are easily eroded, even moderate wind or rainstorms could potentially carry off a lot of material. The variability of these two components leads to great regional differences in geologic erosion rates. Although geologic erosion rates are relatively low, with enough time, spectacular landscapes may result.

It might be hard to believe, but the force of raindrops on the ground packs a powerful punch in detaching material (Figure 9.6). This does not cause a loss of material from the landscape, but the splashed soil particles clog openings at the surface preventing water from flowing down into the soil. Water will begin to pool and flow as **runoff**, carrying the splashed soil particles along with it. This is known as a type of positive feedback, where one process (**rainsplash**) reinforces another process (runoff generation). Runoff can be unconcentrated, where water moves as a sheet across the landscape, but more commonly, runoff becomes concentrated into small channels (or **rills**) that develop in the landscape (Figure 9.7). Rills may ultimately grow into larger features called

Figure 9.6 High speed photograph of raindrop impact (left) and soil dislodged and splashed by raindrops on tobacco crops in North Carolina (right). The cartoon below shows the scale of an individual impact crater and dispersion of material. Heavy rainstorms can dislodge many tons of soil per hectare (Photo: Mike Slattery).

Figure 9.7 A large rill on a field in southern England. This photograph was taken in an area that is believed to be relatively stable in terms of soil loss. The wrong choice in plowing method (up-and-down slope) resulted in some of the worst documented erosion in the country (Photo: Mike Slattery).

gullies. When soil is mobilized and erodes from a particular location, it enters what is known as the **sediment delivery system**. Some ends up lower down individual slopes and on flood plains while the rest is transported by rivers to the sea. However, large volumes of sediment are stored on river bottoms (as channel bars) and on the bottoms of reservoirs.

Vegetation is the critical erosion control. It anchors soil in place with its roots and protects the soil surface from the full erosive force of wind and rain. In arid and semi-arid environments, sparse vegetation coupled with high-intensity, short-duration storms leads to high runoff and removal of soil material. Figure 9.8 shows this generalized relationship between **annual sediment yield** (the amount of sediment leaving a known area) and annual rainfall. As annual rainfall increases, erosion and sediment yield increase dramatically, peaking where there is enough rainfall but not enough vegetation to protect the surface. In more vegetated environments (such as grasslands), the relationship still holds true although sediment yield might not be as large (because the rainfall could encourage further vegetation growth thereby enhancing land stability). Forests represent the most stable surface as the canopy absorbs most of the kinetic energy of raindrops. However, under very high rainfall regimes, erosion would likely remain high due to the large volume of generated runoff.

Human Activity and Soil Degradation

Overall, geologic soil erosion rates are low. However, they can be accelerated greatly with human actions. As humans remove vegetation or uncover soil for any reason, erosion rates greatly increase. On a global basis, soil loss is caused primarily by overgrazing livestock, deforestation, agricultural activities, over-exploitation of land to produce fuel wood, and industrialization (Figure 9.9).

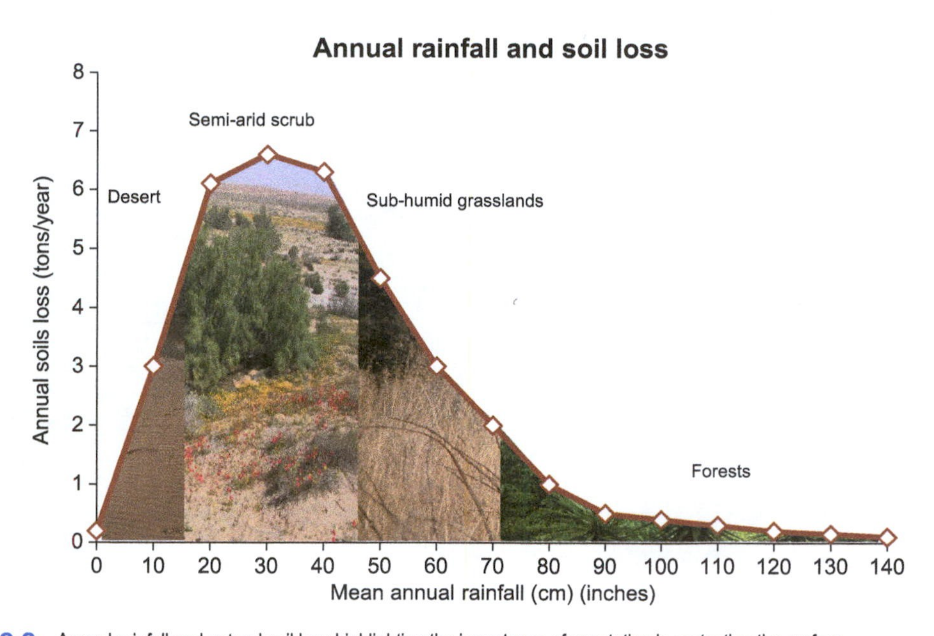

Figure 9.8 Annual rainfall and natural soil loss highlighting the importance of vegetation in protecting the surface.

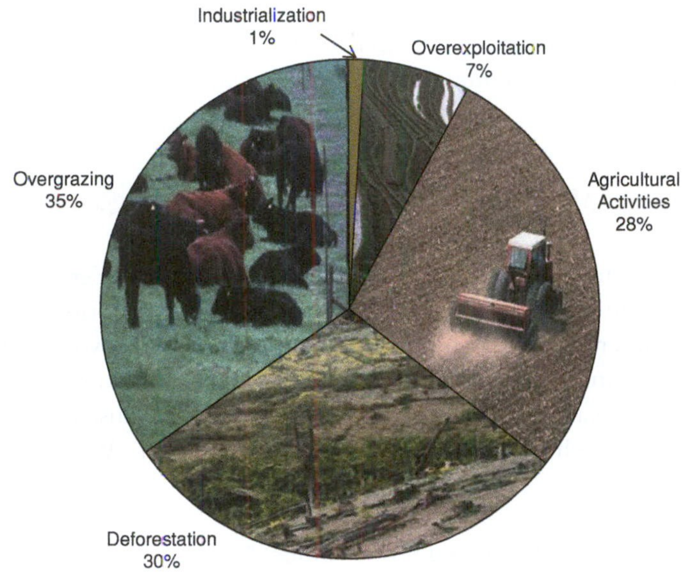

Figure 9.9 Major causes of soil degradation worldwide.

Livestock can do a remarkable amount of damage to a landscape. Besides removing vegetation and exposing soil, they compact soil with their hooves (which increases runoff) and destabilize stream banks as they access water. Even areas with relatively stable soils can quickly become degraded with cattle and sheep grazing. These effects are amplified when livestock is placed on steep, unstable slopes.

Soil erosion rates are profoundly affected by both deforestation and conversion of previously undisturbed lands to croplands. As discussed previously in Chapter 7, deforestation is of global concern, particularly in tropical areas where species diversity is so high. Globally, deforestation accounts for about 30% of all soil degradation. Although erosion increases because of tree removal, another key factor is a rapid decline in soil productivity that follows conversion of forests to agriculture. Most areas of tropical rainforest are underlain by infertile soils, a somewhat surprising fact given the sheer weight of biomass found in tropical forests. The bulk of the nutrients are actually stored within the vegetation and surface litter. If these are removed, a 20- to 30-year fallow period (meaning a period without any agricultural activity) is required for nutrients to be brought up from deep-rooted plants. If no plants remain to access those deep-stored nutrients, rainfall quickly leaches nutrients from the soil body.

Agricultural activities account for about 28% of global soil degradation. The largest contributor to soil loss has been the rapid conversion to mechanized farming coupled with intensive farming methods. For example, widespread planting of row crops (like corn and wheat) on the same fields year after year leaves the soil extremely vulnerable to erosion. Traditional methods of plowing leave soil exposed to rainfall and erosion for an extended period of time before the growing crop is able to provide any sort of protection from erosion. Row crops are especially problematic because runoff becomes concentrated between crops and along vehicle wheelings (areas of compacted soil due to vehicular traffic). The situation is exacerbated when farmers plow upslope and downslope instead of following the natural contours of the land that funnels runoff downslope, as shown in Figure 9.7.

HOW MUCH SOIL ARE WE REALLY LOSING?

We obviously need all the land we have now to produce our food. Yet, with soil loss and soil deterioration, we are losing that resource and, in many parts of the world, losing it very quickly. Although estimates vary, recent research suggests that one-third of the world's arable land has been lost since 1960 due to human activity (Figure 9.10). Each year, an estimated 10 million hectares of cropland worldwide are abandoned due to lack of productivity caused by soil erosion. Worldwide, soil erosion losses are highest in the agroecosystems of Asia, Africa, and South America.[2]

Estimated Size of Global Soil Loss

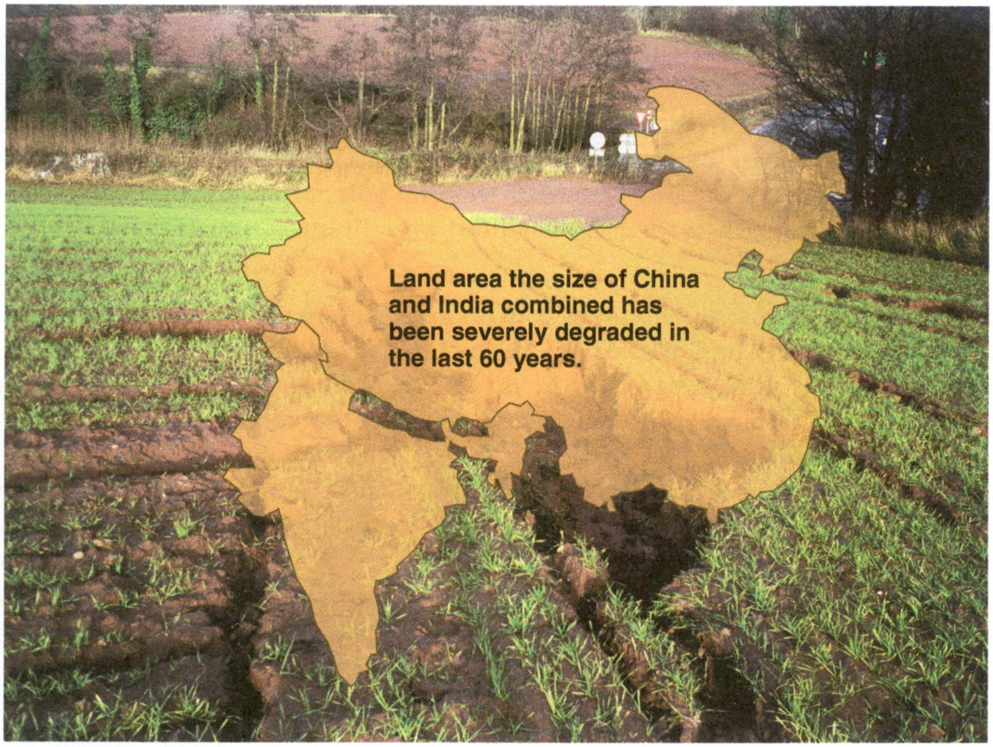

Figure 9.10 Soil, one of our most precious resources, is being lost at an alarming rate.

[2] Source: Pimentel, D. (2006), *Environment, Development, and Sustainability,* Vol. 8: p. 119–137.

In developing countries, soil erosion is particularly severe on small farms that are often located on marginal lands where the soil quality is poor and the topography is frequently steep. In addition, poor farmers tend to raise row crops, such as corn, which are highly susceptible to erosion because the vegetation does not cover the entire soil surface.

In 1982, erosion on U.S. cropland totaled 3.1 billion tons per year, with 1.7 billion tons removed via sheet and rill erosion and 1.4 billion tons lost through wind erosion. By 1992, soil loss had been reduced by one-third to 2.2 billion tons per year (Figure 9.11) because of improved soil conservation measures. Soil loss has continued to decline, averaging about 1.7 billion tons per year over the past decade. That number may seem difficult to conceptualize, but it represent enough soil to fill 150 million dump trucks! On a per area basis, 1.7 billion tons of soil equates to approximately 12 tons of soil being lost each year for every hectare of cropland in the United States. Are such rates something to be concerned about? If they are, why do we not hear more about soil loss being a major environmental concern?

To try to answer these questions, soil scientists have turned to a system in which they define how much soil we can afford lose every year without substantially altering soil productivity. The **Tolerable Erosion Value,** or ***T*-value**, defines an upper or allowable rate of soil loss in a particular region. The magnitude of a particular *T*-value depends on the thickness of the soils within a region. Thin soils, where the productive topsoil is less than 10 inch thick, have low *T*-values, about 2–3 tons per hectare per year (which equates to losing about 0.008 inch of vertical soil horizon per year). Deeper soils, where the topsoil is thicker than 60 inch, can afford to lose more soil and therefore have higher T-values, usually in the range of 11 tons per hectare per year (or about 0.04 inch per year). In the United States, *on average*, 11 tons per hectare per year is considered the

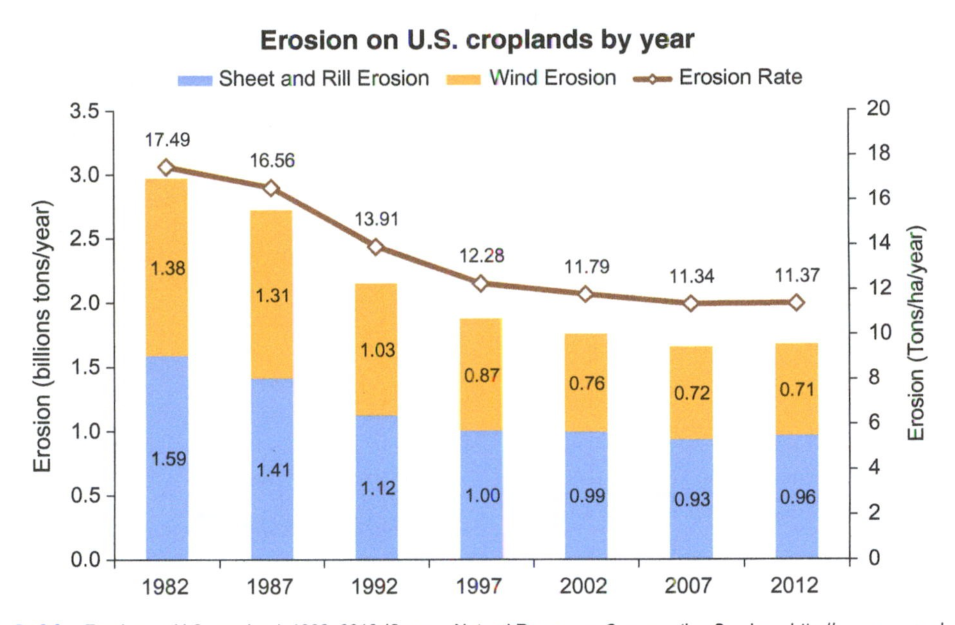

Figure 9.11 Erosion on U.S. cropland, 1982–2012 (Source: Natural Resources Conservation Service—http://www.nrcs.usda.gov/).

maximum allowable limit. Any soil loss that exceeds this amount is deemed excessive. Unfortunately, with an average soil loss rate of 12 tons per hectare per year, we are exceeding the tolerable rate in many areas, although seemingly not by much. Let us try to make 12 tons per hectare per year easier to visualize.

As discussed earlier, we know that soil forms very, very slowly. Soil scientists estimate that it takes at least, *on average*, 500 years to form 1 inch of topsoil under normal agricultural conditions, a rate of about 0.002 inch per year. Successfully growing crops requires approximately 6 inch of topsoil, which means that at least 3,000 years are required to build up a reasonable amount of topsoil. Now, 12 tons of eroded material would equal a depth of about 0.04 inch if spread back out over a hectare (see Figure 9.12). Thus, current erosion rates on U.S. cropland equates to an annual depth of soil loss of 0.04 inch per hectare. Initially, this does not sound like a particularly severe rate of soil loss. However, losing 0.04 inch per year would mean that it would take just 25 years to lose an inch of soil and 150 years to lose the entire 6 inch in the profile. From this, we can say that, *on average* in the United States, soil is being depleted about 20 times faster than is being formed (i.e., 3,000 divided by 150, or 0.04 divided by 0.002). Around the world, soil is being swept and washed away 10–40 times faster than it is being replenished, destroying cropland that are cumulatively the size of Indiana every year!

I have italicized the term *on average* in the above-mentioned paragraphs because it is important to appreciate that soil may be eroding much faster in some areas and less so in others (Figure 9.7, for example). In many parts of the United States, soil is eroding at rates that are 1.5–2 times the annual tolerable rate, which is an unsustainable situation in the long run (Figure 9.13). In some states, such as parts of Missouri, Iowa, and the upper Piedmont of North Carolina, soil is eroding at rates that are 3–4 times the annual tolerable rate. Large areas of West Texas are still losing soil

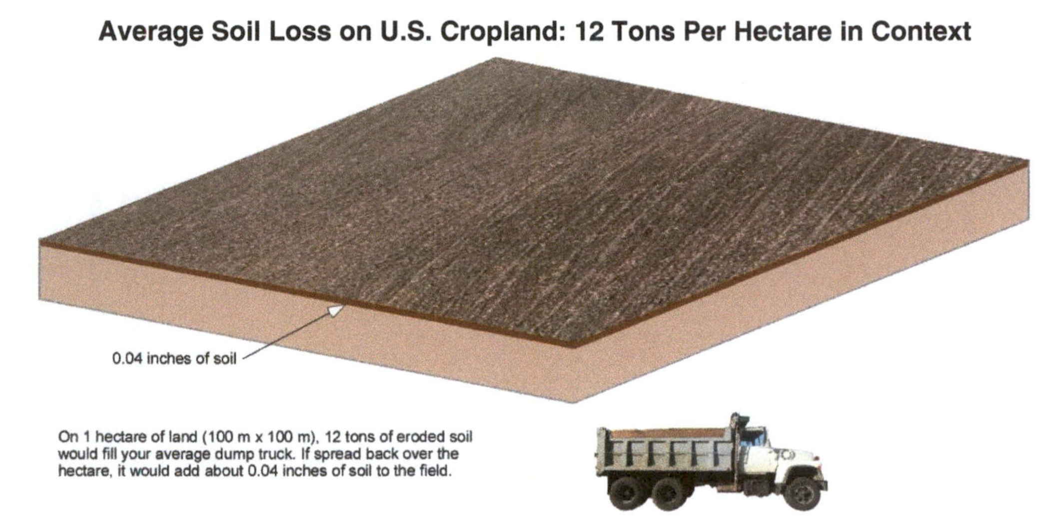

Figure 9.12 Cartoon showing the depth of soil being lost each year (0.04 inches) on 1 hectare at an erosion rate of 12 tons per hectare per year.

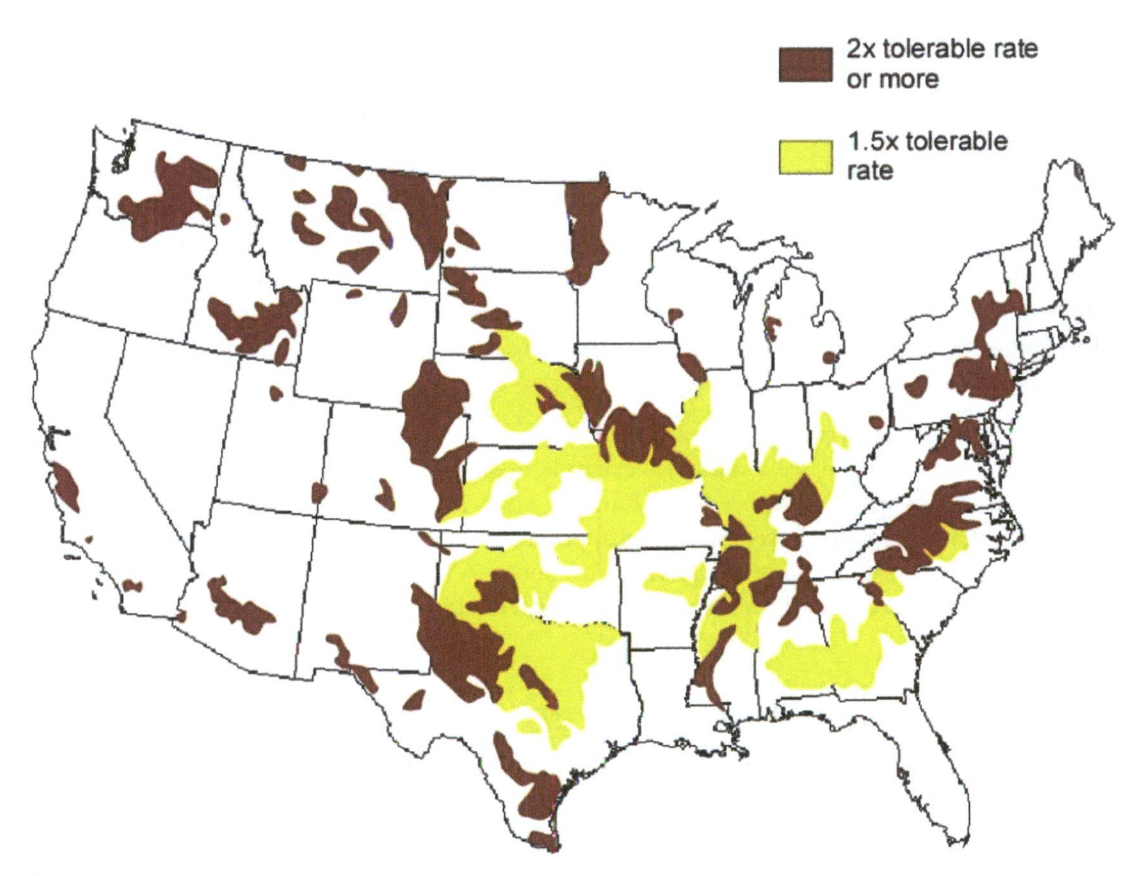

Figure 9.13 Soil loss as a proportion of the tolerable rate, or T-value, in the U.S.

via wind erosion at rates well in excess of the tolerable rate. In many developing countries, reports of soil losses exceeding 100 tons per hectare per year are common, especially on sloping terrain. Gully erosion (a prelude to cropland abandonment) is characterized by erosion rates of roughly 400 tons per hectare per year, a rate that eliminates all topsoil in roughly a decade.

IMPACTS OF SOIL LOSS

According to a report by researchers at Cornell University, damage from soil erosion worldwide is estimated to be $400 billion per year.[3] The economic impact of soil erosion in the United States alone costs the nation about $38 billion each year in productivity losses, although some estimates have put the costs of topsoil loss in the United States as high as $125 billion per year. As you might imagine, these are very difficult calculations to make, since topsoil production rates are so slow,

[3] Source: Pimentel, D. (2006), *Environment, Development, and Sustainability,* Vol. 8: p. 119–137.

and lost topsoil is essentially irreplaceable. Nonetheless, the impacts of erosion are widespread and generally fall into two categories: (1) on-site impacts and (2) off-site impacts.

In terms of on-site impacts, when soil is lost from productive cropland, the remaining soil is generally less valuable. It is thinner and frequently lower in organic matter (OM) content, meaning it loses much of its fertility. OM is the glue that holds soil particles together. Thinner soils also have less water holding capacity, which puts greater stresses on irrigation systems. In areas where wind erosion is prevalent, infertile B horizons commonly become exposed at the surface as topsoil is blown away. The Dust Bowl in the 1930s was primarily caused by blowing away of topsoil when long-rooted grasses in the prairies of Texas, Kansas, Oklahoma, and Colorado were removed during plowing.

Off-site impacts of soil erosion occur as soil is transported down streams and rivers, ultimately reaching the sea. You can see a dramatic example of such sediment output in the Mississippi River as seen from space in Figure 9.14. Ranking sixth in the world in sediment discharge to the oceans, the Mississippi River transports approximately 250 million tons of sediment to the Gulf of Mexico each year. This is equivalent to about 7.5 tons of soil being deposited in the Mississippi delta every second, an amount that would fill about fifteen Ford F-150 pickup trucks, every, single, second!

Figure 9.14 The sediment plume of the Mississippi River is quite visible in this NASA image as it empties into the Gulf of Mexico (Source: www.visibleearth.nasa.gov, provided by the SeaWiFS Project, NASA/Goddard Space Flight Center, and ORBIMAGE).

During its delivery from source (areas of erosion) to sink (areas of deposition), sediment-laden water may flood properties along its route, causing considerable damage. However, the major environmental impact is the siltation of lakes and reservoirs. Dams can potentially capture a river's entire sediment load. As the sediments accumulate in the reservoir, the dam gradually loses its ability to store water. According to the United Nations Environment Program, the world's dams are losing 1% per year of their water-holding capacity due to accumulated silt. If we tried to remove all the sediment in existing lakes and reservoirs that has already accumulated, estimates by the World Bank suggest that the cost would be between \$130 and \$200 billion per year.[4]

The Aswan High Dam in Egypt is a classic example of a large-scale reservoir sedimentation (Figure 9.15). Completed in 1971, the Aswan High Dam created Lake Nasser, which stretches back some 170 miles from the dam wall. It brought approximately 520,000 hectares of new land along the Nile River into production through irrigation and extended year-around cropping to another 285,000 hectares. However, every year, almost all the 134 million tons of silt transported by the Nile above Lake Nasser ends up behind the dam, decreasing its storage capacity. Prior to completion of the dam, the Nile would flood its valley annually during the rainy season, bringing nutrient-rich silt to fields and renewing fertility. However, farmers downstream from the dam now have to use artificial fertilizers and pesticides to keep crop yields high, which further degrades this once pristine floodplain. The originally fertile soil is now considered substandard and a poor quality soil for growing crops.

Numerous reservoirs around the United States are filling with sediment and losing storage capacity at an average rate of 0.2% per year. In California, a number of reservoirs primarily in the Coastal Ranges have already filled or are nearly filled with sediment. Of the approximately 42 million acre-feet[5] of water storage in the state, about 5.1 million acre-feet is currently occupied by sediment.

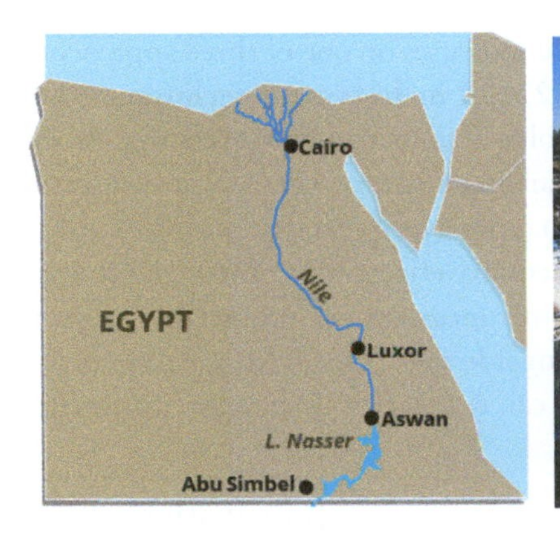

Figure 9.15 The Aswan Dam in Egypt built to control flooding, provide increased water storage for irrigation, and generate hydroelectricity. The dam has starved the lower river of sediment, causing erosion of the delta in many areas.

[4] World Bank (www.worldbank.org).

[5] An acre-foot is a measure of water supply for irrigation. It represents the volume of water required to cover one acre to a depth of one foot (1 acre-foot = 1,233 m^3).

This sediment takes up 12% of the state's water supply (the cost of reservoir storage capacity loss in the United States has been estimated annually at $819 million/year).[6]

As noted earlier, the transport of sediment from eroded fields to the ocean is not a direct and continuous route. Hydrologists estimate that an average of 25% of soil lost through erosion actually makes it to the ocean. The other 75% is deposited within drainage basins, in reservoirs, on river flood plains, or in the river bed itself. This reduced storage capacity makes waterways more prone to flooding and contamination from fertilizers and pesticides in soils.

Finally, and perhaps most importantly, global soil loss is having a profound impact on our ability to meet rising food demand. As outlined in Chapter 2, there is currently about 1.5 billion hectares in crop production, with most of the remaining potential arable land already grazed by livestock. Given the population projections over the next decades, feeding a further 2 billion people will require either significantly increasing crop yields or putting much more land into production. Agronomists generally agree that, while some new land could be brought into cultivation, the competition for land from other human activities makes this a costly and unlikely solution. We could turn to the tropical forests and subtropical grasslands but, as discussed in Chapter 7, farming such marginal lands will produce a short-term return until the land quickly becomes degraded. The ecological cost of bringing such land into production is also prohibitive. The fact that large areas of productive agricultural land are being lost to soil erosion is only confounding the issue of providing long-term food security.

PROGRESS IN SOIL CONSERVATION

The goal of soil conservation is to obtain a sustained level of production from a given area of land while maintaining soil loss below a threshold level. In theory, this permits the natural rate of soil formation to keep pace with erosion. Soil conservation generally focuses on one of three approaches: (1) agronomic measures (i.e., manipulating vegetation), (2) soil management techniques, and (3) mechanical methods. Agronomic measures essentially involve crop management techniques. For example, experiments on Iowa farmland have shown that planting a cover crop (typically a sort of grass) every third year with **minimum tillage** instead of corn crops every year decreases soil loss by about 80%. Using a cover crop, such as fescue grass, is particularly effective in protecting the soil surface from rainsplash and runoff (Table 9.1). Other agronomic measures include **strip cropping** (where crops with a more protective canopy are planted alongside row crops with greater erosion potential), contour farming, and mulch application, which can dramatically increase infiltration and reduce runoff (Figure 9.16).

Soil management techniques focus on preserving or increasing the organic matter content of the soil. A major advance in soil conservation has been the spread of conservation tillage cropland management techniques. These techniques greatly reduce soil erosion, reduce

[6] Source: Minear, J.T. and Kondolf, G.M. (2004), Estimating reservoir sedimentation rates: Long-term implications for California's reservoirs, presented at the Fall 2004 meeting of the American Geophysical Union.

Table 9.1 Soil cover and erosion (based on 14 years data from Missouri Experiment Station, Columbia, Missouri).

Cropping system	Average annual soil loss (tons/ha)	Percent rainfall ending up as runoff
Bare soil (no crop)	41.0	30
Continuous corn	19.7	29
Continuous wheat	10.1	23
Rotation: corn, wheat, and clover	2.7	14
Continuous bluegrass	0.3	12

Figure 9.16 Alternating strips of alfalfa with corn plowed along the contour protects this crop in northeast Iowa from soil erosion (Photo courtesy of the USDA Natural Resources Conservation Service).

farming costs, and conserve water (which increases yields). Techniques like minimum tillage or even no tillage and the practice of leaving the past year's crop residues on the fields have been widely adopted in the United States (Figure 9.17). Over 40 million hectares worldwide are planted under conservation tillage, and soil erosion rates can be reduced by 80–90% with such systems.

Mechanical methods of soil conservation involve increasing soil drainage (by both tile drains and ditches), applying chemical stabilizers to the soil surface, and constructing soil retention structures. Generally, mechanical methods are used in conjunction with agronomic and soil management techniques rather than a stand-alone technological solution to soil erosion. This is because the aforementioned techniques look to prevent directly soil erosion and deterioration, whereas mechanical methods are more geared toward protecting property once runoff and erosion are already occurring.

Major advances have been made worldwide in several aspects of soil conservation. For example, the USDA's Conservation Reserve Program (CRP) has taken much of the nation's most highly erodible croplands out of production (they do compensate the landowners for their lost crops). In four years,

Figure 9.17 Young soybean plants thrive in a residue of what crop. This form of no till farming provides good protection for the soil from erosion and helps retain moisture for the new crop (Photo courtesy of the USDA Natural Resources Conservation Service).

the CRP reduced erosion on participating farms from 9 to 1 tons per hectare. If the CRP reaches its enrollment target of 20 million hectares, it will have saved 0.73 billion tons of topsoil per year. Conservation practices of all kinds have contributed to a 25% reduction in U.S. soil erosion during the past two decades, and we continue to see the benefits today.

CONCLUDING THOUGHTS

The future of the human society is closely linked to the future of its soils yet, alarmingly, relatively little is being done to monitor soil losses and deterioration. Fairly good data exist on the effects of soil loss in developed countries, but, at best, the coverage can be described as patchy. For most developing countries, data on soil loss are still extremely rudimentary.

Erosion is a global problem, and it is heavily affecting our ability to sustain food productivity for our population. Globally, topsoil is eroding faster than it can be replaced on one-third of the world's croplands. Soil degradation is a slow and insidious process that, as one soil scientist put it, literally "nickels and dimes you to death."[7] Yet, controlling soil erosion is really quite simple: the soil can be protected with cover crops when the land is not being used to grow crops.

The pressure on our land is likely to increase in the coming decades. As discussed in Chapter 2, the arable area currently under cultivation is likely to increase only very slowly, if at all. The cost of bringing additional land into production is ecologically high (perhaps even prohibitive). Confounding the issue is the fact that we continue to lose cropland globally at a rate of about 10 million hectares per year. Therefore, it is highly unlikely that the amount of arable land will reach even 2 billion hectares this century, which some suggest is the area necessary to support a population of 9 billion plus people. Farming marginal lands will only provide short-term production benefits as erosion rates are one to two orders of magnitude higher than rates on flat, moist, well-drained bottomlands. If soil losses continue (even at a mere one-tenth of one percent per year), land under cultivation would decline to almost 1.2 billion hectares by 2050. It seems more likely that almost all of the projected increases in food production must come from increased output per hectare (i.e., higher crop yields), rather than from increases in arable land under cultivation.

In the short-term (say 10–20 years), declines in cropland area per capita may not affect us, particularly in the United States. Our consumptive lifestyles and our ability to buy food at any time give us the impression of nutritional security. However, globally, the loss of cropland is a very serious problem because the World Health Organization (WHO) reports that 3.7 billion people are malnourished. If soil conservation is ignored, that number is only likely to grow.

[7] Source: Pimentel, D. (2006), *Environment, Development, and Sustainability*, Vol. 8: p. 119–137.

10

The Water Crisis

"So that's where it goes! Well, I'd like to thank you fellows for bringing this to my attention."

INTRODUCTION

Is it not strange that we call this planet Earth? After all, almost 70% of its surface is covered by water. Of course, most of this is salty water tied up in expansive oceans, saline marshes, and estuaries. Still, there is a perception that our supply of **potable water** (i.e., water of sufficient quality to serve as drinking water) is endless. Vast rivers, like the Mississippi and Amazon, and enormous surface water bodies, such as the Great Lakes, and the countless human-made reservoirs give us the impression that our global water supply is secure and will always be there. Nothing could be further from the truth.

In 2002, the serving EPA Administrator, Christie Todd Whitman, told Congress that "water will be the biggest environmental issue in the 21st century, in terms of both quantity and quality." Although water is relatively abundant in the United States, current trends in population growth and development are already straining our water resources and will continue to do so for years to come. Infrastructure supporting our water systems has deteriorated (the Flint water contaminant crisis in Michigan jumps to mind), and contamination of surface and groundwater resources threatens public and ecological health. For example, New York and other major U.S. cities are distributing water through pipes that are more than a century old. Researchers from the Water and Health Program at the Harvard School of Public Health found that maintenance and repair of the public water infrastructure has been severely neglected, and more than $150 billion must be spent over the next two decades to guarantee the United States will continue to have high-quality water.[1]

Such problems pale in comparison to other parts of the world, where severe droughts and chronic water pollution are an ongoing concern. The statistics are alarming. According to the World Health Organization (WHO), at least 2 billion people use a water source contaminated with feces and almost a million people die each year from diarrhea and dehydration as a result of unsafe drinking water. Tragically, over 90% of these who die are children under the age of five.[2] As a result, several global organizations like the WHO and the United Nations have declared a worldwide water crisis.

THE HYDROLOGICAL CYCLE

If we compressed all of Earth's water into one sphere, its diameter would be about 860 miles (the distance from Salt Lake City, Utah, to Topeka, Kansas) with a volume of 1.4 billion cubic kilometers. This sphere includes all of the water in the oceans, ice caps, lakes, rivers, groundwater, atmospheric water, and even the water in you (Figure 10.1). About 97.5% of this volume is saltwater in the oceans; terrestrial freshwater systems contain just 2.5%. In terms of water supply, we are primarily concerned with available freshwater. Herein lies a surprising fact: approximately 87% of our freshwater is unavailable because it is stored in glacial ice and snow. Of the remaining 13% (i.e., liquid freshwater), 95% is stored as groundwater, often at great depths. What this really amounts to is that water in lakes and streams is just a tiny fraction of the total amount of freshwater on Earth,

[1] The full article can be downloaded from www.ehponline.org/docs/2002/suppl-1/toc.html.

[2] Source: http://www.who.int/news-room/fact-sheets/detail/drinking-water

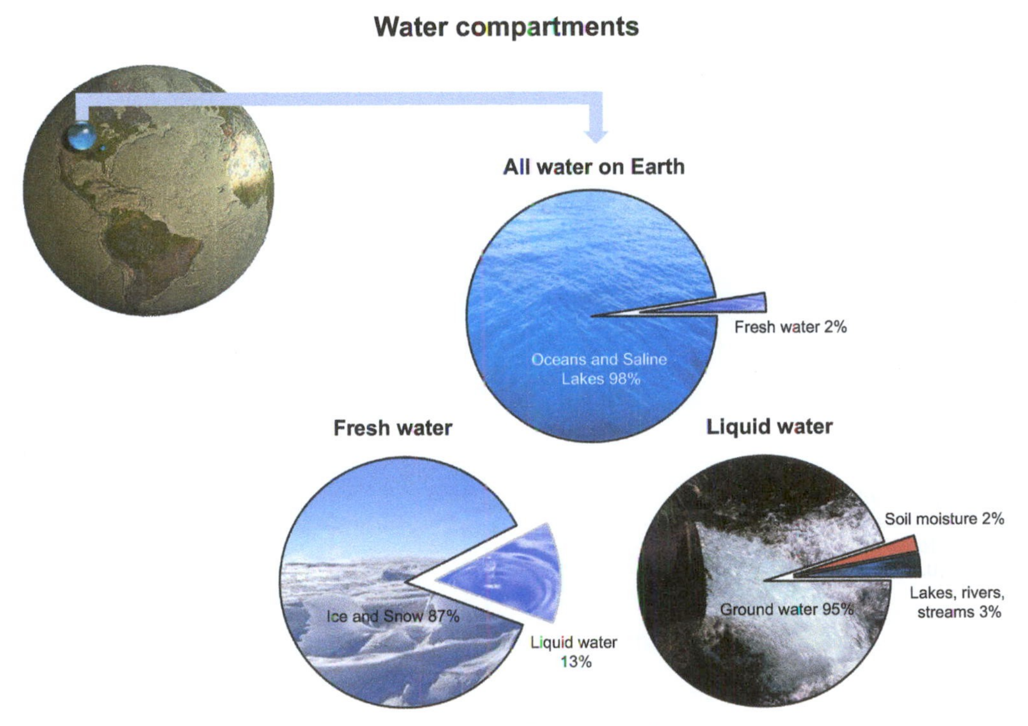

Figure 10.1 Water compartments on Earth (percentages up to whole numbers). The blue sphere on the globe in the upper left of the diagram represents all water on Earth, which would cover an area in the Western U.S. with a sphere with a diameter of 860 miles. The tiny blue dot represents all of Earth's freshwater. Source of images: USGS.

comprising less than 0.5%! If you are having a hard time visualizing this, imagine all of the world's water squeezed into 26 one-gallon milk containers (Figure 10.2). Our freshwater would occupy three-quarters of a single gallon, and readily available freshwater would fill a single teaspoon! Thus, the notion that we have this unlimited supply of freshwater is simply incorrect.

Scientists refer to our planet as a **closed system**, which means that very little material (including water) escapes into outer space. This means that the water that existed on Earth millions of years ago is the same water that exists today. Water is used and reused over and over again. Every glass of water you drink contains water molecules that have been used countless times before. This brings us to a major point: water moves, continually cycling both around and through the Earth as water vapor, liquid water, and ice.

The pathways by which water constantly moves through the Earth's atmosphere system are known as the **hydrological cycle.** A simplified version of this cycle is shown in Figure 10.3. Water is cycled quickly between Earth's surface and the atmosphere via **evaporation** (water changing phase from liquid water into water vapor), **transpiration** (water moving through plants and back into the atmosphere), and **precipitation**. Annual global precipitation is more than 30 times the atmosphere's total capacity to hold water. This means water is continually being transported between Earth and the atmosphere. Interestingly, our atmosphere, which plays such an important role in weather, contains less than 0.001% of the total water volume.

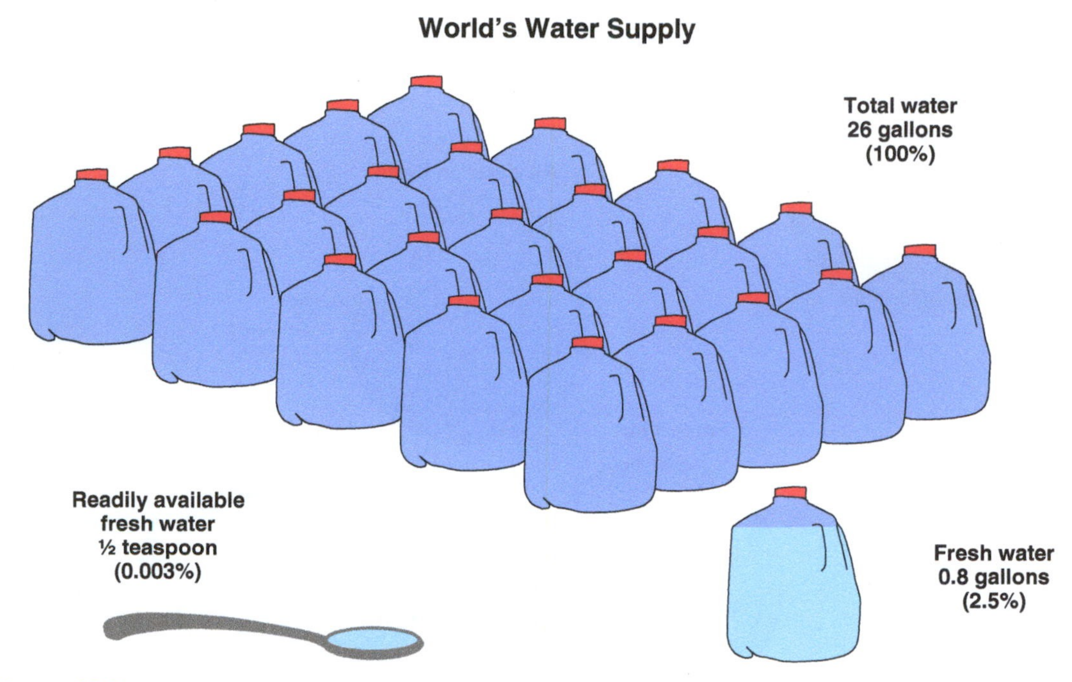

Figure 10.2 The world's water supply expressed in terms of one-gallon containers.

The terrestrial component of the cycle is more complex. Water can be stored within lakes and streams as groundwater (via infiltration through the soil) or as glacier ice. Residence times associated with these pathways can be short (e.g., rivers transporting water quickly back to the ocean) or very long (e.g., water tied up in glaciers). Notice also that there is more evaporation of water from the ocean surfaces (425,000 km^3) than returns via rainfall[3]—a difference of approximately 40,000 km^3. Of course, the oceans are not emptying of water. This excess water is ultimately transported over continental areas by the global winds, where it joins another 71,000 km^3 of water that then falls as terrestrial precipitation. This 40,000 km^3 is critical because it represents **runoff** (the difference between terrestrial rainfall and evapotranspiration) that recharges subsurface water stores (groundwater reservoirs) and surface water bodies, such as lakes and streams, before ultimately returning to the oceans.

The hydrological cycle is a useful conceptual tool that helps visualize water flux at the global scale. However, it does not show the considerable spatial variability in the global distribution of rainfall. As shown in Figure 10.4, the tropics (i.e., between 23.5° North and 23.5° South latitudes) receive abundant rainfall, whereas other parts of the globe are extremely dry. In the United States, the south is humid and wet due to evaporated moisture originating from the warm waters of the Gulf of Mexico. A marked decrease in moisture occurs west of the Mississippi River, punctuated by

[3] 1 km^3 = 264.2 billion gallons of water.

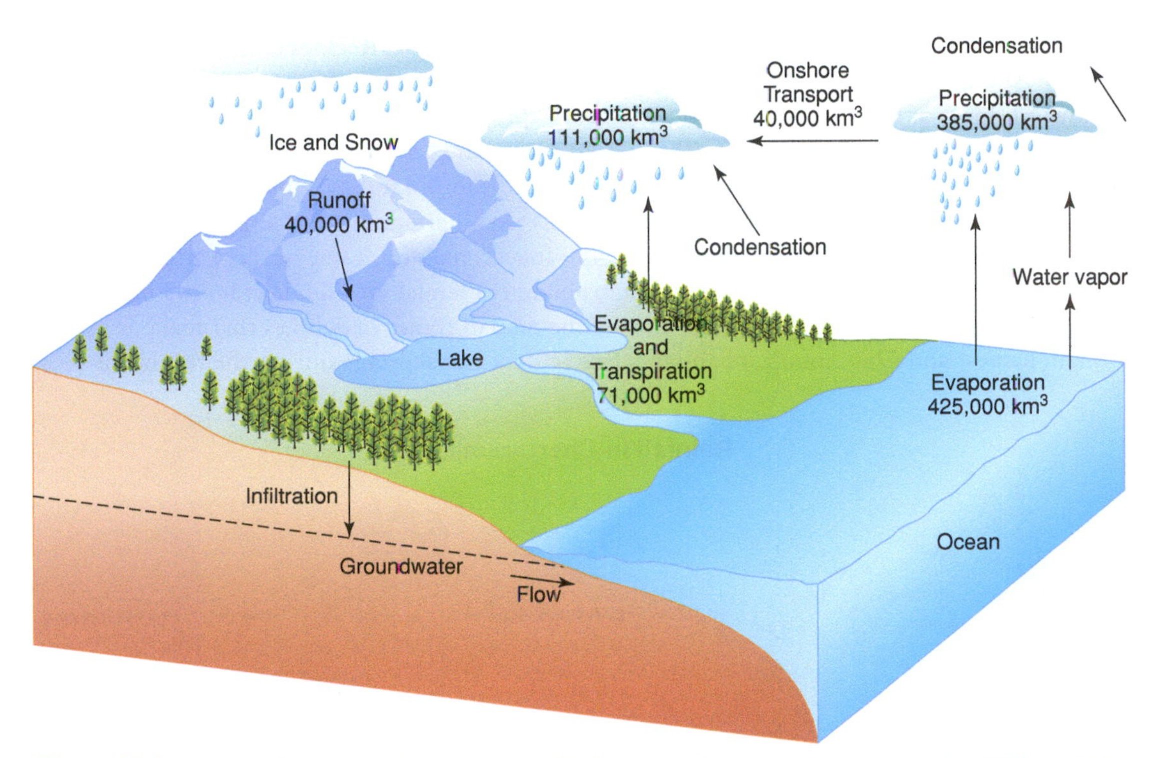

Figure 10.3 Schematic illustration of the hydrological cycle. Note that evaporation over our oceans is greater than rainfall, and rainfall over land is greater than evapotranspiration, by approximately 40,000 km^3. This amount ends up being runoff, the critical component of the cycle that we have to manage.

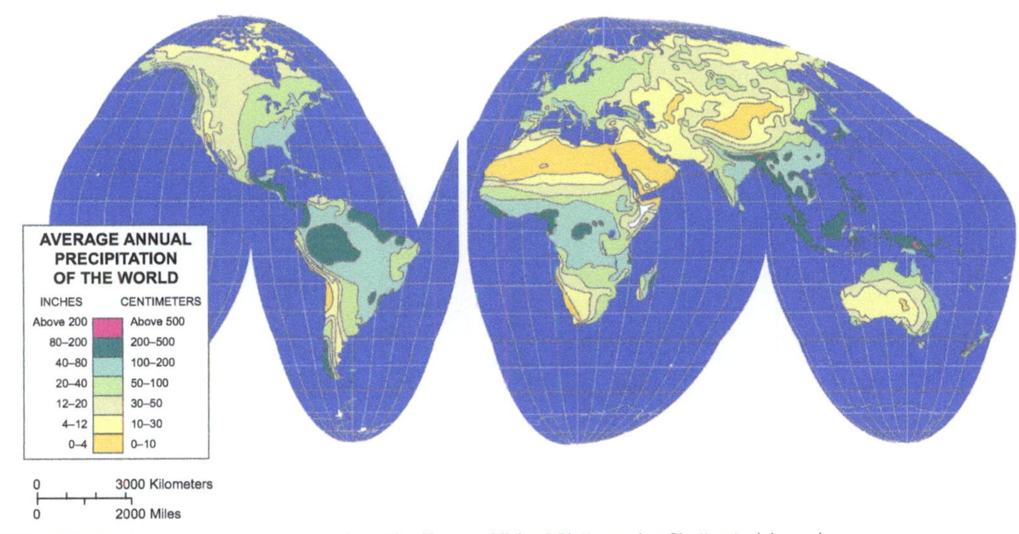

Figure 10.4 Annual precipitation across the globe (Source: Michael Slattery using Shutterstock image).

corridors of rainfall along the great mountain ranges (notably the Rockies and the Pacific coastal ranges). Along the west coast, moist, cooler air originating from the Pacific Ocean is forced up the windward slopes of the mountains producing **orographic rainfall** (or mountain-induced precipitation). This process produces some of the rainiest places on Earth. During South Asia's wet monsoons, moisture-laden air flowing off the Indian Ocean is forced up the Himalayas (Figure 10.5), producing annual rainfall of incredible proportions. In Cherrapunji, India, a small town at an elevation of 4,500 ft, annual rainfall averages 450 inches, making it one of the wettest places on Earth. It holds two world records: the maximum amount of rainfall in a single year (905 inches) and a single month (366 inches). Ironically, most of the rainfall comes in deluges that make it impossible to harvest and Cherrapunji inhabitants face acute water shortage during the rest of the year.

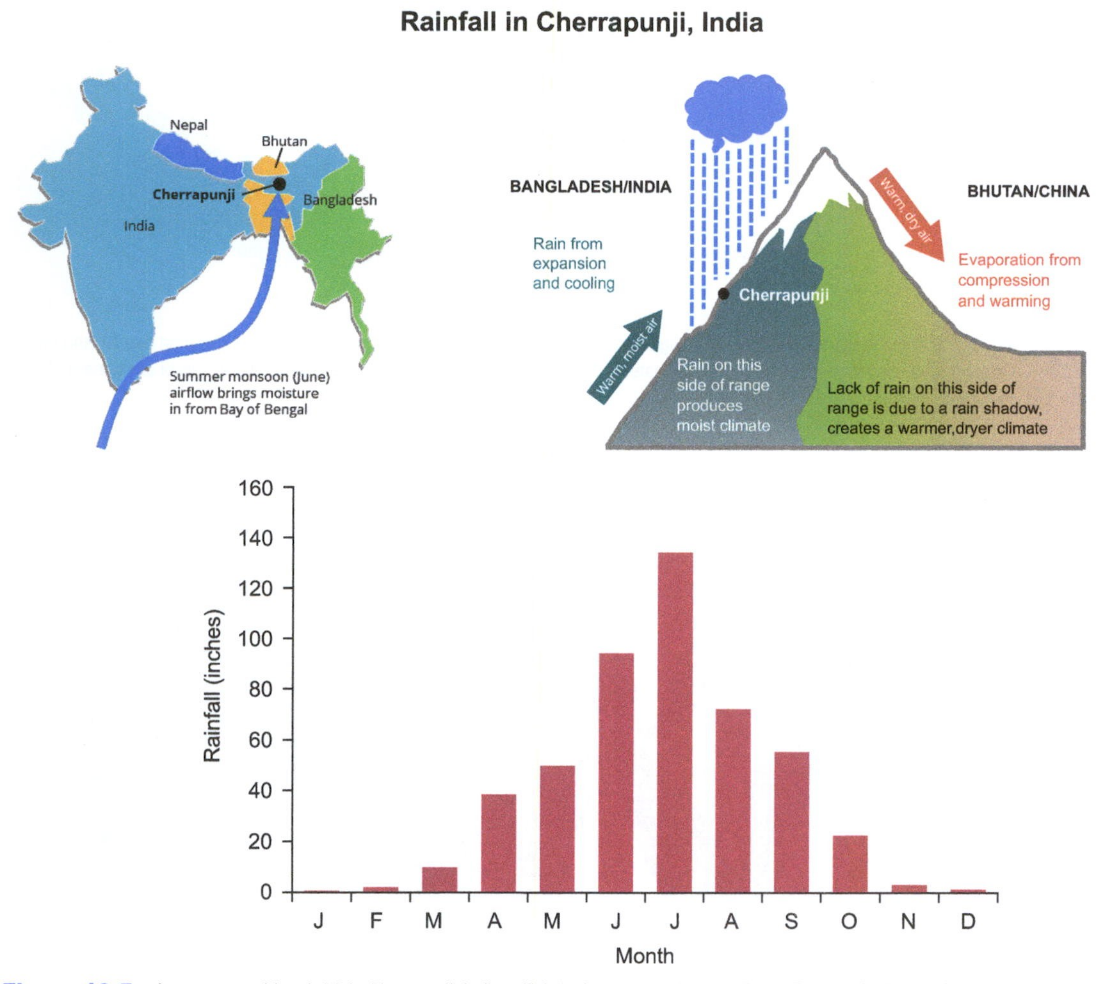

Figure 10.5 Average monthly rainfall in Cherapunji, India, officially the wettest place on Earth. This rainfall is largely orographic in origin, stemming from the Asian Monsoon bringing warm, moist air up the Himalaya Mountains.

GROUNDWATER AQUIFERS

Since groundwater composes a major part of our water supply, let us talk about it a little more in depth (no pun intended!). Once water infiltrates the ground, it is called subsurface water, or **groundwater**. This water can remain in a soil layer near the land surface where gaps between the soil particles (also known as pores) are filled with both water and air. This is called the **zone of aeration** and is the zone where most plants obtain their water (Figure 10.6). As water continues to infiltrate the soil, pore spaces become increasingly saturated. When the water reaches an area with no air within its pores, it has reached the **zone of saturation**. The boundary between the zone of aeration and the zone of saturation is called the **water table**. As the amount of groundwater increases or decreases, the water table rises or falls, accordingly. When the entire area below the ground surface is saturated (i.e., the water table is at the surface), flooding occurs because water can no longer infiltrate into the ground.

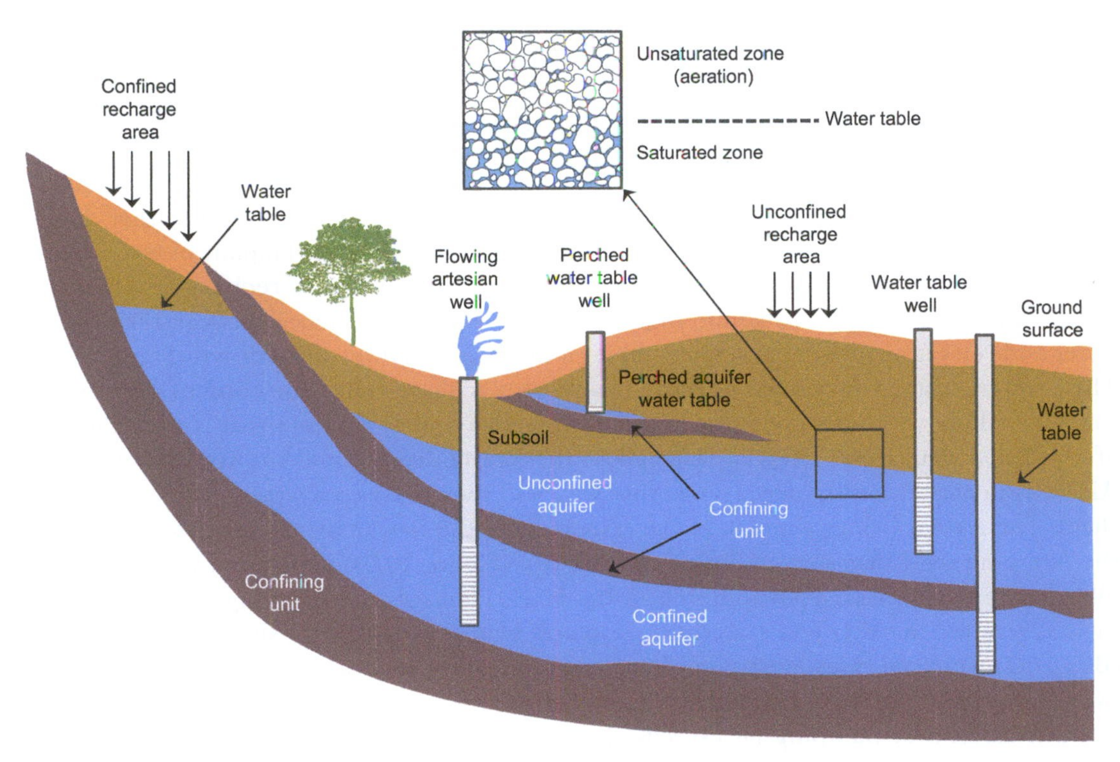

Figure 10.6 Schematic illustration showing the distinction between confined and unconfined aquifers. Notice in the confined aquifer, water is held under pressure and if the pressure is high enough, water can rise and flow at the surface, called an artesian well. In unconfined aquifers, water is recharged directly from infiltrating rainfall. The inset shows the vertical distribution of water through the soil profile. Below the water table, all pore space is filled with water and is saturated. This is groundwater that moves slowly back toward lakes, streams and, ultimately, the sea. Above the water table, soil moisture varies between dry and saturated.

When underground water is held in permeable rock in quantities that make it a significant water store, the formation is said to be an **aquifer**. Aquifers will generally have the following characteristics:

1. A large volume in relation to the amounts of water being removed annually;
2. A moderately high **porosity**. This means that the pore spaces in the rock are large, such as in sandstone, and large amounts of water can be stored within the pores; and
3. A well connected network of pores, fractures, and fissures. This is called **permeability** and speeds up water movement through the geologic unit.

Aquifers are classified as either **unconfined** or **confined**. An unconfined aquifer is one where the water table defines the upper surface of the aquifer. The groundwater within an unconfined aquifer is recharged directly by infiltrating rainfall (Figure 10.6). In a confined aquifer, water is under pressure between two confining layers of very low permeability, called **aquitards** (Figure 10.6). Think of a confined aquifer as a sponge sitting between two pieces of wood. To reach this water, hydrogeologists have to drill wells through the confining layers (i.e., the wood) into the saturated rock (i.e., the sponge). Because the water is under pressure, the water will actually rise up within the well and sometimes even overflow the well (referred to as an **artesian well**). However, most of the time, a pump is necessary to bring the water to the surface. The situation shown in Figure 10.6 is a gross oversimplification of reality. Aquifers are normally highly complex, with multiple confining layers and fractures that let water flow from one layer to the next.

Groundwater resources in the United States have deteriorated over the years, either due to over-pumping and/or contamination. Over-pumping occurs when the amount of water being removed from an aquifer exceeds the amount entering the aquifer as **recharge**. When this occurs, the water table of the aquifer is lowered. The vertical difference between the original water table elevation and the new water table elevation is called the **drawdown** (Figure 10.7). The pumping causes drawdown to occur in a particular shape called the **cone of depression** (imagine taking your finger and pressing it down onto a spongy pillow—that is what the cone of depression would look like). The land above the cone of depression is the **area of influence.** All wells in the area of influence of another well will show a lowered water level, even if they themselves are not being pumped. In the extreme case, the cone of depression is so low around the pumping well that surrounding wells run dry (Figure 10.7). In western parts of Texas, Oklahoma, and Kansas, over-pumping of the High Plains aquifer for irrigation and ranching has resulted in water levels declining in places by more than 100 ft since predevelopment (Box 10.1). In some cases, over-pumping occurs only on a seasonal basis, where aquifers are over-pumped during the dry season but have a net gain in water during other times of the year. In other cases, over-pumping far exceeds recharge, leading the loss of the aquifer as a water source, as is the case with the High Plains aquifer.

Human Impact in the High Plains Aquifer

The High Plains aquifer underlies an area of about 451,000 square kilometers in parts of Colorado, Kansas, Nebraska, New Mexico, Oklahoma, South Dakota, Texas, and Wyoming (shaded area). It consists of several geologic units, the most important of which is the Ogallala Formation, a water-bearing unit consisting mostly of unconsolidated gravel and sand. This formation was deposited by an extensive eastward-flowing system of streams that drained the eastern slopes of the Rocky Mountains about 2 to 6 million years ago. The Ogallala Aquifer provides water resources for the extensive agricultural assets and population of West Texas and the Great Plains. Although the area is characterized as a semi-arid environment, water drawn from the Ogallala Aquifer is used to sustain large-scale irrigated agriculture, livestock production, and rural communities. Simply put, the aquifer supports one of the most agriculturally productive regions in the world. In fact, $20 billion a year in food and fiber depend on the aquifer. However, the problem is that withdrawal of groundwater from the aquifer has now greatly surpassed the aquifer's rate of natural recharge, and there is significant concern about depletion of the aquifer.

Extensive research has been carried out on the Ogallala aquifer at universities throughout the region. For example, at Texas Tech University, researchers have used 42,075 water level measurements from irrigation wells in the aquifer to better understand the characteristics of the aquifer and the relationship between the aquifer and the agricultural landscape. Maps of saturated thickness, rates of aquifer decline, and center pivot irrigations systems have been generated over a fifteen year study period from 1990 to 2004 for 41 counties. These 41 counties cover an area of about 21.4 million acres (33,400 square miles) overlying the Ogallala Aquifer.

The results of this study show that the available storage in the Texas Ogallala Aquifer in 1990 was approximately 403.5 million acre feet and by 2004, it was 354.0 million acre feet. A decline in storage of approximately 49.5 million acre feet or 12% decline was measured during the 15-year period. This means there was slightly less than 1% per year. When a map of center pivot irrigation fields is overlaid on a map of saturated thickness change, it becomes obvious that the most intensive irrigated agriculture has developed where the aquifer is thickest, and these areas closely correspond to the highest rates of aquifer depletion. Many parts of the aquifer have experienced declines of greater than 40 feet over the 15 year period. Scientists also calculated the "time to depletion" or "usable lifetime" of the aquifer, where the aquifer essentially becomes unusable for large-volume irrigation when the saturated thickness drops below 30 feet. Again, the effects of humans on this aquifer are clear: vast areas of the aquifer have less than 30 years worth of usable water, suggesting that the era of irrigated agriculture on the Texas High Plains will probably come to an end within the next generation.

(Continued)

(Continued)

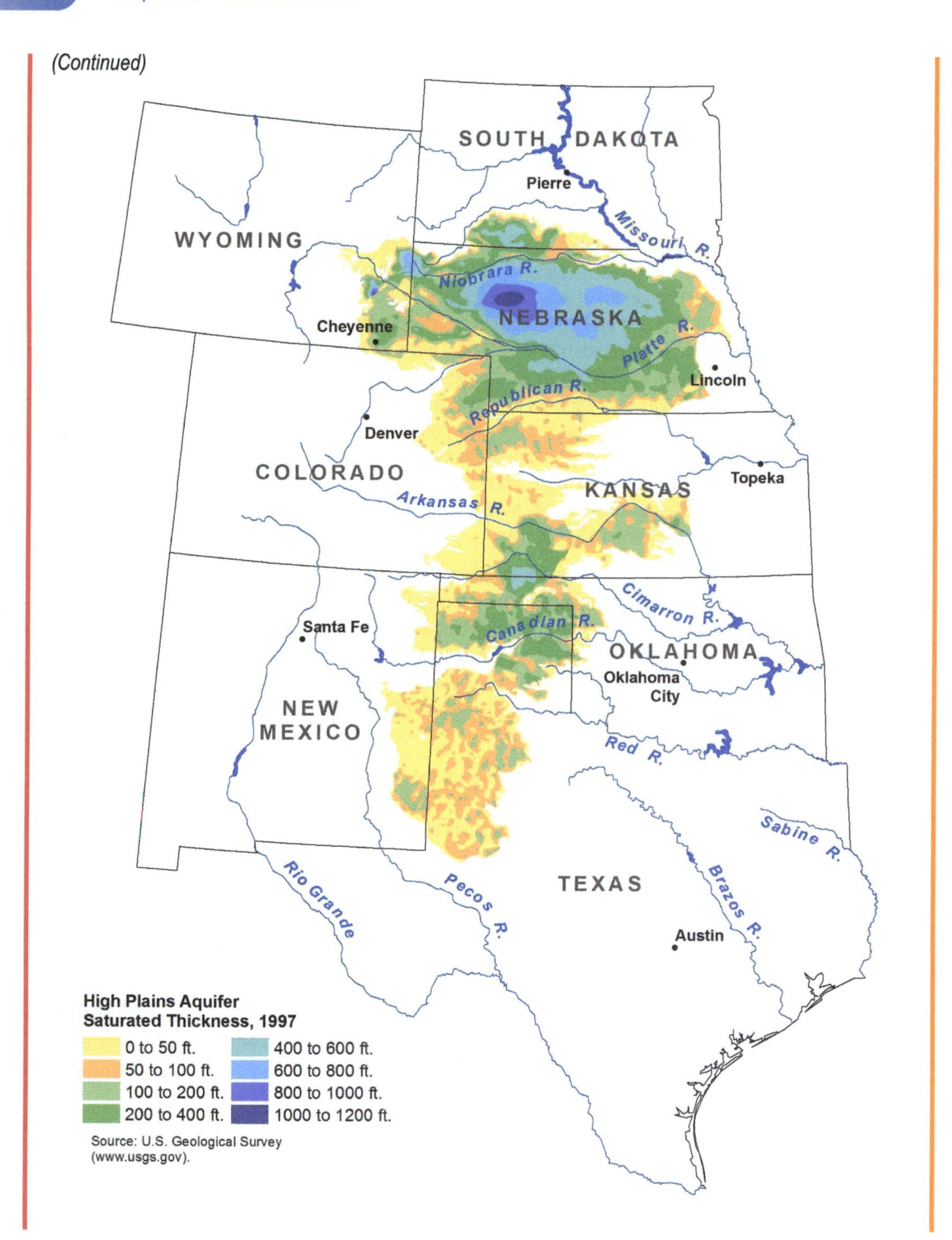
SOUTH DAKOTA
Pierre
Missouri R.
WYOMING
Niobrara R.
NEBRASKA
Cheyenne
Platte R.
Lincoln
Republican R.
Denver
COLORADO
KANSAS
Topeka
Arkansas R.
Cimarron R.
Santa Fe
Canadian R.
OKLAHOMA
Oklahoma City
NEW MEXICO
Red R.
Sabine R.
Rio Grande
Pecos R.
TEXAS
Brazos R.
Austin
High Plains Aquifer Saturated Thickness, 1997
0 to 50 ft.
50 to 100 ft.
100 to 200 ft.
200 to 400 ft.
400 to 600 ft.
600 to 800 ft.
800 to 1000 ft.
1000 to 1200 ft.
Source: U.S. Geological Survey (www.usgs.gov).

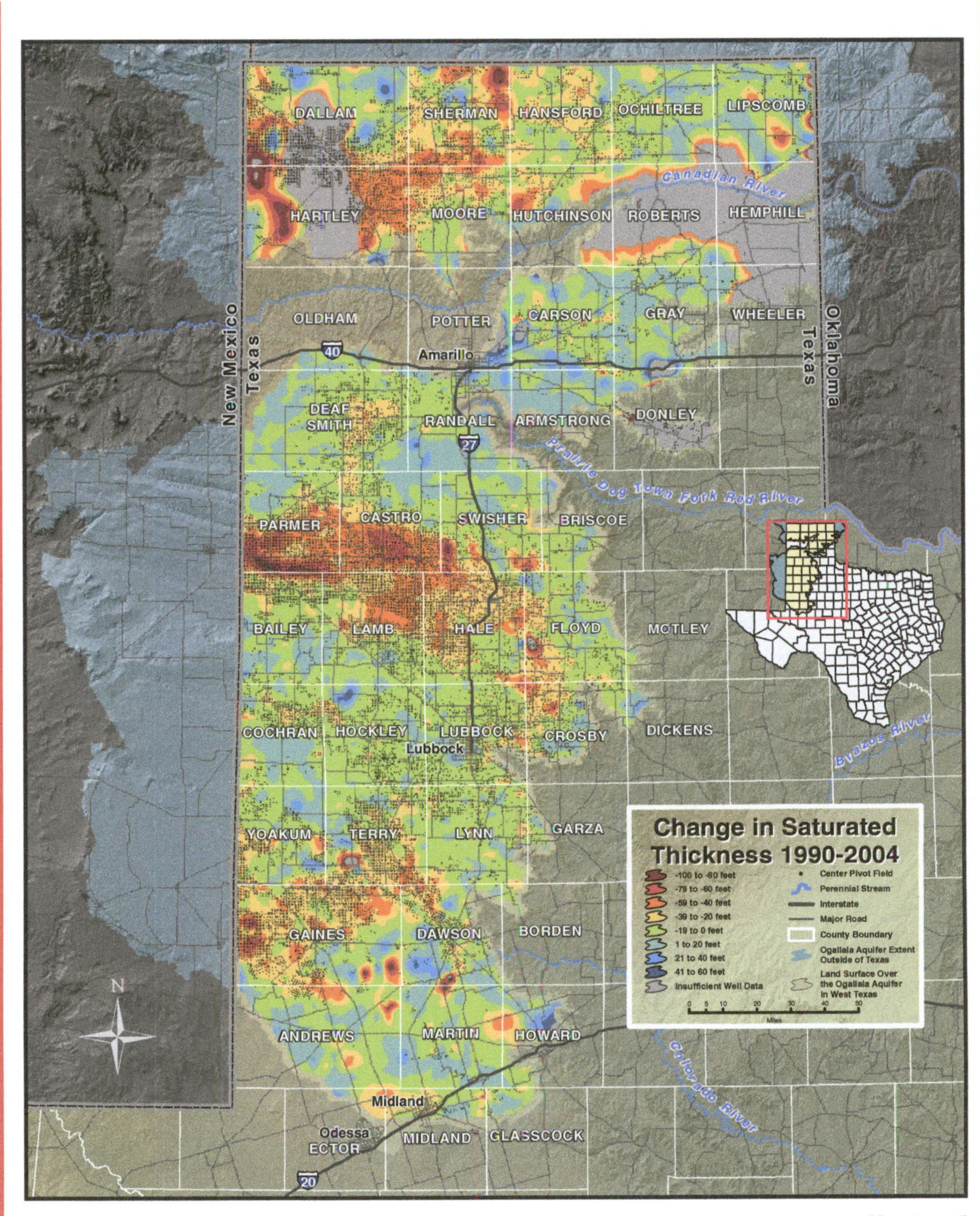
Change in Saturated Thickness 1990-2004
-100 to -80 feet
-79 to -60 feet
-59 to -40 feet
-39 to -20 feet
-19 to 0 feet
1 to 20 feet
21 to 40 feet
41 to 60 feet
Insufficient Well Data
Center Pivot Field
Perennial Stream
Interstate
Major Road
County Boundary
Ogallala Aquifer Extent Outside of Texas
Land Surface Over the Ogallala Aquifer in West Texas
0 5 10 20 30 40 50
Miles
DALLAM
SHERMAN
HANSFORD
OCHILTREE
LIPSCOMB
HARTLEY
MOORE
HUTCHINSON
ROBERTS
HEMPHILL
OLDHAM
POTTER
CARSON
GRAY
WHEELER
Amarillo
DEAF SMITH
RANDALL
ARMSTRONG
DONLEY
PARMER
CASTRO
SWISHER
BRISCOE
BAILEY
LAMB
HALE
FLOYD
MOTLEY
COCHRAN
HOCKLEY
LUBBOCK
Lubbock
CROSBY
DICKENS
YOAKUM
TERRY
LYNN
GARZA
GAINES
DAWSON
BORDEN
ANDREWS
MARTIN
HOWARD
Midland
Odessa
ECTOR
MIDLAND
GLASSCOCK
New Mexico
Texas
Oklahoma
Canadian River
Prairie Dog Town Fork Red River
Brazos River
Colorado River
40
27
20
N

(Continued)

(Continued)

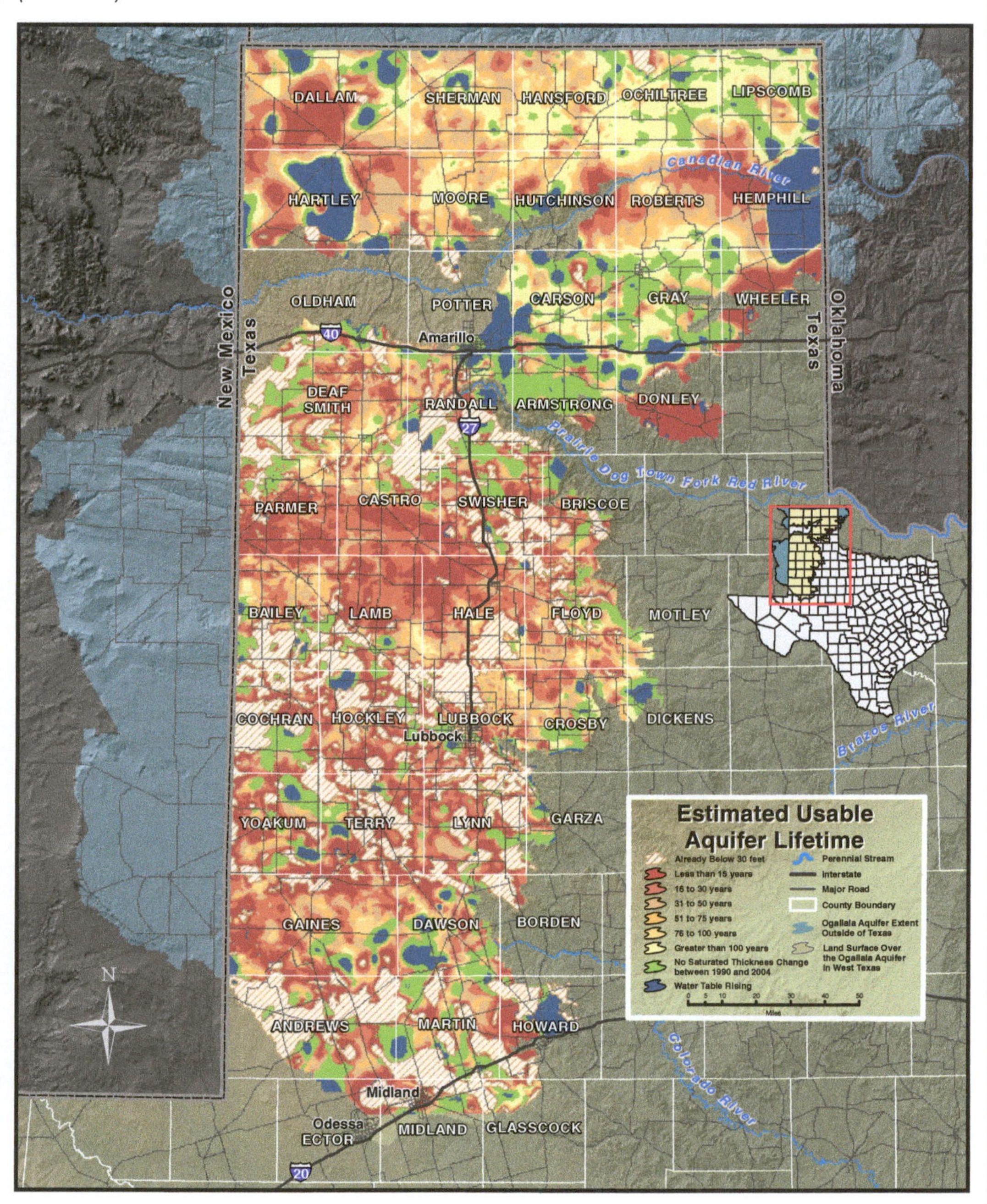

Courtesy of the Center for Geospatial Technology, Texas Tech University.

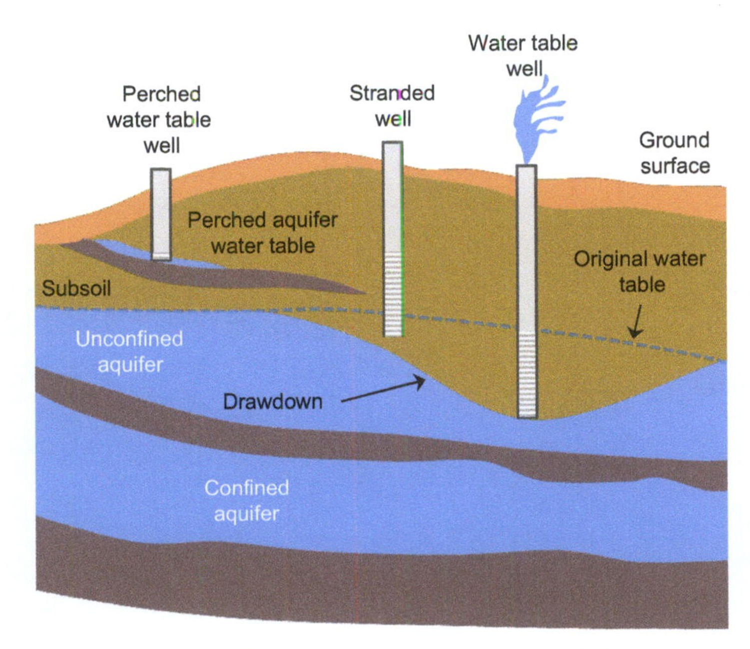

Figure 10.7 A water table being "drawn down" due to pumping from a groundwater well. Notice that the well on the left has become stranded as the water table has fallen below the base of the well.

One major effect of over-puming is **ground subsidence**. This occurs when the reduction in fluid pressure in the pores and cracks of aquifers allows overlying rocks to compact the sediments, lowering the land surface. The Houston area, possibly more than any other metropolitan area in the United States, has been adversely affected by land subsidence, having subsided as much as 9 ft in some areas. In Mexico City, subsidence rates approach 2 ft per year in places, and the total subsidence over the past 100 years is 30 ft. Annual costs in the United States from flooding and structural damage caused by land subsidence exceed $125 million.[4]

Groundwater contamination is also a serious problem, especially when talking about **saltwater intrusion**. Sea water naturally extends under land, but because it is denser than freshwater, it is at much greater depths. However, during pumping, the water level is changed not only at the upper surface of the aquifer (i.e., the drawdown) but also at the lower saltwater/freshwater boundary. Figure 10.8 illustrates the situation that occurs when freshwater is withdrawn from an unconfined coastal groundwater basin. Excessive pumping can lead to saltwater in the well, a situation that is very costly and problematic to reverse once it has occurred. Control of saltwater intrusion in coastal areas occurs by (1) injecting freshwater between the coast and the production wells, (2) constructing recharge basins seaward of the production wells, and (3) placing pumping wells near the coast and discarding the saltwater back into the ocean. Florida, Southern California, and New York are some areas in the United States where saltwater intrusion is particularly problematic, but two-thirds of the United States is underlain by saltwater aquifers. So, the problem is not limited to coastal areas.

[4] https://water.usgs.gov/watuse/

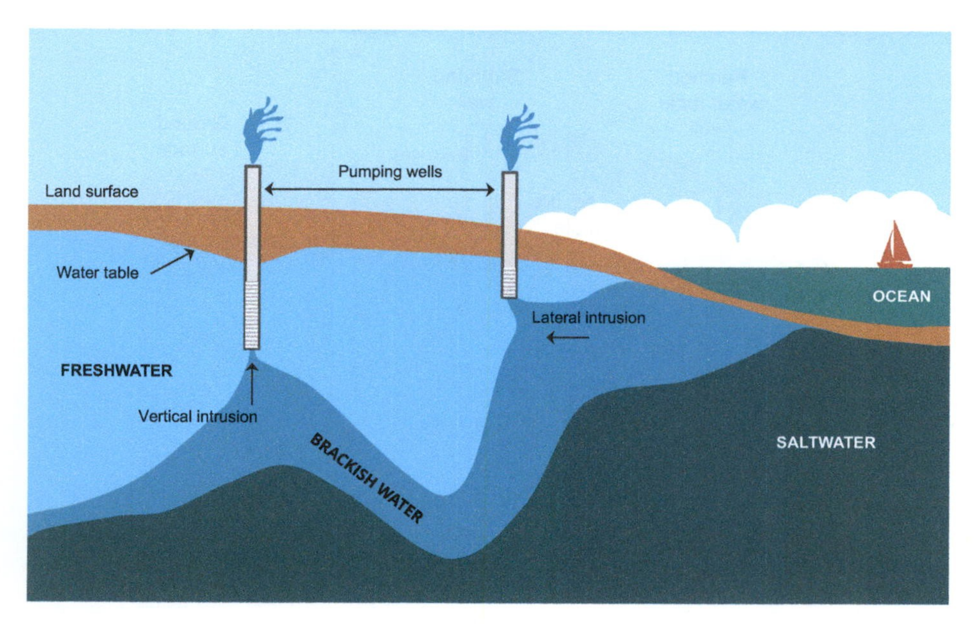

Figure 10.8 Along the coast, freshwater from the land interacts with saltwater forming a "lens" of groundwater. As freshwater is pumped from coastal wells the lens shrinks and saltwater intrudes landward, eventually entering the wells and rendering them useless.

HOW MUCH WATER DO WE USE?

The U.S. Geological Survey (USGS) has been compiling U.S. water-use data every five years since 1950. With the help of local, state, and federal environmental agencies, the USGS can collect site-specific data that includes where water is being used, how it is being used, how much is being used, and even where the water comes from. They can find out how much water is being used to produce power at a fossil-fuel power-generation plant or how much water a farmer uses to irrigate his crops. The data from hundreds of thousands of sites is compiled to produce aggregated water-use information at the county, state, and national levels. The national water-use data system is published on the web[5] and is available to the public.[6]

Total freshwater and saline-water withdrawals in 2015 were estimated to be about 322 billion gallons per day (Bgal/d), equivalent to emptying 490,000 olympic-sized swimming pools each day! Despite this rather large number, our withdrawals in 2015 were actually 9% less than in 2010 (354 Bgal/d) and are continuing a declining trend since 2005 (Figure 10.9). In fact, the 2015 estimates put total withdrawals at the lowest level since before 1970, which is a very encouraging sign in terms of water conservation. The reason for the decline is a significant decrease in thermo-electric power water demand.

[5] http://pubs.usgs.gov/circ/1344/

[6] https://pubs.usgs.gov/circ/1441/circ1441.pdf

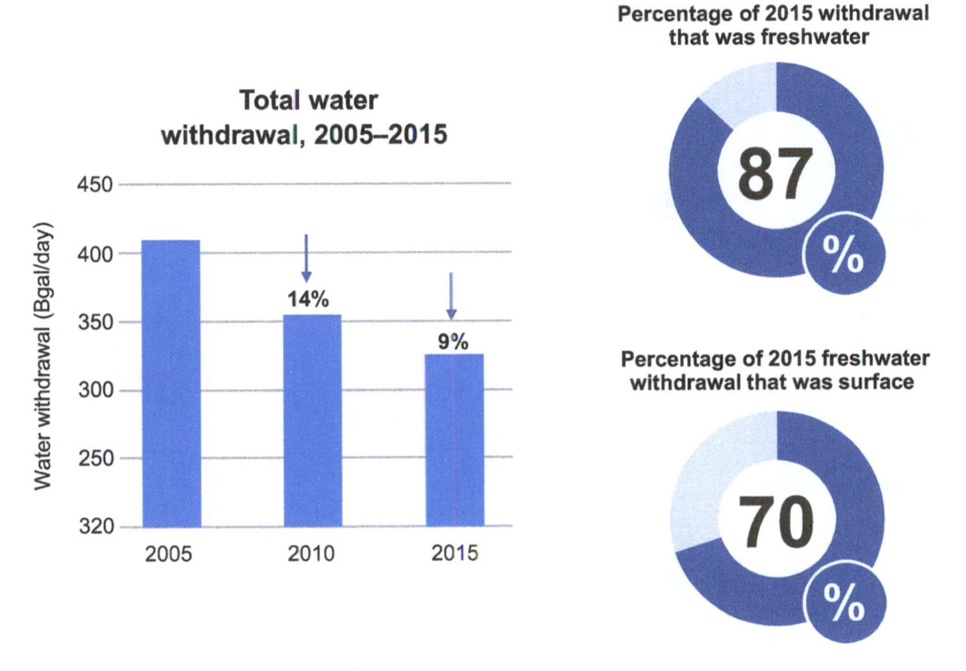

Figure 10.9 Total water withdrawals in the U.S. in 2015 (Source: www.water.usgs.).

Freshwater withdrawals of 281 Bgal/d made up 87% of the total (Figure 10.9) with the remaining 41,000 Bgal/d (13%) saline withdrawals. Most saline-water withdrawals were seawater and brackish (slightly salty, as is the mixture of river water and seawater in estuaries) coastal water used to cool thermoelectric power plants. The majority of our freshwater withdrawals were from surface sources (70%).

In terms of categories of use, about 133 Bgal/d (42% of the total water withdrawal) were used for thermoelectric power (Figure 10.10) which is the lowest level in 45 years. Almost all of this water was derived from surface water and used for cooling at power plants. Irrigation withdrawals totaled 118 Bgal/d for 2015 (37% of total water withdrawal). Historically, more surface water than ground water has been used for irrigation, but this trend is changing. Ground water irrigation withdrawals have increased from 23% in 1950 to almost 50% in 2015.

The geographical distribution of total surface water and total groundwater withdrawals in the United States is shown in Figure 10.11. What this figure primarily shows us is that, despite becoming more efficient, we are still a highly consumptive country with regard to water. Four states (California, Texas, Idaho, and Florida) account for 26% of all water withdrawals in the nation. California accounted for 9% of all withdrawals in the United States. Nearly three-fourths of the freshwater withdrawn in California was for irrigation alone. Withdrawals in Texas accounted for about 7% of the national total and were primarily for thermoelectric power and irrigation. Idaho's withdrawal for for irrigation were the second largest nationwide.

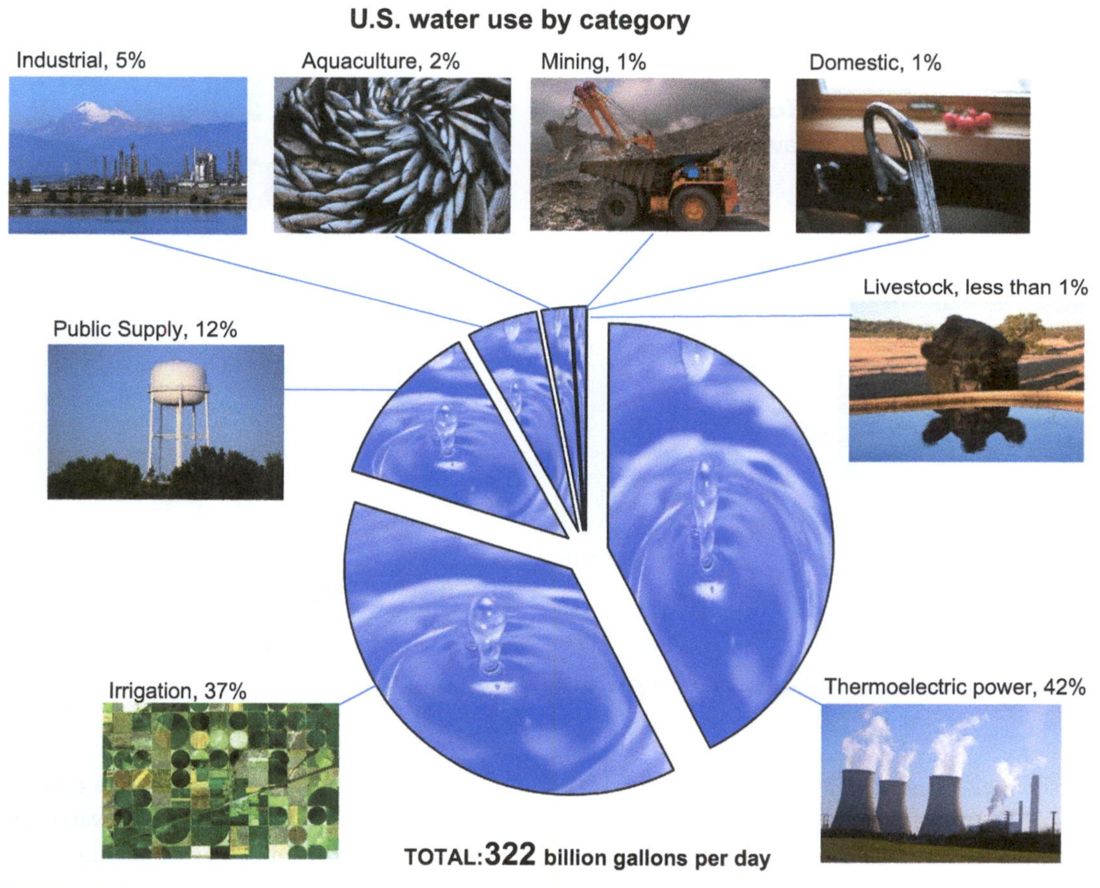

Figure 10.10 Total water withdrawals in the U.S. by category, 2015 (Source: www.water.usgs.).

To set these water data into a broader (and perhaps more digestible) context, a water withdrawal rate of 322 Bgal/d equates to 1,001 gallons of water withdrawn for every person each day,[7] equivalent to filling over 10 large bathtubs to the brim! Of course, this reflects withdrawal for all uses including irrigation and power generation. Nevertheless, estimates suggest that each person in the United States uses about 150 gallons of water per day, on average, in and around the home (called domestic use), for drinking, food preparation, washing clothes and dishes, flushing toilets, watering lawns and gardens, and washing cars.[8] By world standards, that much water is a luxury (Figure 10.12). Furthermore, most of us get our water delivered to our homes and can almost always count on it being sanitary. In the United States alone, almost 34 billion gallons of water are treated every day at water facilities. We simply take it for granted. However, in developing countries, people (primarily women and children) have to walk an average of four miles a day to get water that may or may not be clean. Although it is difficult to estimate the amount of water

[7] This is computed using the 2015 water use data and a 2015 U.S. population of 322.3 million (https://www.census.gov/popclock/)

[8] http://pubs.usgs.gov/circ/1344/pdf/c1344.pdf

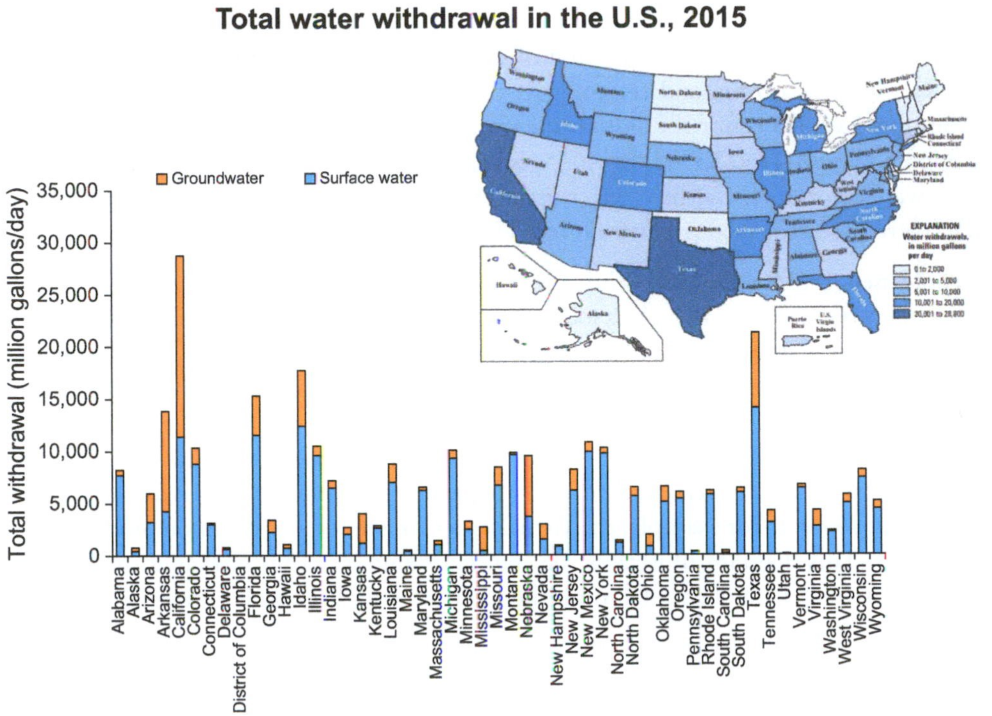

Total water withdrawal in the U.S., 2015
Groundwater
Surface water
Total withdrawal (million gallons/day)
35,000
30,000
25,000
20,000
15,000
10,000
5,000
0
Alabama
Alaska
Arizona
Arkansas
California
Colorado
Connecticut
Delaware
District of Columbia
Florida
Georgia
Hawaii
Idaho
Illinois
Indiana
Iowa
Kansas
Kentucky
Louisiana
Maine
Maryland
Massachusetts
Michigan
Minnesota
Mississippi
Missouri
Montana
Nebraska
Nevada
New Hampshire
New Jersey
New Mexico
New York
North Carolina
North Dakota
Ohio
Oklahoma
Oregon
Pennsylvania
Rhode Island
South Carolina
South Dakota
Texas
Tennessee
Utah
Vermont
Virginia
Washington
West Virginia
Wisconsin
Wyoming

Figure 10.11 Total water withdrawal by state (groundwater and freshwater) in 2015. California, Florida, Idaho, and Texas are the thirstiest states. This is due primarily to water use for agriculture and power generation (Source: www.water.usgs.gov).

Average Water Use Per Person Per Day
Average American
151 gallons
of water per day
55,115 gallons annually
France
76 gallons
Germany
34 g
China
15 g
Japan
99 gallons
India
24 g
Nigeria
6 g
Rwanda
3 g
Brazil
33 g
The world's poorest live on less than 2.5 gallons of water per day

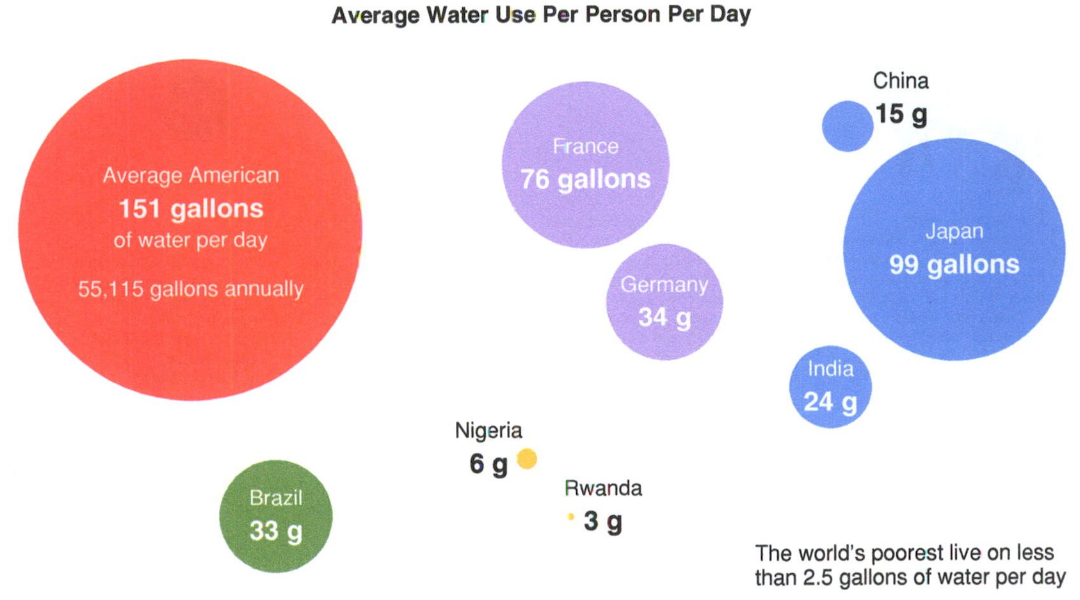

Figure 10.12 Average water use per person per day. The U.S. population has the highest per capita demand of any country (Source: Human Development Report 2006, United Nations Development Program (UNDP).

needed to maintain acceptable or minimum living standards, the World Health Organization suggests that 5 gallons is the minimum amount of water needed to meet a person's daily needs—for drinking, sanitation, bathing, and cooking. At the start of the 21st century, 20% of the world population from over 50 countries did not meet this standard. Major nations on the list include India, Ethiopia, Nigeria, and Kenya. By 2025, two out of three people in the world will experience significant water shortages. By then, water use is expected to have increased 40% and 17% more water will be required for food production to meet the needs of growing populations.[9]

While people use lots of water for drinking, cooking, and washing, we actually use even more in the production of everyday products such as food, paper, and cotton clothes. This indirect use of water is called the **virtual water content** of a product (a commodity, good, or service) and is defined as the volume of freshwater used to produce the product, measured at the place where the product was actually produced. It refers to the sum of the water use in the various steps of the production chain. The adjective "virtual" refers to the fact that most of the water used to produce a product is not contained in the product. The real-water content of products is generally negligible if compared to the virtual-water content. For example, the virtual water content of four common products is shown in Figure 10.13. This diagram illustrates the enormous amount of water that is used to produce 1 kg of beef relative to 1 kg of corn (it might be useful to refer back to Figure 2.18 at this point).

As with many environmental problems, the issue of water supply and use seems a little overwhelming. There is no doubt that we are enormously wasteful with our water supply. The world's golf

Hidden (Virtual) Water in Everyday Items

Beef
3,960 gallons/kg

In an industrial beef production system, it takes an average of three years before the animal is slaughtered to produce about 200 kg of boneless beef. The animal consumes nearly 1,300 kg of grain, 7,200 kg of roughages, 6,340 gal of water for drinking and 1,850 gal of water for servicing. To produce 1 kg of boneless beef, we use about 6.5 kg of grain, 36 kg of roughages, and 41 gal of water. This excludes vast amounts of water need to produce the grains.

Chicken
1,030gallons/kg

By industrial standards it takes ten weeks before the chicken is slaughtered. It will produce 1.7 kg of meat. A chicken consumes about 3.3 kg of grains and 8 gallons of water for drinking and servicing the farmhouse. Producing these grains takes 1,690 gallons of water on average.

Corn
238 gallons/kg

Corn consumes about 145 trillion gallons of water annually, which is 8% of the global water use for crop production. About 10 billion gallons of water is used for producing corn for export.

Tea
32 gallons/kg

To produce one kilogram of fresh tea leaves, 634 gallons are required. One kilogram of fresh tea leaves gives 0.26 kg of made tea, so that one kilogram of made tea (black tea as we buy it in the shop) requires 2,430 gallons of water. A standard cup of tea requires three grams of black tea, so that one cup of tea has a water footprint of 8 gallons.

Figure 10.13 The virtual water content of beef, chicken, corn and tea (Source: www.virtualwater.eu).

[9] World Health Organization: Towards a Healthier Future (report available at http://www.who.int/gb/ebwha/pdf_files/EB117/B117_16-en.pdf).

courses require 2.5 billion gallons of water a day for irrigation. U.S. lawns and landscapes alone claim 7.9 billion gallons of water a day, much of which evaporates or runs off along streets. Now, I am not suggesting that we shut down the golf industry (it is, after all, my favorite pastime), nor am I suggesting that we turn our backyards into concrete wastelands. The simple fact is that there is more than enough water on Earth, but so far, the political will and financial commitments to provide the poor with reliable access to clean water have not been sufficient.

WATER QUALITY

Water quality is a term used to describe the chemical, physical, and biological characteristics of water, usually with respect to its suitability for a particular purpose. Water quality is a complex subject because there is no simple property that tells whether water is good or bad. We frequently hear about microorganisms that have gotten into drinking-water supplies or chemical pollutants that have been detected in streams or seeped into the ground. Water quality has become a very big issue today, partly because of the tremendous growth of the world's population and sprawling urban development but also because of the scale and intensity of agricultural operations.

In 1972, the U.S. Congress passed the **Clean Water Act** (CWA). The CWA established the basic structure for protecting surface water quality, employing a variety of regulatory and non-regulatory tools to sharply reduce and manage pollutant discharges in waterways and runoff. It does not deal directly with groundwater or with water quantity issues. These tools are employed to achieve the CWA's broad goal of restoring and maintaining the chemical, physical, and biological integrity of the nation's waters for "the protection and propagation of fish, shellfish, and wildlife and recreation in and on the water."

At the time of passage of the CWA:

- One-third of the nation's waters were deemed safe for swimming and fishing.
- Wetland losses approximated 460,000 acres annually.
- Agricultural runoff caused three billion tons of topsoil to be lost annually.
- Sewage treatment plants served 85 million people.

Today:

- More than two-thirds of the nation's waters are deemed safe for swimming and fishing,
- Wetland losses approximate 80,000 acres annually.
- Agricultural runoff causes less than one billion tons of topsoil loss annually.
- Sewage treatment plants serve over 180 million people.

Despite such enormous progress, about one-third of our waters are still unsafe for swimming and fishing. In a country as wealthy and technologically advanced as the United States, these numbers are not so great. According to *The State of the Nation's Ecosystems* (a recent report published by the

John Heinz III Center for Science, Economics, and the Environment), the United States may have no streams left that are free from chemical contamination.[10] In this important report, ecological indicators are used to assess the nation's environmental health. One indicator focused on using the concentration of phosphorus, a vital plant nutrient that can lead to huge algal blooms when in excess. About half of all river sites tested had phosphorus concentrations that exceeded the EPA's recommended level.

What, then, should we be concerned about with regard to water quality? What must we do to restore our rivers, lakes, and coastal areas as originally envisaged in the Clean Water Act?

Major Pollutants and Areas of Concern

A key distinction in terms of water quality is that between **point source pollution** and **nonpoint source pollution** (Figure 10.14). Point source pollution refers to effluent being released from a single outlet, meaning it is generally easier to monitor and control. Nonpoint source (NPS) pollution refers to pollution that cannot be linked directly to one specific source, such as when runoff picks up pollutants from agricultural fields and deposits them into rivers, lakes, coastal waters, or even groundwater. Pollutants that are generally NPS include fertilizers, herbicides, and insecticides from agricultural lands and residential areas; toxic chemicals from urban runoff; sediment from

Point and Non-Point Source Pollution

(a)

(b)

Figure 10.14 Examples of point and non-point source pollution. (A) untreated sewage discharging from a pipe (Photo: iStockphoto.com/Nancy Nehring); (B) runoff from a heavy rain carries topsoil from unprotected, highly erodible soils on this field in south-central Iowa (Source: www.nrcs.usda.gov).

[10] Source: http://www.heinzctr.org/ecosystems/index.shtml

improperly managed construction sites, farms, and eroding stream banks; and acid drainage from abandoned mines.

Following the passage of the Clean Water Act, efforts focused on regulating discharges from traditional point source facilities, such as municipal sewage plants and industrial facilities. The CWA made it unlawful for any person to release any pollutant from a point source unless a permit was obtained under the Act's provisions. The CWA was later amended to address critical problems posed by NPS pollution. Monitoring and controlling NPS pollution are far more challenging, as discharge is widespread and often seasonal. NPS pollution remains the nation's largest source of water quality problems and is the reason that so many of our water bodies are not clean enough for uses such as fishing or swimming.

The EPA has identified more than 200 types of pollutants that impair water quality. The number one pollutant is actually the introduction of sediment into waterways. Sediments are loose particles of sand, clay, and/or silt (all less than 2 mm in size) that result from natural erosional processes where soil is removed by wind or water. However, sediment pollution in our streams, lakes, and estuaries is usually the product of accelerated erosion, which, as discussed in Chapter 9, can come from any activity involving significant earth disturbance (planting crops, road, or building construction, etc.). The cost of controlling sedimentation is enormous. The U.S. Army Corps of Engineers alone spends over $1.5 billion each year to dredge sediment from navigation channels at coastal inlets.

The introduction of too much sediment into a stream causes the water to become turbid (cloudy) and can lead to detrimental effects on native aquatic life. Excessive sediment can smother fish eggs, newly hatched fish (called fry), and other aquatic organisms that fish rely on for food. Most game fish (like white bass, smallmouth bass, and salmon) require relatively clear streams for spawning. One study in the Southern Appalachians examined the effects of clay particles on reproductive success in the tricolor shiner, *Cyprinella trichroistia*, a small (less than 3 inch long) fish native to the Southeast. Spawning adults were placed in tubs with varying amounts of suspended clay (the more clay, the cloudier the water). The numbers of spawns were inversely proportional to sediment concentrations, meaning that as the number of clay particles increased, the numbers of spawns decreased.[11] Besides affecting the reproductive success of many fish, a sediment-clogged stream also means less light will penetrate the water, which leads to less photosynthesis of algae and aquatic plants (Figure 10.15).

Aside from sediment behaving as an actual pollutant, it can also act as a storage unit for other pollutants. The pesticide DDT (dichlorodiphenyltrichloroethane) was first released into the environment in the 1940s and has since have been banned in the United States. However, chemicals such as these can remain in our soils for years, attached to individual sediment particles. As water flows through soil or as soil is washed away, these legacy pollutants will detach from the soil and compromise water quality. Simply imagine these pollutants attached to all the sediment shown in the photograph in Figure 10.15!

Overall, it has been estimated that 10% of the sediment underlying our nation's surface water is significantly contaminated. Of the 300 million cubic yards of sediment that are dredged each year

[11] Source: Burkhead, N.M. and Jelks, H.L. (2001), *Transactions of the American Fisheries Society*, Vol. 130: p. 959–968.

Figure 10.15 Sediment chokes this stream due to many years of erosion on nearby unprotected farmland (Source: www.nrcs.usda.gov).

to deepen harbors and clear shipping lanes in the United States, roughly 3–12 million cubic yards are so contaminated that they require special, costly handling. This pollution can affect and even kill small creatures such as worms, crustaceans, and insect larvae that inhabit the beds of water bodies (also known as the benthic environment). These **benthic** organisms can uptake and ingest some of the pollutants in a process called **bioaccumulation**. When larger animals feed on these contaminated organisms, the pollutants are taken into their bodies. The movement and concentration of pollutants within a food chain is called **biomagnification**. When contaminants accumulate in trout, salmon, ducks, and other food sources, they pose a threat to human health.

Other major pollutants include chemical nutrients, such as nitrogen (Box 10.2), which are chemical elements or compounds used in an organism's metabolism. Nutrients in polluted runoff can come from a variety of sources such as agricultural fertilizers, septic systems, home lawn care products, and yard and animal wastes. In a process called **eutrophication**, excess nutrients can leach from soil into groundwater and surface waters, leading to excessive growth of aquatic plants called **algal blooms**. Subsequent decay of these aquatic plants by bacteria can result in foul odors and **hypoxic conditions**[12] (reduced dissolved oxygen levels. **Dead zones** are large areas of hypoxia

[12] *Hypoxia* means "low oxygen." In estuaries, lakes, and coastal waters low oxygen usually means a concentration of less than 2 parts per million. In many cases, hypoxic waters do not have enough oxygen to support fish and other aquatic animals. (http://toxics.usgs.gov/definitions/hypoxia.html).

The Nitrogen Cycle

Nitrogen is the key nutrient in plant growth. Large quantities of nitrogen, in the form of N_2 gas, reside in the atmosphere above the surface of the Earth. However, this form of nitrogen cannot be used by plants. The N_2 must first be changed by microorganisms into organic forms then ionic forms. A simplified version of the nitrogen cycle is shown in the diagram below. In the first step, N_2 gas is taken from the air and changed into forms used by the plants, a process called **fixation**. This is accomplished by microorganisms either in the soil or in root nodules of plants. The first conversion process is **mineralization**, where organic nitrogen is changed into the ammonium form (NH_4^+). Bacteria then convert NH_4^+, first into nitrite (NO_2^-), which occurs slowly and is highly toxic, and then quickly into nitrate (NO_3^-) via **nitrification**. Plants can absorb both NH_4^+ and NO_3^-; however, the latter is highly soluble and mobile in soils which means plants can easily get at nitrate nitrogen. Nitrate can also be converted back to gaseous nitrogen (N_2) through bacteria in the soil and returned to the atmosphere via **denitrification**. Anthropogenic additions to the cycle occur via fertilizer

The Nitrogen Cycle

(Continued)

(Continued)

which can be applied in a variety of nitrogen-based forms. If nitrogen fertilizer is added as ammonium nitrate (NH_4NO_3), the nitrate ions are immediately available for plant use whereas the ammonium ions nitrify to nitrate via nitrification. If nitrogen fertilizer is applied as anhydrous ammonia (NH_3), it reacts with water to produce NH_4^+, which then undergoes nitrification. Nitrogen losses occur primarily through NO_3^- leaching in percolating water and with eroding soil. Leaching losses are increased when plant growth is not sufficient to absorb the nitrates which often occurs because of over-fertilization of agricultural fields.

where oxygen levels are so low that fish and other aquatic organisms cannot survive. The largest of these dead zones occurs every summer in the Gulf of Mexico, covering up to 20,000 square kilometers of one of the nation's most important commercial and recreational fisheries (Figure 10.16). Researchers at Louisiana State University have been mapping the size and location of the dead zone since 1985 by measuring the amount of **dissolved oxygen** in bottom waters at nine off shore stations (see http://www.gulfhypoxia.net/). NASA satellites are also used to monitor the health of the oceans and spot conditions that lead to the dead zone.

The map in Figure 10.16 shows the extent of the hypoxic zone in 2017.[13] It is caused by nutrient enrichment from the Mississippi River, particularly nitrogen and phosphorous. Most of the nitrogen input comes from major farming states in the Mississippi River Valley, including Minnesota, Iowa, Illinois, Wisconsin, Missouri, Tennessee, Arkansas, Mississippi, and Louisiana. Nitrogen and phosphorous enter the river through upstream runoff of fertilizers, soil erosion, animal wastes, and sewage. Figure 10.17 shows just how much nitrate is carried to the ocean each year by the Mississippi River, which drains more than 40% of the lower 48 states. This nitrate load has increased over the past several decades. It carries roughly 15 times more nitrate than any other U.S. river, and the amount of nitrate it carries has approximately doubled since the 1950s.

Why is this important? Well, the hypoxic zone forms in the middle of one of the most important commercial and recreational fisheries in the coterminous United States. The Gulf of Mexico is a major source area for the seafood industry, supplying 72% of U.S. harvested shrimp, 66% of harvested oysters, and 16% of commercial fish. These Gulf fisheries generate about $2.8 billion annually. Consequently, if the hypoxic zone continues or worsens, fishermen and coastal state economies will be greatly impacted. The key to minimizing the Gulf dead zone is to address it at the source. Solutions include using fewer fertilizers in upstream watersheds, controlling animal wastes so that they are not allowed to enter into waterways, and careful industrial practices such as limiting the discharge of nutrients, organic matter, and chemicals from manufacturing facilities.

While the discussion here has focused on just two major pollutants in our waters, it must be recognized that other contaminants enter our streams, rivers, reservoirs, and oceans every day. Such contaminants include pesticides, polychlorinated biphenyls (PCBs), volatile organic compounds (VOCs), and potentially toxic trace elements. Plastic pollution is also becoming a major environmental concern (see Box 10.3).

[13] Data courtesy of http://water.epa.gov/type/watersheds/named/msbasin/zone.cfm

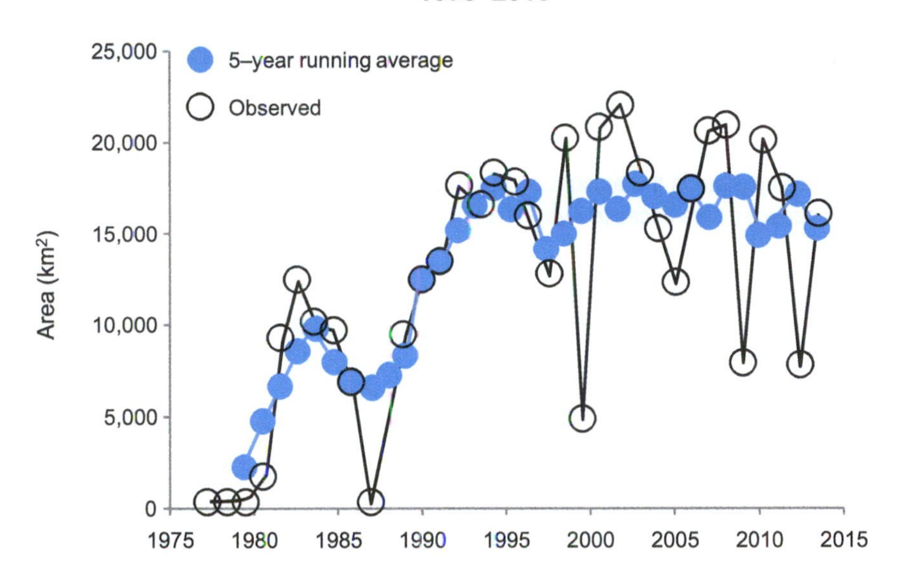

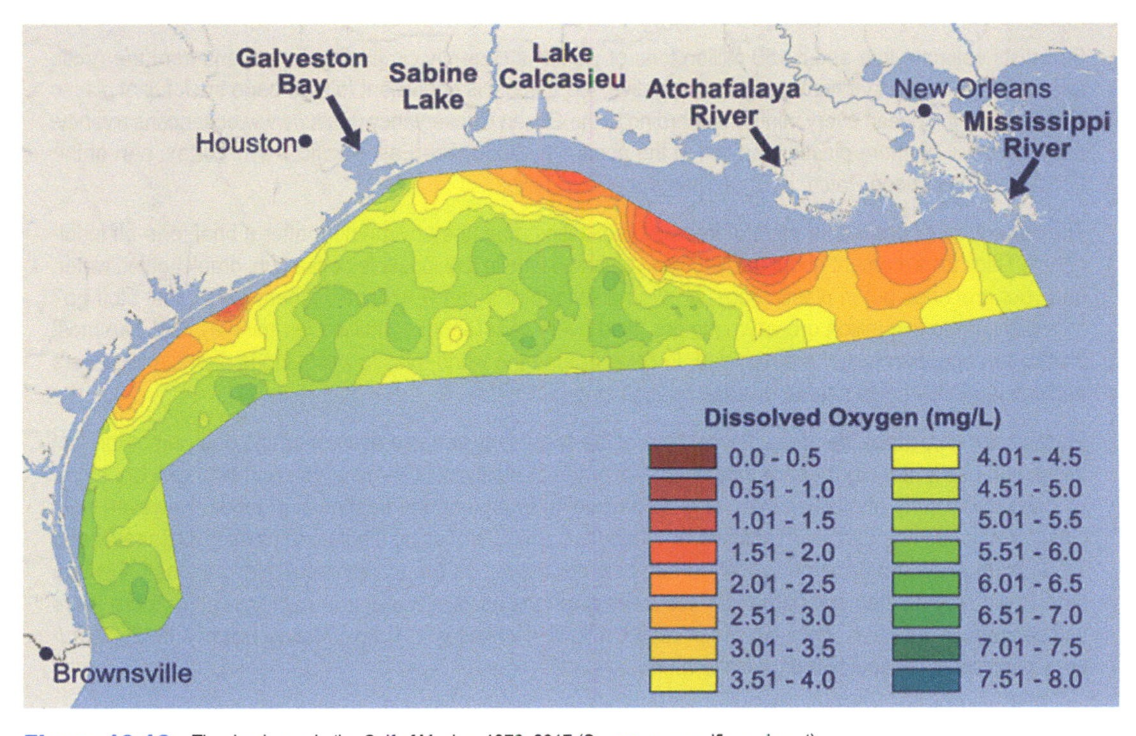

Figure 10.16 The dead zone in the Gulf of Mexico, 1978–2017 (Source: www.gulfhypoxia.net).

BOX 10.3

Plastic Buildup in our waters and Environment

A major water quality issue is the amount of plastic in our environment. This topic could cover an entire chapter, but I introduce it here because a vast amount of the plastic we use ends up in ocean waters through rivers and streams.

Humans produce hundreds of millions of tons of plastic each year. In fact, scientists estimate that over nine billion tons of plastic has been produced worldwide, with about 80% staying in either landfills or the marine and land environment. The problem is that while plastics can break down into smaller pieces, they almost never biodegrade (i.e., break down through natural decomposition into organic matter with common elements such as carbon and nitrogen, elements that can be utilized by living organisms). This has led to a huge buildup of plastic in the environment.

Living organisms, particularly marine mammals, can be harmed through entanglement in plastic objects, such as the plastic rings used in six packs of beer and other soft drinks, or by ingesting plastic waste. Studies have shown that the world's wildlife is drowning in plastic, with a reported 60% of seabirds containing plastic debris in their bodies.

Scientists estimate that about 150 million tons of plastic is currently circulating our oceans from the Arctic to Antarctica. To put that number into perspective, the amount is equivalent to a garbage truck full of plastic dumping into the ocean every minute. According to the Ocean Conservancy (https://www.oceanconservancy.org), there will be more plastic than fish in the oceans by 2050. The East Pacific Trash Vortex, part of the Great Pacific Garbage Patch, is larger than the size of Texas.

The tragedy of all this is that most of the world's plastic enters the environment after a brief, one-off indulgence. Think for a moment about the last time you had a latte in a disposable coffee cup, drank bottled water, had your groceries put in plastic bags, or even disposed of a candy wrapper. Where did all that stuff go? Incredibly, up to five trillion grocery bags are used every year (about 10 million per minute), and, like most plastic garbage, hardly any are recycled. In cities around the world, plastic waste clogs drains and sewers and consequently helps spread disease through stagnant water.

Plastics certainly make life easier for us. One of the first things you use every morning and one of the last things you use every night—your toothbrush—is made of plastics. Every time you visit your grocery store, you see many sorts of plastics that serve as packaging to prolong the freshness of foods. However, anti-plastic sentiment has gathered momentum among the public. Not using single-use plastics like straws and six-pack rings can go a long way toward reducing our impact on the environment. Nevertheless, this is a drop in the ocean when compared with the millions of tons used every year in packaging. For more ways you can reduce your use of plastic, check out the following websites: (1) https://4ocean.com/ and (2) https://www.plasticpollutioncoalition.org/environment/.

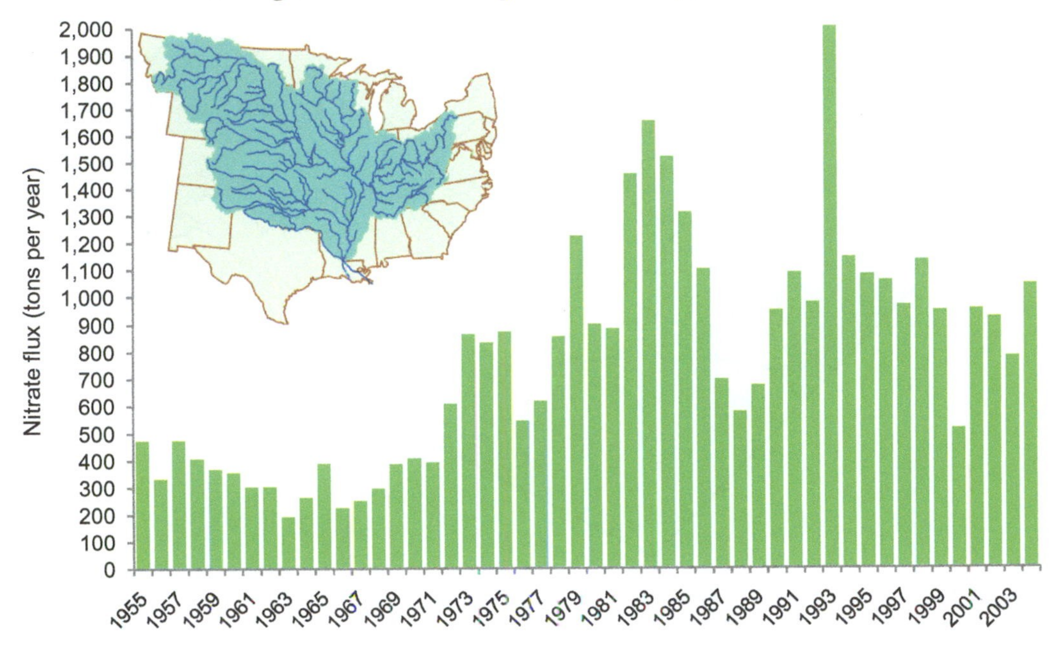

Figure 10.17 Total nitrate delivered by the Mississippi River, 1955–2004 (Source: http://ks.water.usgs.gov/Kansas/pubs/fact-sheets/fs.135-00.html).

A NEW WAY FORWARD: FOCUSING ON WATERSHEDS

Thanks to pollution-regulating laws like the CWA, many of our rivers and waterways are much cleaner than they were 30 years ago. However, NPS pollution remains a serious challenge. The most effective framework to address today's water resource challenges is through what hydrologists call the **watershed approach**. A watershed approach looks not only at the water body itself but also the entire area that drains into it (Figure 10.18).

Right now, wherever you are, you are within a watershed, whether it be a small, tributary watershed in your own neighborhood or the giant Mississippi River Basin, in which thousands of lakes, tributaries, large rivers, wetlands, and estuaries feed fresh water into the northern Gulf of Mexico. It is a coordinating framework that focuses public and private sector efforts to address the highest priority problems within hydrologically defined geographic areas, taking into consideration both ground and surface water flow. This allows communities to focus on the most serious water concerns within an area. The key term in this kind of approach is the word *community*. A watershed approach involves all stakeholders and not just federal scientists and agency officials.

Many public and private organizations are joining forces to create multidisciplinary and multijurisdictional partnerships addressing water quantity and quality problems. These watershed approach programs will result in significant restoration, maintenance, and protection of water resources in the United States. Supporting them is a high priority for the EPA's National Water

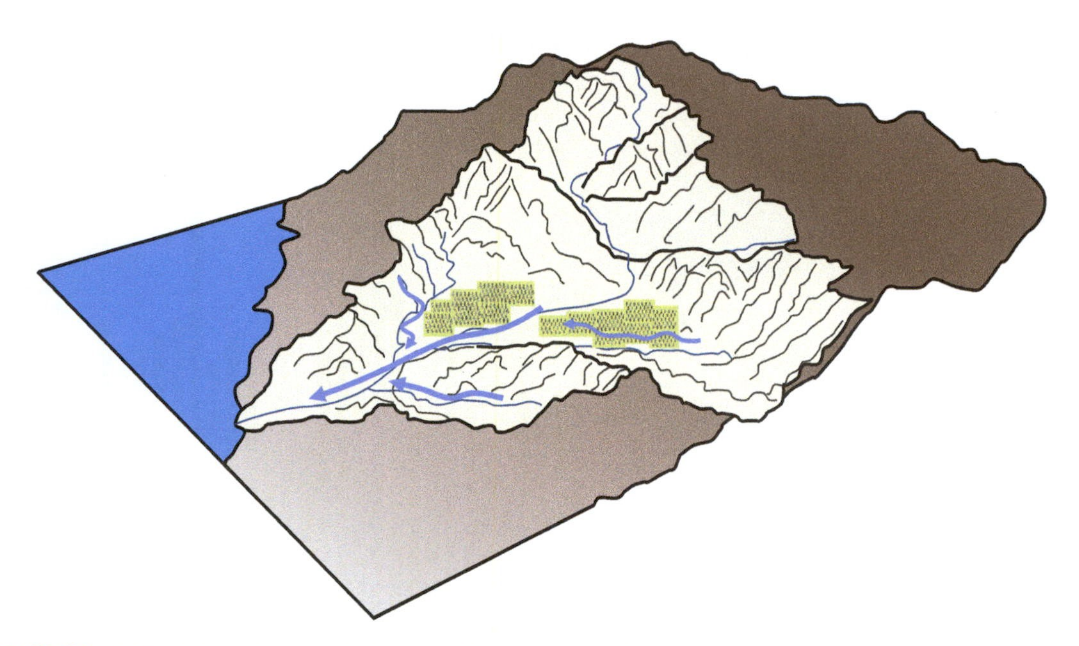

Figure 10.18 Schematic illustration of a watershed. Large blue arrows indicate the overall direction of flow from headwater areas to the basin outlet. Light green shaded areas represent agricultural land.

Program.[14] There is already reason to be optimistic: over 4,000 locally based organizations are involved in community watershed protection efforts. More citizens are practicing water conservation and participating in stream walks, cleanups, and other environmental activities. Such efforts ensure that more of our rivers, lakes, and coastal waters will become safe for swimming, fishing, drinking, and aquatic life.

CONCLUDING THOUGHT

Water is a key environmental issue because it integrates and connects ecosystems and carries political and social connotation. There is a finite amount of water that can be used, and as the global population continues to grow, demand will only increase. The question is how do we deal with this? How do we plan for the future, balancing the water needs of society and the needs of ecosystems? Simply put, we have to become smarter and use our water more efficiently which, as our 2015 water use data show, is exactly what we are doing! Monitoring is critical, and this is where you can really make a difference. For example, in Texas only about 10% of the 200,000 river miles are monitored by water resource professionals. This creates a huge information gap that can only realistically be filled by citizen volunteer monitors. Anybody can become a volunteer monitor, as long as he or she has the desire and is available once a month to help take water samples. Volunteering gives people the opportunity to take an active role in the management of our water resources. Through such active and broad involvement, we can build a sense of community, reduce conflicts, and increase commitment to taking the necessary actions that will ultimately promote long-term sustainability of our water resources.[15]

[14] http://water.epa.gov/resource_performance/planning/

[15] A very useful resource is the EPA's *Handbook for Developing Watershed Plans to Restore and Protect Our Waters* which is designed to help anyone undertaking a watershed planning effort, but should be particularly useful to persons working with impaired or threatened waters (see http://www.epa.gov/owow/nps/watershed_handbook/#contents to download the handbook).

11

There Is No Planet B

"And may we continue to be worthy of consuming a disproportionate share of this planet's resources."

INTRODUCTION

Over the past 70 years, we have been witness to an unprecedented transformation of the human relationship with the natural world.

The Anthropocene, the concept that the Earth has moved into a new geological epoch characterized by human domination of the planetary system, is becoming a widely accepted framework for understanding this transformation. I have no doubt that if we continue with the Great Acceleration in its present form, we will set in motion irreversible climate change and other significant changes to Earth's life-support system. Scientists overwhelmingly agree that we need to adopt a different development model, one that will undoubtedly require transitions to new energy systems. However, one of the great challenges of the 21st century is that there is also a growing disparity between the wealthy and the poor, and, through modern communication, a growing awareness of this gap by the poor is leading to heightened desire of material goods. At the same time, many of the ecosystem services upon which human well-being depends are depleted or degrading with looming (though yet clearly defined) thresholds.

For example, we do not really know how sensitive the global climate is to increases in CO_2. The 40% rise in CO_2 since the Industrial Revolution raises concerns of potentially abrupt and irreversible changes in the planetary environment as a whole (the so-called "Anthropocene rupture" outlined in Chapter 1 and Figure 1.4). There are also many **positive feedbacks** within our Earth-atmosphere system, the consequences of which are difficult to quantify. For example, thawing of **permafrost** soils, such as the Yedoma in Eastern Siberia, will most likely accelerate as the climate continues to warm. Current research estimates that permafrost alone stores the equivalent of roughly twice the carbon in the atmosphere. Projected releases of greenhouse gas emissions from the melting of the Yedoma permafrost are about 2.0 to 2.8 Gt C per year.[1] By comparison, fossil fuel emissions today are roughly 8.8 Gt C per year (Box 6.1, Chapter 6). Thus, we would potentially be adding the equivalent of an additional 30–40% *more* CO_2 into the atmosphere on top of increasing emissions due to global growth.[2] This would likely further accelerate climate warming. The key point is either we turn around many of these trends—the CO_2 trend, deforestation, and so on—or we allow them to continue unabated and potentially push beyond Earth's limits. We need to find ways of identifying what those thresholds are so that we can continue to support humanity without completely ruining the environment we depend upon. As the title of this chapter implies, we only have one planet.

HOW MUCH GROWTH CAN OUR PLANET SUPPORT?

In the early 1970s, a study was undertaken by a team of Massachusetts Institute of Technology scientists on the topic of global sustainability. The study used a computer model of the world economy to show that population and economic growth rates at that time could not continue

[1] Source: Richardson, K. et al. (2011), *Climate Change: Global Risks, Challenges and Decisions.*

[2] Permafrost also stores methane, the powerful GHG that would accelerate warming even further. Although scientists have many more questions than answers regarding methane, estimates suggest that methane from permafrost could eventually equal 35% of today's annual human GHG emissions.

indefinitely on a planet with limited natural resources and limited ability to deal with pollution. They predicted a "rather sudden and uncontrollable decline in both population and industrial capacity"—a prediction that quickly became known as The Doomsday Scenario.[3] The study was published in a very famous 1972 book called *The Limits to Growth.*

Almost five decades on, some scholars and thinkers say that we have already exceeded the carrying capacity of the planet. Others say the Earth can hold billions more. Well, which is it? Is the Earth already overpopulated? Are our resources on the verge of running out? Are we oscillating about some critical threshold or have we moved beyond it?

In ecological terms, the carrying capacity of an ecosystem is the size of the population of a given species that can be supported indefinitely upon the available resources and services of that ecosystem.[4] However, such a definition cannot simply be applied to human beings. Rather, human carrying capacity should be defined as the maximum load that can safely be imposed on the environment by people. In this sense, human load is a function not only of numbers (i.e., the population) but also of per capita resource consumption that can be sustained indefinitely without damaging the functionality and productivity of the biosphere. This definition may be written as:

$$\text{(Total human impact on the ecosphere)} = \text{(Population)} \times \text{(Per capita impact)} \qquad (11.1)$$

Another way of expressing carrying capacity is to provide an estimate of natural capital requirements in terms of productive landscape, called **biocapacity**. So, instead of asking how many people a particular region can support sustainably, the question becomes how much productive land (and water area) is required to support the region's population at their consumption levels? We call this the **ecological footprint** or the area required to maintain any given population (more on this below).

History tells us that humans can exceed the carrying capacity of their particular environment. Easter Island, located about 2,300 miles west of South America and 1,100 miles from the nearest island, is one of the most isolated inhabited islands on the planet. When the Polynesians first arrived on this 63 square mile island, somewhere between 400 and 600 A.D., it was covered with trees with a large variety of food types. Resources must have seemed inexhaustible to the inhabitants, the Rapa Nui. However, they quickly began to cut trees for wood to make fires, houses and, eventually, for the rollers and lever-like devices used to move and erect the Moai, the massive statues that surround the island. Pollen records suggest that the last forests were destroyed by A.D. 1400.[5] With the loss of the forests, the land began to erode. The small amount of topsoil quickly washed into the sea. Crops began to fail, and the clans began to fight over the increasingly scarcer resources.

When Dutch Admiral, Jacob Roggeveen, landed on the island on Easter Day in 1722, he reported 2,000–3,000 inhabitants with very few trees (estimates put the population as high as 10,000 only a century earlier). It was a desolate place, and by 1877, only 100 or so inhabitants remained on the

[3] Source: Meadows et al. (1972), *The Limits to Growth*, p. 23.
[4] Source: http://www.sustainablemeasures.com/node/33
[5] Source: Fenley, J.R. et al. (1991), *Journal of Quaternary Science*, Vol. 6: p. 85–115.

island. As Jared Diamond describes in his best-selling book, *Collapse*, Easter Island is the "clearest example of a society that destroyed itself by overexploiting its own resources." Once the Rapa Nui started clearing trees, they did not stop until the whole forest was gone. Diamond called this self-destructive behavior "ecocide" and points to the civilization's demise as a model of what can happen if human appetites go unchecked.

The ecological collapse of Easter Island is widely attributed to overpopulation, unchecked consumption of natural resources and the introduction of European disease and invasive species. For whatever reasons, the question of how many humans the island could comfortably support never seemed to have come up. As such, scholars often view Easter Island as a metaphor for Earth: If humans, isolated on the planet and unaware of any life beyond, treat our world as the Rapa Nui treated theirs, will we suffer the same fate they did?[6]

Of course, our Earth is not a small, isolated island like Easter Island, and the question of carrying capacity and sustainability is certainly more complicated at the global scale. Still, we must at least attempt to define the ecological boundaries. Some things are certain: another two and a half billion more people in the next 40 to 50 years will need to be fed. They will need access to potable water. They will consume resources, particularly things made of wood. They will almost certainly want to develop and improve their standard of living. And, like all organisms, they will seek out carbon to survive which, in human terms, means a continued reliance on carbon-based fossil fuels.

IS POPULATION REALLY THE DRIVER OF ENVIRONMENTAL DEGRADATION?

Environmentalists have long debated whether human population growth is the *fundamental cause* of our environmental problems. This is a touchy issue, because if population growth is truly the principal cause of environmental degradation, then surely population control should be a (or, perhaps, *the*) central strategy in any environmental protection program.

In 1968, biologist and influential environmentalist, Paul Ehrlich, wrote a best-seller entitled *The Population Bomb*, in which he contended that birth rates in developing countries would probably not decrease enough to avoid environmental collapse due to overpopulation. He observed that developed countries have already transitioned from high birth and death rates to low birth and death rates because of economic and social development (a process called the **demographic transition**, see Figure 11.1). Furthermore, he asserted that developed countries must take the lead in promoting birth control to avert a worldwide ecological disaster, and he promoted this goal by founding the Zero Population Growth organization (now called the Population Connection).[7] This view was (and still is) supported by many environmentalists. At its core is the belief that the

[6] For an interesting look at an alternative theory to Easter Island's collapse (i.e., due to the introduction of rats), see https://www.npr.org/sections/krulwich/2013/12/09/249728994/what-happened-on-easter-island-a-new-and-even-scarier-scenario

[7] http://www.populationconnection.org

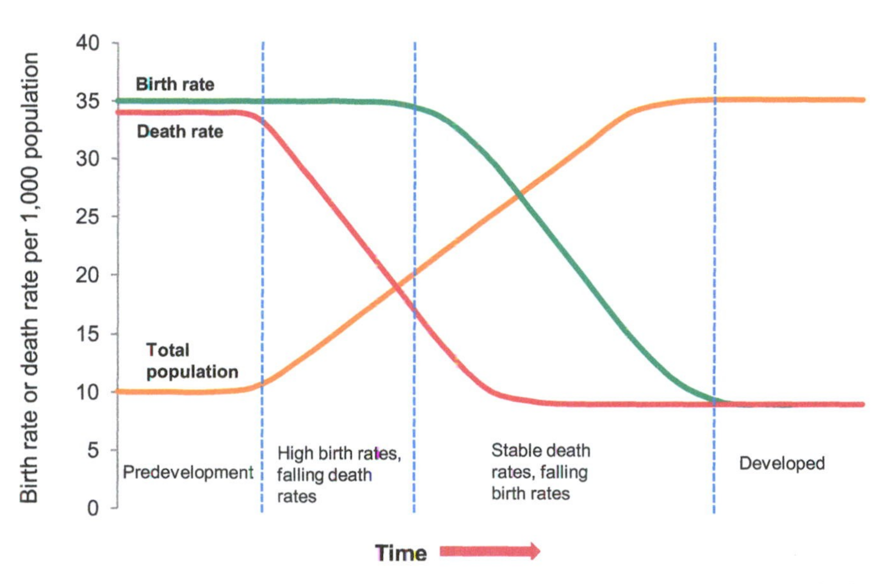

Figure 11.1 The demographic transition, showing the transition from high birth and death rates to low birth and death rates as a country develops from a pre-industrial to an industrialized economic system. In stage two, that of a developing country, the death rates drop rapidly due to improvements in food supply and sanitation, which increase life spans and reduce disease. These changes usually come about due to improvements in farming techniques, access to technology, basic healthcare, and education. In stage three, birth rates fall due to access to family planning, increases in wages, a reduction in subsistence agriculture, and an increase in the status and education of women. During stage four, there are both low birth rates and low death rates. Birth rates may drop to well below replacement level as has happened in countries like Germany, Italy, and Japan. Most developed countries are in stage 3 or 4 of the model; the majority of developing countries have reached stage 2 or stage 3.

well-being and survival of humanity depends on the attainment of a balance between population and the environment, even though we do not really know what that balance is.

If a large and growing population is not the root cause of our environmental problems, then what else may cause environmental degradation? Many scholars, like Barry Commoner (a biologist and retired college professor who incidentally ran for President in 1980), have argued that pollution and other environmental and social problems result from ill-conceived programs of industrial development, whose short-term, profit-oriented designers and practitioners have ignored the ecological consequences of their actions. This view is not an anti-industrial development stance. Indeed, Commoner and those who share his views believe that population control is likely to be achieved in developing countries only after poverty has been reduced through appropriate development. Appropriate development is another term that is difficult to define, but ideally, it represents development where resources are used for social good rather than simply for profit.

A more alternative view on Earth's population capacity is held by a group of scholars referred to as ecological and environmental optimists. The late Julian L. Simon, a professor of economics and business at the University of Maryland, was arguably the most well-known of the group, and he contended that environmental conditions and standards of living are likely to improve as global populations increase. In his controversial 1981 book, *The Ultimate Resource*, Simon states that virtually all measures of human well-being have improved since the 18th century, and no reason exists why such trends should not continue into the indefinite future (see Figure 11.2). He argues that pollution is (and always has been) a problem, but on average, we now live in a less dirty and more healthy environment than in earlier centuries. The central premise of this view is that human resourcefulness and enterprise will continue to respond to impending shortages and existing problems: if a particular resource becomes scarce, either new resources will be discovered, people will learn to do more with less, or resource substitutes will be utilized.

What, then, should we do about the growing numbers in the developing world? Do we continue to let the population grow, hoping that one day we will all achieve a level of development that will solve our environmental problems (i.e., the larger the population, the better)? Or do we attempt to control population growth by lowering birth rates? Of course, there are ethical considerations surrounding developing countries' right to grow, both in population and economically, even if it is causing more destruction. Why should "they" pay the price for the mess largely made by developed countries, i.e., "us?"

Perhaps we should adopt the **lifeboat ethics theory** developed by Garret Hardin, professor of biology and human ecology at the University of California, Santa Barbara. Hardin, an influential

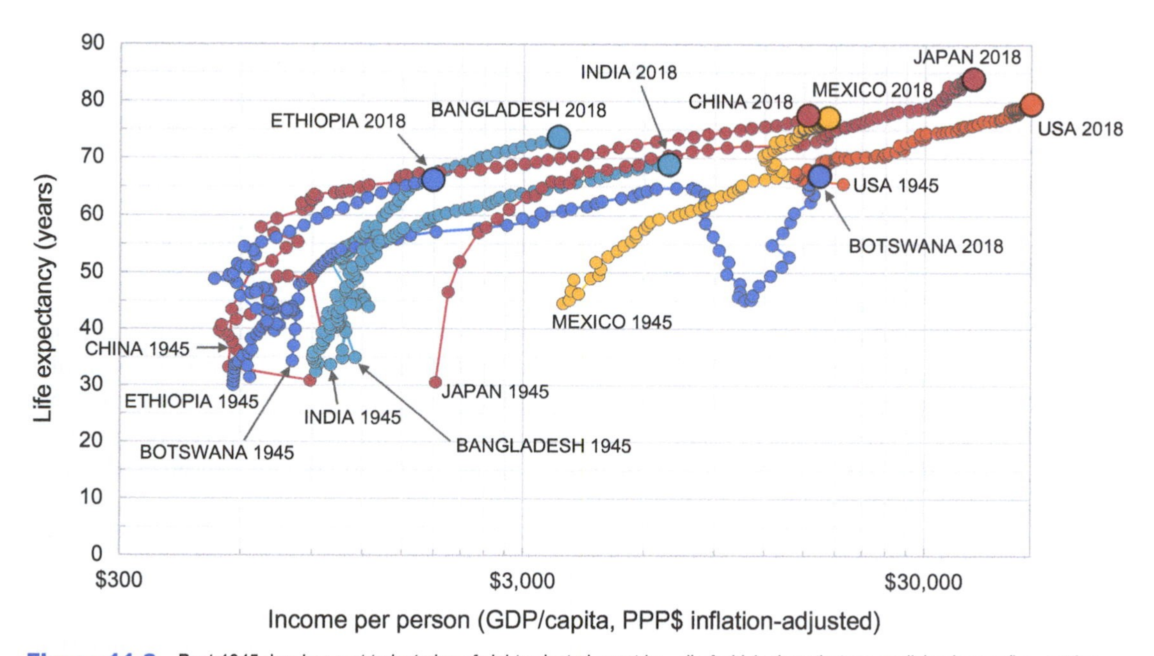

Figure 11.2 Post-1945 development trajectories of eight selected countries, all of which show that we are living longer (i.e., getting healthier) and getting richer, with some interesting trends. For example, Mexico's decline in income per person post-1981 was the result of the worst recession since the 1930s, a period known as La Década Perdida, "the lost decade." The Great Acceleration in Japan is profound following World War II with life expectancy now the highest in the world. Life expectancy in Bangladesh is now the same as it was in the U.S. in 1979. The significant dip in life expectancy in Botswana after 1991 is due to the HIV/Aids crisis in the country, with infection rates approaching 25% of the population. Since 1997, the government has been significantly more proactive in combating the epidemic.

and outspoken critic of the ecological and environmental community, proposes a metaphor in which developed, resource-controlling nations are in a lifeboat and struggling developing nations are floundering in the surrounding ocean. He concludes that it is folly to try to rescue all the swimmers and the only option to prevent the entire boat from sinking is for those in the lifeboats to decide who to rescue and who to let drown. I would argue that this harsh analysis, which has won praise from many pragmatic environmentalists, is, at the very least, untenable on ethical grounds. Would you be able to sit idly by and watch millions of people potentially die because we are trying to preserve "the greater good" on a planet that may very well be able to cope with 9 or 10 billion people (perhaps more) in the long run? And, aside from the ethical problems, it accepts a premise that those in the "lifeboat" do not have intricate interconnections with those outside the lifeboat, which we know is not the case.

It seems we have yet to find a definitive answer to the dispute about Earth's carrying capacity and whether or not we have already exceeded some upper, absolute limit. Ultimately, it is a tough question to put numbers on. The relationships among population, resources, consumption patterns, and the environment are complex, perhaps too complex, to simply be reduced to some static, quantitative "capacity threshold." Still, we should be wondering whether we are beginning to approach a limit, beyond which irreversible changes will alter the world, as we know it.

CALCULATING HUMANITY'S ECOLOGICAL FOOTPRINT

You have probably heard of the **Ecological Footprint,** the metric that allows us to calculate human pressure on the planet and come up with facts, such as "If everyone lived the lifestyle of the average American, we would need five planets" Statements such as these sound alarming and give the impression that humanity is on a collision course with the planet. Of course, humanity needs what nature provides, but how do we know how much we are using and how much we have to use? The Ecological Footprint[8] helps address these questions. It has become one of the most widely used and accepted measures of humanity's demand on nature. Technically, the Ecological Footprint measures how much land and water area a human population requires to produce the resource it consumes and to absorb its carbon dioxide emissions, using prevailing technology. Essentially, it is a measure of how much nature your lifestyle requires.

Footprints are not bad or good *per sé*. Every living entity possesses an ecological footprint, but it is the size that differs. For example, land consumed by urban areas is typically huge. Every city draws on the material resources and production of a vast and increasingly global hinterland of an ecologically productive landscape many times the size of the city itself. A simple mental exercise may help to illustrate the ecological reality behind this. Imagine what would happen to any urban region if it were enclosed in a glass or plastic dome completely closed to material flows. At some point, the city would cease to function, and its inhabitants would perish quickly. The population

[8] See http://www.footprintnetwork.org

and economy contained by the capsule would have been cut off from both vital resources and essential waste sinks leaving it to starve and suffocate at the same time. In other words, the ecosystems contained within our imaginary human terrarium would have insufficient carrying capacity to service the ecological load imposed by the contained population.

On a global scale, humanity's entire ecological footprint can be compared to the total capital and services that nature provides. When humanity's footprint is within the annual regenerative capabilities of nature, its footprint is sustainable. In 2014,[9] the biosphere had an estimated 12.2 billion global hectares (gha) of biologically productive area (or **biocapacity**), about one-quarter of the planet's surface. When the total ecologically productive land area (including forests, oceans, and cropland) is divided by the human population, there are about 1.6 gha available for each person.[10] Humanity's total ecological footprint worldwide in 2014 was estimated at 20.6 gha. With the world population at 7.6 billion in 2018, the average person's footprint was approximately 2.7 gha, about 60% greater than Earth's biocapacity. This overshoot means that, in 2014, humanity used the equivalent of 1.7 Earths to support its consumption (Figure 11.3).

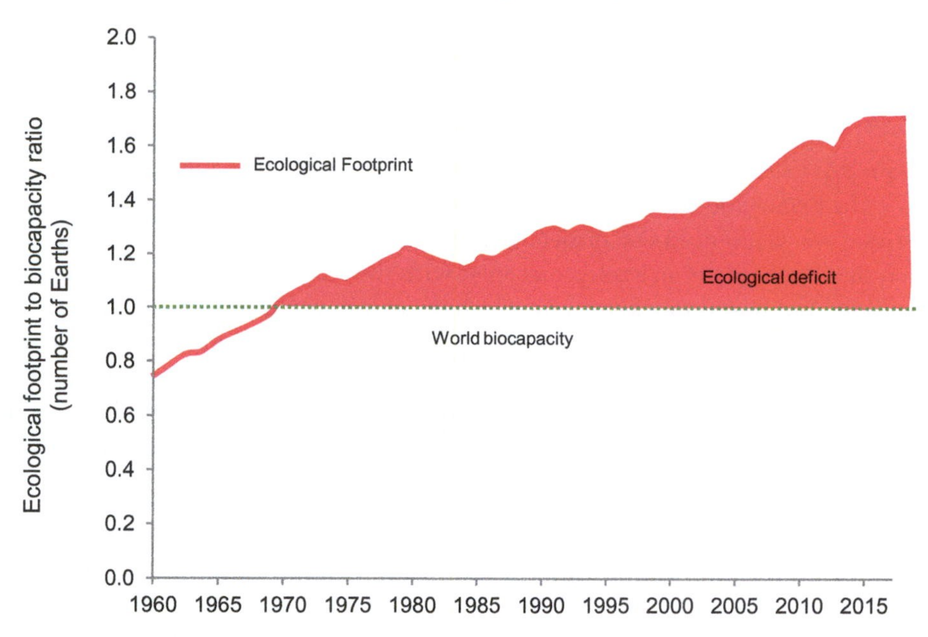

Figure 11.3 Humanity's ecological footprint through time in relation to biocapacity (Source: www.footprintnetwork.org).

[9] National footprint accounts are updated annually based on the latest complete data sets available, which usually entails a time lag of about three years. So the data discussed here and shown in Figures 11.3, 11.4 and 11.5 are for 2014, the latest available.

[10] To delve into the data for each country, go to http://data.footprintnetwork.org/#/. Because the global biocapacity does not change much from year-to-year, we use the 2014 data here divided by the 2018 global population to keep the statistics as up-to-date as possible.

We can calculate Earth's overshoot day by simply taking the global biocapacity and dividing by the global footprint, then multiplying by the number of days in the year, as follows:

$$\text{Earth Overshoot Day} = 365 * (1.6/2.7) = 227\text{th day in the year} \tag{11.2}$$

The 227th day is August 15. What this means is that on August 15th, we have used our total carrying capacity.

Scientists involved in refining the global footprint estimates now suggest that by the early 2030s, we will need the equivalent of two Earths to support us. Of course, we only have one! There is no Planet B! In short, we are depleting the very resources on which human life and biodiversity depend. The results, as we have seen in the preceding chapters, are diminishing forest cover, depletion of soil and fresh water systems, and the buildup of pollution and waste, which creates problems like global climate change.

Figure 11.4 shows the top ten largest countries ranked in terms of their total ecological footprint. Figure 11.5 tracks the ecological footprint and biocapacity in four countries that I have mentioned frequently throughout this book. In Figure 11.6, I highlight Costa Rica, a unique country in terms of not only its biodiversity but also its legacy of environmental degradation juxtaposed to some of the most progressive conservation laws in the world. Incidentally, Qatar's overshoot day is February 9 while Costa Rica's is September 2.

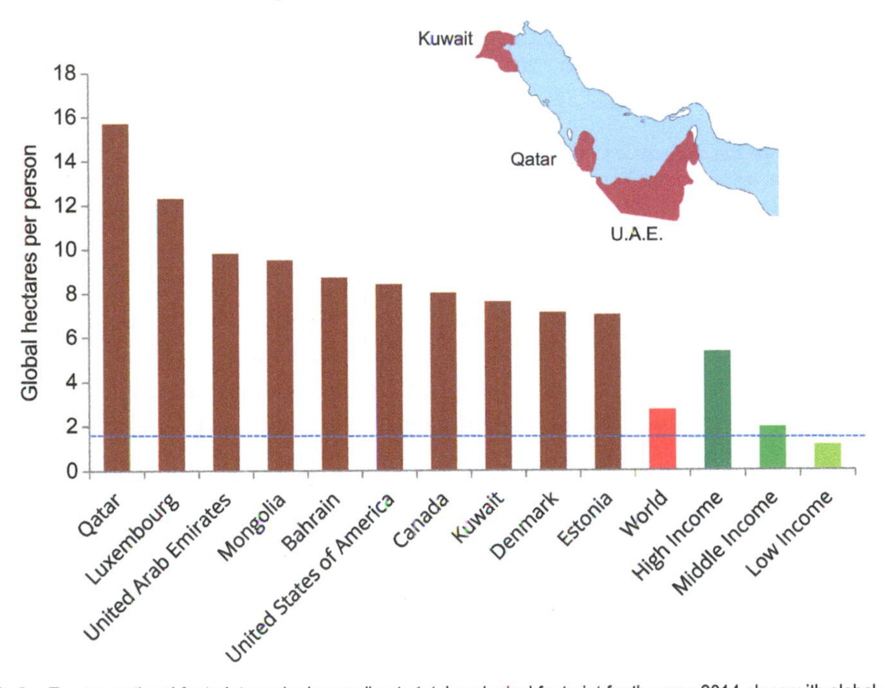

Figure 11.4 Top ten national footprints ranked according to total ecological footprint for the year 2014 along with global and income-related averages. The horizontal dashed line shows the worldwide biocapacity per person of 1.6 global hectares per person (Source: www.footprintnetwork.org).

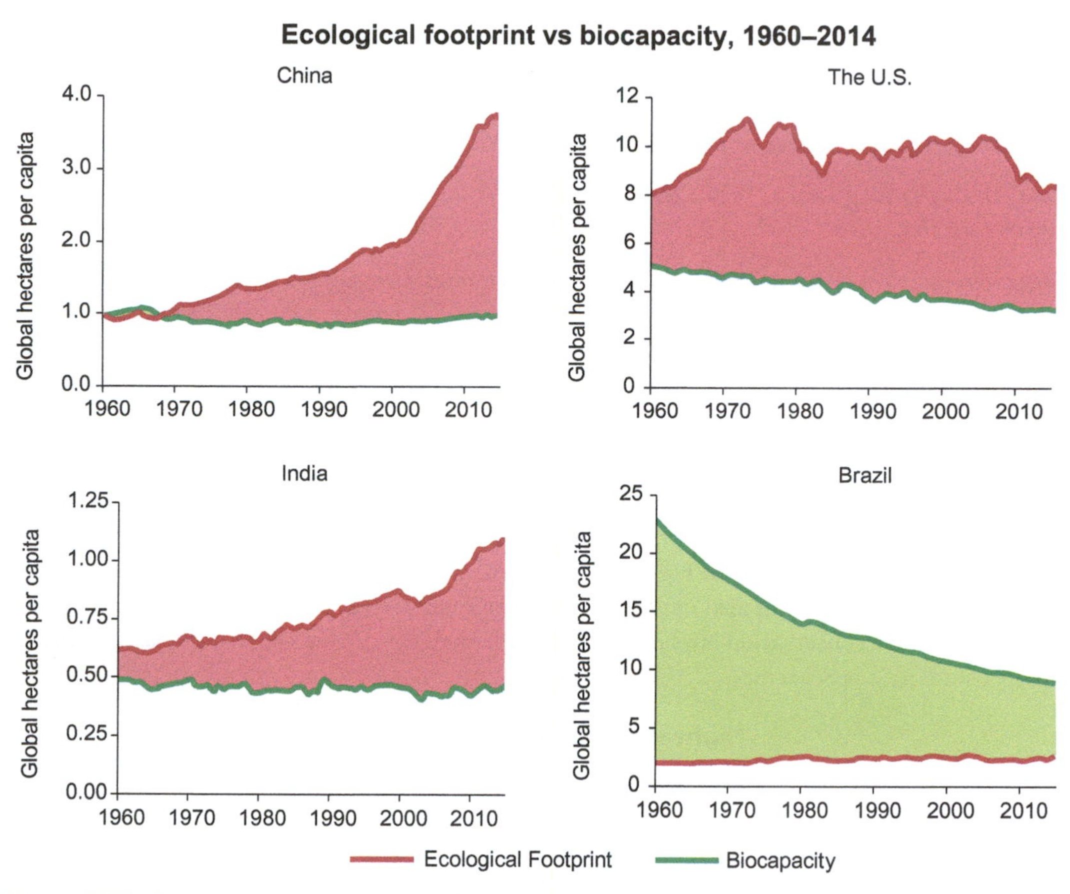

Figure 11.5 The per-person resource demand ecological footprint and biocapacity for China, the United States, India, and Brazil. Biocapacity varies each year with ecosystem management, agricultural practices (such as fertilizer use and irrigation), ecosystem degradation, weather, and population size. Footprint varies with consumption and production efficiency. Note the difference in scales on the *y*-axes (Source: www.footprintnetwork.org).

Not surprisingly, high-income countries dominate the rankings, with countries in Western Europe, North America, and the Middle East having the highest per capita footprints of any region on Earth. Much of this is due to their carbon footprint which, in almost all cases, accounts for more than 50% of the total footprint. The top three (Qatar, Luxembourg and the United Arab Emirates) have carbon footprints that account for between 70 and 85% of their total footprint. These three countries also have the world's largest ecological deficits due to their extremely low biocapacity. In effect, Qatar and the UAE do not have any cropland, grazing land, or forested land to speak of with only fishing grounds offering any real bio-significance. Almost everything has to be shipped into Qatar and the UAE which greatly increases their carbon output. Landlocked Luxembourg, of course, has no fishing grounds at all, and has a dense population with little supporting hinterland. Compare that to Brazil's current biocapacity of 8.9 gha per person (Figure 11.5), and you begin to appreciate the challenge many of these countries face in terms of sustainable development.

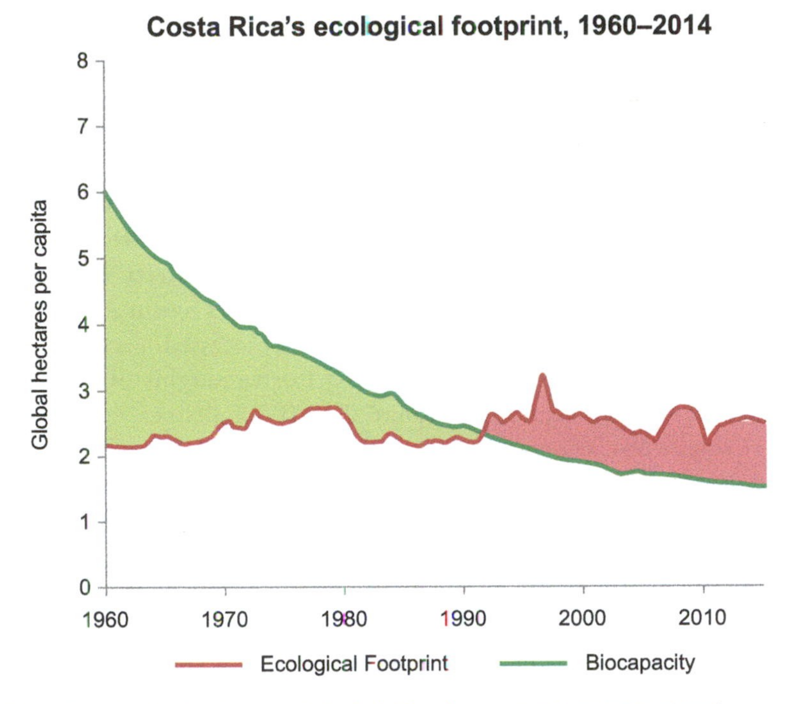

Figure 11.6 The ecological footprint and biocapacity for Costa Rica (Source: www.footprintnetwork.com).

At 8.4 global hectares per person, the United States has the sixth largest per capita ecological footprint on the planet. As shown in Figure 11.5, our footprint has remained fairly static since the mid-1980s with evidence of a decline since the mid-2000s. The good news is that this trend is likely to continue as we move toward an increasingly cleaner energy future. India and China, on the other hand, have been steadily increasing their ecological footprints though without any significant degradation to their biocapacity. China, in particular, has seen a dramatic acceleration since the mid-2000s, and this trend too is likely to continue given the predictions for China's future energy demand (refer back to Figures 3.4 and 3.11 for context). I included Costa Rica here (Figure 11.6) because I have referred to this country several times throughout this book. Costa Rica's biocapacity has been in constant decline not just because of deforestation (see Figure 7.16) but primarily because of the expansion of monocultures such as palm oil and unsustainable land use practices. In addition, consumption and development has been increasing and, along with it, pollution fuelled, quite literally, by rising oil demand of the transportation sector. Estimates show that a 27% cut in carbon emissions would ensure that Costa Rica could achieve a balanced ecological footprint, but achieving such a goal is not going to be easy because nearly three-quarters of the energy demand is dependent on oil.[11]

[11] http://www.estadonacion.or.cr/

I encourage you at this point to go ahead and complete the footprint quiz, (www.footprintnetwork.org) but do not be discouraged by your results. There are some portions of the footprint calculation that are really out of your control. For example, each resident of a city is "responsible" for a portion of the city's infrastructure, such as roads, schools, and government offices, regardless of whether or not the resident actually uses those services. In addition, some options that could make your footprint smaller may not be available to you due to local policy decisions, such as reliable and efficient public transport or city recycling programs. I have included my ecological footprint in Figure 11.7. As you can see, it hardly makes for pretty reading! However, my footprint is significantly affected by a combination of factors: (1) I live in a large city where goods and services are expensive; (2) The majority of my food is processed, packaged, and not locally grown; (3) Although I live within five miles of work, I drive roughly 300 miles each week on personal business; (4) I fly a lot on business (more than 100 hours per year); and (5) My home, though just 1,247 square feet, energy efficient, and without any lawn to water, still requires air conditioning and heat in the variable southern climate. I clearly have some work to do to begin to reduce my consumptive lifestyle!

Admittedly, the ecological footprint approach cannot capture all of humanity's impacts on the environment, and we therefore consider it a generally conservative estimate. For example, toxic pollutants and species extinction are not explicitly incorporated into the footprint model. Moreover, the value of nature extends far beyond the goods and services that humans take from it, and footprinting does not (and cannot) take account of this intrinsic value. However, footprinting does offer a fairly robust set of metrics that can help inform, educate, and point the way toward a more sustainable path. Moreover, the simplicity of the concept enables people to easily understand it.

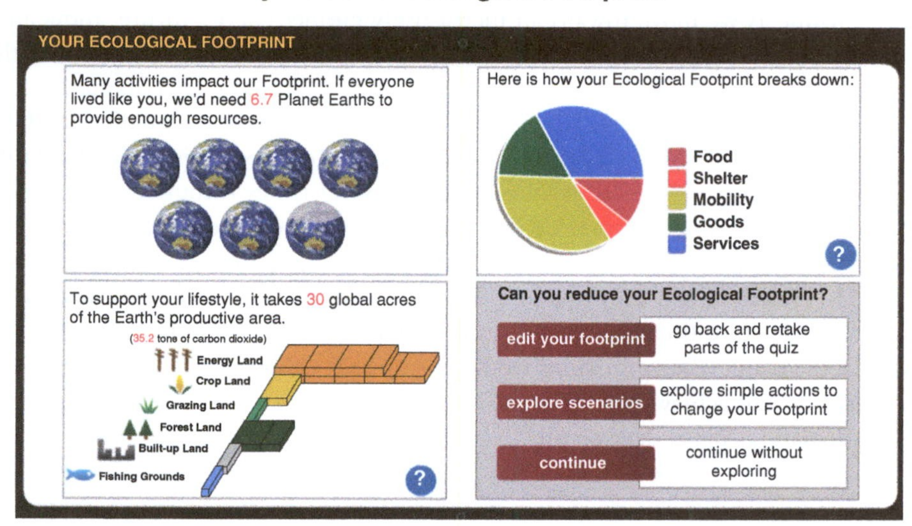

Figure 11.7 My personal ecological footprint. Note that the output from this online quiz is given in global acres, so my 26.6 global acres = 10.8 global hectares.

CONCLUDING THOUGHTS

Our Earth is a vast and complex set of interconnected systems. Defining the ecological boundaries or the carrying capacity of the planet is a complicated and perhaps even futile task. We may not end up like the Rapa Nui on Easter Island, but at the same time, we know that the world cannot sustain an unlimited number of people consuming non-renewable resources. Furthermore, while studies may single out population size and growth as the primary cause of global environmental degradation, this may not be strictly valid. Certainly, it is in our best interest to slow population growth while recognizing how consumption, lifestyles, and technology also impact our environment. While population may play a central role, it will ultimately be the *combination* of population, lifestyles, and technology that will determine whether we maintain a habitable planet and achieve a sustainable society.

Ecological footprints are the result of individual choices and the activities and policies of government, corporations, and civic institutions. This means that actions that lead to changes in personal lifestyle are as important as changes in policy and the ways products are produced. The first step toward reducing our ecological impact is to recognize that the environmental problems outlined in this book are largely behavioral and societal problems that can be resolved only with the help of behavioral and societal solutions. We must recognize Earth's ecological limits and keep them in mind while we make decisions and use human ingenuity to find new ways to live within the Earth's bounds. As the authors of the Global Footprint Network note, this means investing in technology and infrastructure that will allow us to operate in a resource-constrained world. It means taking individual actions and creating the public demand for businesses and policymakers to engage. Using tools like the Ecological Footprint to manage our ecological assets is essential for humanity's survival and success. Knowing how much nature we have, how much we use, and who uses what is the first step. This allows us to track our progress as we work toward our goal of sustainable lifestyles.

In this book, we have explored some of the environmental issues that will undoubtedly shape the future. They are likely to affect the political and economic choices made by world leaders profoundly. Some, like climate change, could lead to a dramatically altered world order during this millennium. Edward O. Wilson, the intellectual giant whom I have referred to several times in this book, sees our future as follows:

> *I do believe we now have a very Faustian choice upon us: whether to accept our corrosive and risky behavior as the unavoidable price of population and economic growth or to take stock of ourselves and search for a new environmental ethic.*[12]

Many positive forces for global transformation are at work today and some trends have begun to reverse. For example, globally, human fertility has halved in the past few decades as women's rights and status in society have improved along with progressively greater access to family planning and maternal health services. Economic development and urban lifestyles has also led to people wanting

[12] Source: Wilson, E.O. (1998), *Consilience: The Unity of Knowledge*, p. 277–278.

fewer children because of the financial burden. In time, this should see the global population stabilize. Agricultural intensification is slowing in some areas, and forests are starting to expand in some regions such as Vietnam and parts of Central America. Large areas of our natural heritage are continually being conserved by non-governmental organizations and numerous philanthropic individuals. The carbon intensity of industry has begun to decline with companies finding ways of doing business that are more frugal with energy than before and saving money in the process. Stratospheric ozone depletion appears to have been arrested, and projections suggest a path toward full recovery is possible by the middle of this century. In the U.S., we are using water at levels not seen since the 1970s. In short, there are signs that some drivers of global environmental change are slowing or changing. It is not all doom and gloom!

Still, the world faces some daunting challenges. The human race now produces about *37 billion* tons of carbon dioxide annually, with the majority of it produced by the burning of fossil fuels. In some countries, people are consuming far too much, including carbon, water, and other resources embodied in trade. Globally, the highest income earners are responsible for three times the level of emissions (and hence environmental impact) compared with lowest income earners, but the growing middle classes of many developing or transitional countries are developing consumption habits that mimic ours thereby adding to the burden on the Earth system. Meanwhile, more than a billion people lack access to clean water and electricity.

At times, the task of taking care of our environment seems hopeless, as we appear to continue to repeat our behavior, frequently with known and often dire consequences. The brilliant British novelist and screenwriter, Ian McEwan, summed it up best in an essay titled "Let's talk about climate change":

> *The pressure of our numbers, the abundance of our inventions, the blind forces of our desires and needs are generating a heat—the hot breath of our civilization—whose effects we comprehend only hazily.*
>
> *We are fouling our nest, and we have to act decisively and against our immediate inclinations.*
>
> *Can we avoid what is coming at us, or is there nothing much coming at all? Are we at the beginning of an unprecedented era of international co-operation, or are we living in an Edwardian summer of reckless denial?*

There is much cause for optimism, but we must accompany that optimism with a conscientious awareness of the tasks ahead. The time to get started is now.

Glossary

Acid deposition The accumulation of acids or acidic compounds on the surface of the Earth, in lakes or streams, or on objects or vegetation near the Earth's surface, as a result of separation from the atmosphere; occurs in a wet or dry process.

Acidification See *acid deposition*.

Acute poverty A severe lack of material possessions or money. In 2005, the World Bank defined extreme poverty as living on less than US $1.25 a day.

Age structure Categorization of the population of communities or countries by age groups which depicts growth trends, allowing demographers to make projections of the growth or decline of the particular population.

Aichi Biodiversity Targets A series of targets set out in 2010 at the Convention on Biological Diversity; for example, one target required that at least 10% of each of the world's ecological regions to be effectively conserved.

Albedo The fraction of solar energy (shortwave radiation) reflected from the Earth back into space. It is a measure of the reflectivity of the Earth's surface.

Algal bloom A rapid increase or accumulation in the population of algae (typically microscopic) in an aquatic system. See *eutrophication*.

Alternative energy Energy, such as solar, wind, or nuclear energy, that can replace or supplement traditional fossil-fuel sources, such as coal, oil, and natural gas.

Anthropocene A proposed term for the present geological epoch (from the time of the Industrial Revolution onwards), during which humanity has begun to have a significant impact on the environment.

Anthropogenic climate change See *global warming*.

Aquifer Water-bearing porous soil or rock strata that yield significant amounts of water to wells. Aquifers may either be confined (the aquifer lies between two layers of much less permeable material) or unconfined (where the aquifer is recharged directly from infiltrating rainfall).

Aquitard A water-saturated sediment or rock whose permeability is so low it cannot transmit any useful amount of water.

Area of influence The area covered by the drawdown curves of a given pumping well or combination of wells at a particular time.

Arithmetic growth The situation where a population increases by a constant number of persons (or other objects) in each period being analyzed. Also known as *linear growth*.

Artesian well A well in which water is under pressure; especially one in which the water flows to the surface naturally.

Atomic chlorine See *free chlorine*.

Background extinction The ongoing extinction of individual species due to environmental or ecological factors such as climate change, disease, loss of habitat, or competitive disadvantage in relation to other species. Background extinction occurs at a fairly steady rate over geological time and is the result of normal evolutionary processes, with only a limited number of species in an ecosystem being affected at any one time.

Benthic Of, relating to, or occurring at the bottom of a body of water.

Bioaccumulation The accumulation of a substance, such as a toxic chemical, in various tissues of a living organism in a trophic level.

Biocapacity The capacity of a given biologically productive area to generate an on-going supply of renewable resources and to absorb its spillover wastes.

Biodiversity The degree of variation of life forms within a given species, ecosystem, biome, or an entire planet; a measure of the health of ecosystems relative to latitude. For example, in terrestrial habitats, tropical regions are typically rich in species whereas polar regions support fewer species.

Biofuels (first, second, third generation) Fuel, such as methane, produced from renewable biological resources such as plant biomass and treated municipal and industrial waste.

Biological corridors The concept focused on preserving the physical connections between protected areas with important biodiversity with the aim of preventing the fragmentation of natural habitats. Today these corridors are being promoted as an innovative way to promote sustainable development as well as conservation.

Biomagnification The increasing concentration of a substance, such as a toxic chemical, in the tissues of organisms at successively higher trophic levels in a food chain. As a result, organisms at the top of the food chain generally suffer greater harm from a persistent toxin or pollutant than those at lower levels.

Biomass The amount of living matter in a given habitat, expressed either as the weight of organisms per unit area or as the volume of organisms per unit volume of habitat.

Biotic potential The maximum reproductive capacity of a population if resources are unlimited.

Black lung disease The common name for coal workers' pneumoconiosis (CWP) or anthracosis, a lung disease of older workers in the coal industry, caused over many years by inhalation of small amounts of coal dust.

Buffer zone An area located beyond a natural reserve area, either as another forest area, free state land, or as land with certain rights, which is required and able to preserve the integrity of a natural reserve area.

Buffering capacity The relative ability of a system to resist pH change upon addition of an acid or a base.

Cap-and-trade A regulatory system that is meant to reduce certain kinds of emissions and pollution and to provide companies with a profit incentive to reduce their pollution levels faster than their peers. Under a cap-and-trade program, a limit (or "cap") on certain types of emissions or pollutions is set, and companies are permitted to sell (or "trade") the unused portion of their limits to other companies that exceed their limit and are struggling to comply.

Carbon capture and sequestration (CCS) Refers to technology attempting to prevent the release of large quantities of CO_2 into the atmosphere from fossil fuel used in power generation and other industries by capturing CO_2, transporting it and ultimately, pumping it into underground geologic formations to securely store it away from the atmosphere.

Carrying capacity The maximum, equilibrium number of organisms of a particular species that can be supported indefinitely in a given environment.

Catalytic cycle A series of reactions in which a chemical family or a particular species is depleted, leaving the catalyst unaffected. One example is the destruction of stratospheric O_3 by CFCs.

Chain reaction A self-sustaining reaction in which the fission of nuclei of one generation of nuclei produces particles that cause the fission of at least an equal number of nuclei of the succeeding generation.

Chloroflurocarbons (CFCs) Any of various halocarbon compounds consisting of carbon, hydrogen, chlorine, and fluorine; until the Montreal Protocol, once used widely as aerosol propellants and refrigerants which casue depletion of the atmospheric ozone layer.

Clean Air Act (CAA) First enacted in 1970, it authorized the establishment of federal and state regulations that limit emissions stationary (point) and mobile (nonpoint) sources of air pollutants.

Clean coal Technologies that achieve significant reductions in air emissions of sulfur dioxide and nitrogen oxides, two pollutants that contribute to the formation of acid rain.

Clean Water Act (CWA) Came into effect in 1972 as the primary legislation concerning water pollution and its regulation. It establishes a permit system that must be used by point sources of pollution such as industrial facilities, government facilities, and agricultural operations; regulates several kinds of water pollutants including toxins, biochemical oxygen demand (BOD) pollutants, total suspended solids (TSS), fecal coliform, oil, grease, and pollutants that alter pH.

Clearcutting Felling and removing all trees in a forest area.

Climate change A long-term change in the Earth's climates resulting from an increase in the average atmospheric temperature. This can be a natural occurrence or anthropogenic.

Climate sensitivity A measure of how responsive the temperature of the climate system is to a change in the radiative forcing.

Closed system A term used in population dynamics which describes an isolated system that has no interaction with its external environment.

Concentrating solar power (CSP) A category of solar power that uses reflective surfaces or lenses to direct a concentrated amount of the solar radiation to either a photovoltaic surface or to a device used to first heat liquids, then produce electrical power by means of a generator.

Cone of depression The depression in the water table around a well defining the area of influence of the well.

Confined aquifer See *aquifer*.

Core area The central area in a protected reserve off-limits to any development; designed to protect the most important species.

Correlation A statistical measurement of the relationship between two variables. Possible correlations range from +1 to –1. A zero correlation indicates that there is no relationship between the variables. A correlation of –1 indicates a perfect negative correlation, meaning that as one variable goes up, the other goes down. A correlation of +1 indicates a perfect positive correlation, meaning that both variables move in the same direction together. Correlation does not necessarily mean causation.

Criteria air pollutants A group of six pollutants that the EPA uses to define air quality; they are CO, NO_2, SO_2, PM, O_3 and Pb.

Critical habitat Areas of habitat that are crucial to the survival of a species.

Crop rotation The practice of growing different crops in succession on the same land chiefly to preserve the productive capacity of the soil.

Crude birth rate *(b)* The number of births per 1,000 people in a given population.

Crude death rate *(d)* The number of deaths per 1,000 people in a given population.

DDT A colorless contact insecticide, toxic to humans and animals when swallowed or absorbed through the skin. It has been banned since 1972 in the United States.

Dead zone A hypoxic (low-oxygen) area in the world's oceans and large lakes, caused by excessive nutrient pollution from human activities like agriculture coupled with other factors that deplete the oxygen required to support most marine life in bottom and near-bottom water. A primary example is the mouth of the Mississippi River where it flows into the Gulf of Mexico.

Deforestation The removal of a forest or stand of trees where the land is thereafter converted to a nonforest use.

Demographic transition The change that typically takes place as a country develops; the birth and death rates of its population both eventually tend to fall as per capita income rises.

Denudation The long-term sum of processes that cause the wearing away of the Earth's surface leading to a reduction in elevation and relief of landforms and landscapes.

Directional drilling A drilling method involving intentional deviation of a wellbore which is often used in shale fracking.

Dissolved oxygen The amount of oxygen dissolved (and hence available to sustain marine life) in a body of water such as a lake, river, or stream.

Doubling time (*Td*) The amount of time for a given population to double, based on the annual growth rate; calculated by the *rule of 70*.

Drawdown A lowering of the water level in a reservoir or other body of water, especially as the result of over-pumping.

Ecological economics A field of academic research that aims to address the interdependence and coevolution of human economies and natural ecosystems over time and space.

Ecological footprint The amount of productive land appropriated on average by each person (in the world, a country, etc.) for food, water, transport, housing, waste management, and other demands.

Ecological overshoot See *overshoot*.

Economically recoverable reserves A term used in natural resource industries to describe the amount of resources identified in a reserve that is technologically or economically feasible to extract. A new reserve can be discovered, but if the resource cannot be extracted by any known technological methods, then it would not be considered part of recoverable reserves. Recoverable reserves is also often called proved reserves.

Ecosystem diversity The variety of habitats that occur within a region or the mosaic of patches found within a landscape.

Edge effects The effect of an abrupt transition between two quite different adjoining ecological communities on the numbers and kinds of organisms in the marginal habitat.

Effective chlorine The sum of hydrochloric acid (HCl) and chlorine nitrate ($ClONO_2$); these are the two most important breakdown products of chlorine from CFCs.

El Niño/Southern Oscillation (ENSO) A quasiperiodic climate pattern that occurs across the tropical Pacific Ocean roughly every five years. *El Niño* refers to a warming in the temperature of the surface of the tropical eastern Pacific Ocean; *Southern Oscillation* refers to changes in air surface pressure across the tropical Pacific. Mechanisms that cause the oscillation are not fully understood and remain under study; the phenomenon causes worldwide changes in global climates such as floods and droughts.

Electromagnetic spectrum (EM) The entire range of wavelengths or frequencies of electromagnetic radiation extending from short gamma rays to the longest radio waves and includes visible light.

Endangered species A population of organisms which is facing a high risk of becoming extinct because it is either few in numbers or threatened by changing environmental conditions or predation parameters.

Endemic species A species which is exclusively found in a given region, location, or unique ecosystem and nowhere else in the world; especially likely to develop in geographically isolated regions; more vulnerable to introduced exotic species.

Energy balance Earth's energy balance describes how the incoming energy from the sun is used and returned to space. If incoming and outgoing energy are in balance, the Earth's temperature remains constant; quantified by the equation $I\text{-}O = \Delta S$, or input (I) minus output (O) equals ΔS (change in storage).

Environmental ethics The part of philosophy which studies the moral relationship of human beings to the environment.

Environmental Protection Agency (EPA) An independent federal agency established to coordinate programs aimed at reducing pollution and protecting the environment.

Environmental resistance The influences of regulating environmental factors which put pressure on continuing the increase in numbers of individuals within a community thereby limiting further growth of a population.

Environmental stewardship An ethic within environmental philosophy which emphasizes the responsible management of our planet, in particular, planning and managing the use of natural resources in a sustainable way.

Equitability The evenness with which individuals are distributed among species in a given community.

Eutrophication The process by which a body of water acquires a high concentration of nutrients, especially phosphates and nitrates. These typically promote excessive growth of algae. As the algae die and decompose, high levels of organic matter and decomposing organisms deplete the available oxygen making the water hypoxic, causing the death of other organisms, such as fish.

Evaporation The change by which any substance is converted from a liquid state into a gas state and carried off in the form of vapor; specifically, the conversion of a liquid into vapor in order to remove it wholly or partly from a liquid.

Evapotranspiration (ET) Describes the sum of evaporation and plant transpiration from the Earth's land surface to the atmosphere.

Exponential growth Also known as *geometric growth*. Development at an increasingly rapid rate in proportion to the growing total number or size; a constant rate of growth applied to a continuously growing base over a period of time.

Feedback mechanism The return of some of the output of a system as input so as to exert some control in the process; a negative feedback cycle occurs when the return exerts an inhibitory control; a positive feedback cycle occurs when it exerts a stimulatory effect.

Fishbone pattern A pattern found in deforested areas which typically follows road and highway construction, resulting in cleared land that resembles the skeleton of a fish. Further deforestation is facilitated with greater penetration into the forested area.

Food security When all people at all times have access to sufficient, safe, nutritious food to maintain a healthy and active life.

Forest degradation A reduction in forest quality, including the density and structure of the trees, the ecological services supplied, the biomass of plants and animals, the species diversity, and the genetic diversity.

Forest fragmentation See *fragmentation*.

Fossil fuels Hydrocarbon deposits used for fuel such as petroleum, coal, or natural gas, which formed over geologic time from the remains of living organisms.

Fracking See *hydraulic fracturing*.

Fragmentation A form of habitat fragmentation, occurring when forests are cut down in a manner that leaves relatively small, isolated patches of forest known as *forest fragments* or *forest remnants*.

Free chlorine A single negatively charged chlorine atom capable of destroying thousands of ozone molecules. Also known as *atomic chlorine* or *effective chlorine*.

Genetic diversity The combination of different genes found within a population of a single species and the pattern of variation found within different populations of the same species.

Geologic time The period of time (estimated at 4.6 billion years) covering the physical formation and development of Earth, especially the period prior to human history.

Geometric growth See *exponential growth*.

Global warming An average increase in the temperature of the atmosphere near the Earth's surface and in the troposphere, which can contribute to changes in global climate patterns. In common usage, global warming refers to the warming that can occur as a result of increased emissions of greenhouse gases from human activities or an anthropogenic acceleration of the greenhouse effect.

Global Warming Potential (GWP) A relative measure of how much heat a greenhouse gas traps in the atmosphere. It compares the amount of heat trapped by a certain mass of the gas in question to the amount of heat trapped by a similar mass of carbon dioxide. A GWP is calculated over a specific time interval, commonly 20, 100 or 500 years. GWP is expressed as a factor of carbon dioxide (whose GWP is standardized to 1). For example, the 20 year GWP of methane is 72, which means that if the same mass of methane and carbon dioxide were introduced into the atmosphere, that methane will trap 72 times more heat than the carbon dioxide over the next 20 years.

Greenhouse effect The warming of the surface and lower atmosphere of a planet (as Earth or Venus) that is caused by conversion of incoming shortwave solar radiation into heat in a process involving selective transmission of short wave solar radiation by the atmosphere, its absorption by the planet's surface, and reradiation as infrared which is absorbed and partly reradiated back to the surface by atmospheric gases.

Greenhouse gas (GHG) Any of the atmospheric gases that contribute to the greenhouse effect by selectively absorbing and transmitting infrared radiation produced by solar warming of the Earth's surface. They include carbon dioxide (CO_2), methane (CH_4), nitrous oxide (NO_2), and water vapor (H_2O).

Ground subsidence The collapse or sinking of the ground surface due to over-pumping groundwater. See *cone of depression.*

Ground-level ozone See *tropospheric ozone.*

Groundwater The water beneath the surface of the ground, consisting largely of surface water that has seeped down. All pore space is completely filled in the groundwater zone.

High-level radioactive waste (HLW) Radioactive waste material, such as spent nuclear fuel, initially having a high activity and thus needing constant cooling for several decades by its producers before it can be reprocessed or treated.

Hindcasting A method of testing a mathematical model by using data from a past event.

Holocene Events occurring in the most recent epoch of the geologic time, which might include sedimentary deposits or rock series; this epoch began at the end of the last Ice Age (and the conclusion of the Pleistocene) about 11,000 years ago and also characterized by the development of human civilizations.

Horizontal drilling The deviation of the borehole at least 80 degrees from vertical so that the borehole penetrates a productive formation in a manner parallel to the formation; this method is often used in shale fracking.

Hotspot (biodiversity) Areas featuring exceptional concentrations of highly vulnerable species and often experiencing exceptional loss of habitat.

Humus A brown or black organic substance consisting of partially or wholly decayed vegetable or animal matter that provides nutrients for plants and increases the ability of soil to retain water; this is found predominantly in the O and A horizons of a soil profile.

Hydraulic fracturing An unconventional procedure of creating micro fractures in rocks and rock formations by injecting a mixture of sand and water into the cracks to force rock to open further, which then allows the natural gas, or other substance, to be extracted. Often referred to as *fracking*.

Hydrocarbons Any of numerous organic compounds, such as benzene and methane, that contain only carbon and hydrogen, such as petroleum, coal, and natural gas.

Hydrological cycle The continuous process by which water is circulated throughout the Earth and its atmosphere. The Earth's water enters the atmosphere through evaporation from bodies of water and from ground surfaces. Plants and animals also add water vapor to the air by transpiration. As it rises into the atmosphere, the water vapor condenses to form clouds. Rain and other forms of precipitation return it to the Earth's surface, where it flows into bodies of water and into the ground, beginning the cycle again. Also called *water cycle.*

Hypoxic conditions A synonym for oxygen depletion, an environmental phenomenon where the concentration of dissolved oxygen in the water column decreases to a level that can no longer support living aquatic organisms.

Ice cores Cylinders of ice obtained by drilling into a glacier. Since the different layers of ice are formed over geologic time through buildup of snow, ice cores provide information on climate and atmospheric composition from different periods (up to almost one million years ago) that can be used for research.

Infant mortality rate (*imr*) The number of number of children that die before one year of age divided by the total number of live births that year.

Infrared radiation (IR) Invisible radiation in the part of the electromagnetic spectrum characterized by wavelengths just longer than those of ordinary visible red light and shorter than those of microwaves or radio waves. Characterized by heat.

Instrumental temperature record Actual temperature measurements taken by temperature monitoring stations around the world.

Instrumental value The value of things as means to further some other ends

Interglacial A period of comparatively warm climate between two glaciations.

Intergovernmental Panel on Climate Change (IPCC) A scientific intergovernmental body establsihed in 1998 which is tasked with evaluating the risk of climate change caused by human activity.

Intrinsic value The inherent worth of something, independent of its value to anyone or anything else.

Island biogeography theory A field within biogeography that examines the factors that affect the species richness of isolated natural communities.

Keeling Curve A graph which has plotted the ongoing change in concentration of carbon dioxide in Earth's atmosphere since 1958. It is based on continuous measurements taken at the Mauna Loa Observatory in Hawaii under the supervision of Charles David Keeling.

Keystone species A species whose presence and role within an ecosystem has a disproportionate effect on other organisms within the system and is crucial to the structure of an ecological community.

Kyoto Protocol An international agreement first adopted in 1997 that aims to reduce carbon dioxide emissions and the presence of greenhouse gases. Countries that ratified the Protocol are assigned maximum carbon emission levels and can participate in carbon credit trading. Emitting more than the assigned limit will result in a penalty for the violating country in the form of a lower emission limit in the following period.

Latent heat The quantity of heat absorbed or released by a substance undergoing a change of state, such as ice changing to liquid water or liquid water changing to ice, at constant temperature and pressure. The latent heat absorbed by air when water vapor condenses is ultimately the source of the power of thunderstorms and hurricanes.

Latitudinal diversity gradient (LDG) The latitudinal diversity gradient (LDG) is a pattern in ecology where localities at lower latitudes generally have more species than localities at higher latitudes.

Lifeboat ethics A metaphor for resource distribution proposed by the ecologist Garrett Hardin in 1974. Hardin's metaphor describes a lifeboat bearing 50 people, with room for ten more. The lifeboat is in an ocean surrounded by a hundred swimmers. The "ethics" of the situation stem from the dilemma of whether (and under what circumstances) swimmers should be taken aboard the lifeboat.

Linear growth See *arithmetic growth*.

Logical fallacy A flaw in reasoning.

Longwave radiation The energy leaving the Earth as infrared radiation at low energy. It is a critical component of the Earth's radiation budget and represents the total radiation going to space emitted by the atmosphere. Generally defined as radition with wavelengths longer than 5.0μm.

Lurking variable A variable that is not included as an explanatory variable in an analysis but can affect the interpretation of relationships between variables.

Malnourished The condition that develops when the body does not get the right amount of the vitamins, minerals, and other nutrients it needs to maintain healthy tissues and organ function.

Malthusian growth Of or relating to the theory of the English economist, the Reverand Thomas Robert Malthus (1766–1834), stating that increases in population occur exponentially and therefore tend to exceed increases in the means of subsistence which occurs arithmetically and that therefore sexual restraint should be exercised.

Mass extinction The extinction of a large number of species within a relatively short period of geological time, thought to be due to factors such as a catastrophic global event or widespread environmental change that occurs too rapidly for most species to adapt.

Mercury (Hg) The metallic element that is poisonous to humans.

Methylmurcury An organic ion containing mercury. This form of mercury is most easily bioaccumulated in (and is toxic to) organisms.

Minimum tillage A production system in which soil cultivation is kept to the minimum necessary for crop establishment and growth, thereby reducing labor costs and fuel costs, and damage to the soil structure.

Molecular chlorine A molecule comprised of two chlorine atoms Cl_2.

Mollisol An order of fertile soils having dark or very dark, friable, thick A horizons and are high in humus and bases such as calcium and magnesium.

Monoculture plantation A farming system given over exclusively to a single crop. Its advantages are the increased efficiency of farming and a higher quality of output. Disadvantages include a greater susceptibility to price fluctuations, climatic hazards, the spread of disease, and also discourages any biodiversity among all local organisms.

Montreal Protocol (MP) The treaty signed on Sept. 16, 1987 by 25 nations; 168 nations are now parties to the accord. The Protocol set limits on the production of chlorofluorocarbons.

Mountaintop removal A surface mining practice involving the removal of mountaintops to expose coal seam, and disposing of the associated mining overburden in adjacent valleys.

National Ambient Air Quality Standards (NAAQS) A set of air quality standards set by the EPA (as required by The Clean Air Act) for widespread pollutants considered harmful to the public and environment.

Negative feedback cycle See *feedback mechanism.*

Nitric acid (HNO_3) Transparent, colorless to yellowish corrosive liquid acid that is a highly reactive oxidizing agent; a component of acid rain.

Non-attainment An area of a state that does not meet the National Air Quality Standard for a criteria pollutant for a period of time. For areas that are designated nonattainment, states must submit a plan (called a State Implement Plan) which outlines the specific strategies it will use to get areas back into attainment. There are also specific deadlines that states must meet to submit their plans and achieve compliance.

Non-point source pollution Pollution discharged over a wide land area, not from one specific location. These are forms of diffuse pollution caused by sediment, nutrients, organic and toxic substances originating from land-use activities, which are carried to lakes and streams by surface runoff.

Nuclear fuel cycle The progression of nuclear fuel through a series of differing stages. It consists of steps in the *front end*, which are the preparation of the fuel, steps in the service period in which the fuel is used during reactor operation, and steps in the *back end*, which are necessary to safely manage, contain, and either reprocess or dispose of spent nuclear fuel.

Nuclear meltdown An informal term for a severe nuclear reactor accident that results in core damage from overheating.

Oil sands Sand and rock material which contains crude bitumen (a heavy, viscous form of crude oil). Considered to be an unconventional source.

Organic matter (OM) Usually carbon-based substances derived from living things. Examples include: everything we grow and eat, wood, dung and the humus content of soil. It can be added to soil to make it more fertile.

Orographic rainfall Precipitation which results from the lifting of moist air over an orographic barrier such as a mountain range; strictly, the amount so designated should not include that part of the precipitation which would be expected from the dynamics of the associated weather disturbance, if the disturbance were over flat terrain.

Overburden The sedimentary rock material that covers coal seams, mineral veins, etc.

Overshoot The biological phenomenon used by ecologists to describe a species whose numbers exceed the ecological carrying capacity of the place where it lives.

Ozone layer See *stratospheric ozone.*

Ozone precursors Chemical compounds, such as carbon monoxide (CO), which in the presence of solar radiation react with other chemical compounds to form ozone, mainly in the troposphere.

Pandemic An epidemic of infectious disease that has spread through human populations across a large region, multiple continents, or even worldwide.

Parent material The disintegrated rock material that is unchanged or only slightly changed that generally gives rise to the true soil by the natural process of soil developmen; known as the C-horizon of a soil profile.

Paris Climate Agreement A landmark agreement to combat climate change and to accelerate and intensify the actions and investments needed for a sustainable low carbon future. Signed in Paris in 2015.

Peak oil A hypothetical date referring to the world's peak crude oil production, whereby following this day, production rates will begin to diminish.

Permafrost A permanently frozen layer at variable depth below the surface in frigid regions of a planet. When melted, large amounts of methane are released.

Permeability The property of rocks that is an indication of the ability for fluids (gas or liquid) to flow through rocks.

Photochemical smog Air pollution containing ozone and other reactive chemical compounds formed by the action of sunlight on nitrogen oxides and hydrocarbons, especially those in automobile exhausts.

Photovoltaic (PV) A method of generating electrical power by converting solar radiation into direct current electricity using semiconductors.

Phreatic zone See *zone of saturation*.

Point source pollution Pollution discharged through a pipe or some other discrete source from municipal water-treatment plants, factories, confined animal feedlots, or combined sewers.

Polar stratospheric cloud (PSC) Clouds that form in the winter polar stratosphere at altitudes of 15,000–25,000 meters (50,000–80,000 ft).

Polar vortex The large-scale cyclonic circulation in the middle and upper troposphere centered generally in the polar regions.

Population growth The increase in the number of people who inhabit a territory or state.

Porosity A measure of the empty spaces in rock, a fraction of the volume of voids over the total volume. Porosity lies between 0 and 1, or as a percentage between 0% and 100%.

Positive feedback cycle See *feedback mechanism*.

Potable water Water reserved or suitable for drinking.

Poverty The state or condition of having little or no money, goods, or means of support; condition of being poor.

Precipitation Any of all of the forms of water particles, whether liquid or solid, that fall from the atmosphere and reach the ground. The forms of precipitation include: rain, drizzle, snow, snow grains, snow pellets, hail, and ice pellets.

Primary forest An old-growth forest (also termed *virgin forest* or *primeval forest*) that has attained great age without significant disturbance, and thereby exhibiting unique ecological features.

Primary pollutants An air pollutant emitted directly from a source, like nitrogen dioxide (NO_2), carbon monoxide (CO), and volatile organic compounds (VOCs).

Primary standards Air quality standards that provide public health protection, including protecting the health of "sensitive" populations such as asthmatics, children, and the elderly.

Primeval forest See *primary forest*.

Principal air pollutants See *criteria pollutants*.

Proxies Records which use other phenomena (e.g., tree-ring width) to estimate temperature.

Radiation The process in which energy is emitted as particles or waves.

Radioactive decay The spontaneous transformation of an unstable atomic nucleus into a lighter one, in which radiation is released in the form of alpha particles, beta particles, gamma rays, or other particles. The rate of decay of radioactive substances such as carbon 14 or uranium is measured in terms of their half-life.

Rainsplash Erosion process in which soil particles are knocked into the air by raindrop impact.

Rate of natural increase (*r*) The crude birth rate minus the crude death rate in a given population.

Recharge A hydrologic process where water moves downward from surface water to groundwater.

Replacement fertility rate The total fertility rate—the average number of children born per woman—at which a population exactly replaces itself from one generation to the next, without migration. This rate is roughly 2.1 children per woman.

Rills Small channels primarily on agricultural land that concentrate runoff and cause erosion. They can be removed by plowing.

Rule of 70 A way to estimate the number of years it takes for a certain variable to double. The rule of 70 states that in order to estimate the number of years for a variable to double, take the number 70 and divide it by the growth rate of the variable.

Runoff Water that moves over the soil surface to the nearest surface stream frequently transporting soil.

Saltwater intrusion The movement of saline water into freshwater aquifers. Most often, it is caused by groundwater pumping from coastal wells, from construction of navigation channels, or oil field canals.

Scrubbers An apparatus for removing impurities released during the combustion of certain types of fossil fuels (especially gases).

Secondary forest A forest or woodland area which has re-grown after a major disturbance such as fire, insect infestation, or timber harvest, until a long enough period has passed so that the effects of the disturbance are no longer evident. It is distinguished from an old-growth forest (*primary* or *primeval* forest), which have not undergone such disruptions.

Secondary pollutants A pollutant which is not directly emitted from a source but forms once other primary pollutants react or interact in the atmosphere. A prime example is tropospheric ozone.

Secondary standards Air quality standards that provide public welfare protection, including protection against decreased visibility and damage to animals, crops, vegetation, and buildings.

Sediment delivery system A collection of processes (rainsplash, runoff, river flow) that erode and transport sediment from the land surface to the ocean.

Sediment yield The amount of sediment lost from a surface area per unit time.

Selective absorption The absorption by a substance of only certain wavelengths of radiation with the coincident exclusion or transmission of others.

Selective gas A gas with the unique property of selective absorption. Examples include carbon dioxide and water vapor.

Selective logging A method of cutting timber that takes only selected trees from a stand. Usually applies to cutting trees marked by a forester for removal under a forest-management program.

Shale A sedimentary rock composed of layers of claylike, fine-grained sediments.

Shifted cultivators A land-use system, especially in the tropics, in which a tract of land is cultivated until its fertility diminishes, at which point it is abandoned until this is restored naturally.

Shortwave radiation A term used to describe radiant energy with wavelengths in the visible (VIS), near-ultraviolet (or UV), and near-infrared (NIR) spectra. There is no standard cut-off for the near-infrared range; therefore, the shortwave radiation range is also variously defined. It may be broadly defined to include all radiation with a wavelength between 0.1μm and 5.0μm.

Simpson's Index of Diversity A measure of species diversity. In ecology, it is often used to quantify the biodiversity of a habitat. It takes into account the number of species present, as well as the abundance of each species.

Slash-and-burn A form of agriculture in which an area of forest is cleared by cutting and burning and is then planted, usually for several seasons, before being left to return to forest.

Soil degradation Occurs when soil deteriorates because of human activity and loses its quality and productivity. Characterized by loss of nutrients or organic matter, breakdown of the soil structure, or high toxicity from pollution.

Soil development A suite of processes that add, remove, transform, and translocate material leading to the formation of soil horizons.

Soil erosion Removal of topsoil faster than the soil formation processes can replace it; due to natural, animal, and human activity (over-grazing, over-cultivation, forest clearing, mechanized farming, etc.); results in land infertility and leads to desertification and often devastating flooding.

Soil formation factors Five factors that lead to the formation of soils: climate, parent material, topography, organic matter, and time.

Solar irradiance The amount of solar energy that arrives at a specific area at a specific time.

Species diversity The variety and abundance of different types of organisms which inhabit an area; this takes both *species richness* and *equitability* into consideration. This can be measured with the Simpson's Index of Diversity.

Species evenness See *equitability*.

Species richness The number of species compared with the number of individuals in the community; measures the diversity of species within a community.

Spent fuel Nuclear reactor fuel that has been irradiated to the extent that it can no longer effectively sustain a chain reaction because its fissionable isotopes have been partially consumed and fission-product poisons have accumulated in it.

Stabilization triangle The difference between the business-as-usual scenario of CO_2 emissions and a flat path of emissions for the next 40 years.

Stabilization wedges An existing technology or societal practice implemented to counteract the current rise in CO_2 emissions and achieve stabilization.

Stratosphere The part of the Earth's atmosphere which extends from the top of the troposphere to about 30 miles (50 kilometers) above the surface and in which temperature increases gradually to about 32° F (0° C).

Stratospheric ozone Naturally occurring atmospheric ozone that is concentrated in the lower stratosphere in a layer between 9–18 miles (15–30 kilometers) above the Earth's surface; plays a critical role for the biosphere by absorbing a large proportion of the damaging ultraviolet radiation. Also known as the *ozone layer*.

Strip cropping The growing of a cultivated crop, such as cotton, and a sod-forming crop, such as alfalfa, in alternating strips following the contour of the land, in order to minimize erosion.

Strip mining An open mine, especially a coal mine, whose seams or outcrops run close to ground level and are exposed by the removal of topsoil and overburden.

Subsistence farming Farming whose products are intended to provide for the basic needs of the farmer, with little surplus for marketing; brings little or no profit to the farmer, allowing only for a marginal livelihood.

Sulfuric acid (H_2SO_4) A colorless, odorless, extremely corrosive, oily liquid; a component of acid rain.

System regulator (forests) Describes how forests modify the environment; examples include: production of oxygen and reduction of runoff and erosion.

Taxonomic kingdom The highest taxonomic classification into which organisms are grouped, based on fundamental similarities and common ancestry.

The Doomsday Scenario A term coined in the early 1970s to describe the collapse of the planet due to rapid population growth and the depletion of resources.

Thermal expansion The increase in volume of a material as its temperature is increased, usually expressed as a fractional change in dimensions per unit temperature change.

Thermohaline circulation A part of the large-scale ocean circulation that is driven by global density gradients created by surface heat and freshwater fluxes. The adjective *thermohaline* derives from *thermo-* referring to temperature and *-haline* referring to salt content, factors which together determine the density of sea water.

Threatened species Any species (including animals, plants, fungi, etc.) which are vulnerable to endangerment in the near future.

Tolerable erosion value (T) The amount of soil a region could afford to lose given the balance between rates of soil formation and soil loss; varies depending on the thickness of the soil.

Topsoil Surface soil usually including the organic layer in which plants have most of their roots and which the farmer turns over in plowing; known as the A-horizon of a soil profile.

Total fertility rate (*tfr*) The average number of children a woman would bear during her lifetime, assuming her childbearing conforms to her age-specific fertility rate every year of her childbearing years (typically, age 15 to 44).

Transpiration The passage of watery vapor from a living body through a membrane or pores into the atmosphere.

Trophic level A group of organisms that occupy the same position in a food chain.

Troposphere The lowest region of the atmosphere where weather forms; located between the Earth's surface and the tropopause; characterized by decreasing temperature with increasing altitude.

Tropospheric ozone A molecule made up of three oxygen atoms; formed in the lower ground level as a result of chemical reactions between oxides of nitrogen (NOx) and volatile organic compounds (VOCs) in the presence of sunlight. Emissions from industrial facilities and electric utilities, motor vehicle exhaust, gasoline vapors, and chemical solvents are some of the major sources of NOx and VOCs.

Ultraviolet light (UV) Radiation lying in the ultraviolet range of the electromagnetic spectrum and not visible to the human eye (wavelengths shorter than visible light but longer than X-rays); this type of radiation is necessary to catalyze many chemical reactions that take place in the atmosphere.

Unconfined aquifer See *aquifer*.

Unconventional sources A type of petroleum that is produced or obtained through techniques other than traditional oil well extraction. See *hydraulic fracking*.

Unsaturated zone See zone of aeration.

Uranium (U) A metallic element that is used as nuclear fuel and is highly toxic and radioactive.

Vadose zone See *zone of aeration*.

Virgin forest See *primary forest*.

Virtual water content The volume of freshwater used to produce the product, measured at the place where the product was actually produced. It refers to the sum of the water use in the various steps of the production chain.

Volatile organic compound (VOC) Any organic compound which is unstable and evaporates readily to the atmosphere; involved in tropospheric ozone production.

Waldsterben The symptoms of tree decline in central Europe from the 1970s, considered to be caused by atmospheric pollution.

Water cycle See *hydrological cycle*.

Water table The top zone of soil and rock in which all voids are saturated with water. The level of the water table varies with topography and climate.

Wavelength (λ) The distance between one peak or crest of a wave of light, heat, or other energy and the next corresponding peak or crest.

Weathering Any of the chemical or mechanical processes by which rocks exposed to the weather undergo chemical decomposition and physical disintegration. Although weathering usually occurs at the Earth's surface, it can also occur at significant depths, for example through the percolation of groundwater through fractures in bedrock. It usually results in changes in the color, texture, composition, or hardness of the affected rocks.

Yield gaps The difference between the crop yield observed at any given location and the potential crop yield at the same location, given current agricultural practices.

Zero population growth (*zpg*) The point when a population stops growing. It occurs when the birth rate equals the death rate and the TFR = 2.1.

Zone of aeration The subsurface sediment above the water table containing air and water. Also known as *unsaturated zone, vadose zone, zone of suspended water*.

Zone of saturation A subsurface zone in which water fills the interstices and is under pressure greater than atmospheric pressure. Also known as *phreatic zone, saturated zone*.

Index

B

C

D

O

P

T

U